ADVANCED TEXTS IN ECONOMETRICS

General Editors

Periodicity and Stochastic Trends In Economic Time Series

Philip Hans Franses

OXFORD UNIVERSITY PRESS
1996

Oxford University Press, Walton Street, Oxford OX2 6DP
Oxford New York
Athens Auckland Bangkok Bombay
Calcutta Cape Town Dar es Salaam Delhi
Florence Hong Kong Istanbul Karachi
Kuala Lumpur Madras Madrid Melbourne
Mexico City Nairobi Paris Singapore
Taipei Tokyo Toronto
and associated companies in
Berlin Ibadan

Oxford is a trade mark of Oxford University Press

Published in the United States
by Oxford University Press Inc., New York

British Library Cataloguing in Publication Data
Data available

Library of Congress Cataloguing in Publication Data
Franses, Philip Hans, 1963–
Periodicity & stochastic trends in economic time series / Philip Hans Franses.
p. cm.—(Advanced texts in econometrics)
Includes bibliographical references and index.
1. Econometrics. 2. Time-series analysis. 3. Stochastic analysis. 4. Cycles. I. Title. II. Series.
HB139.F723 1996 330'.01'5195—dc20 96–420

ISBN 0–19–877454–0 (Pbk)
ISBN 0–19–877453–2

Typeset by Alliance Phototypesetters
Printed in Great Britain
on acid-free paper by
Biddles Ltd, Guildford & King's Lynn

Foreword

The last decade has witnessed a renewed interest in problems of seasonality. In general it has been accepted that seasonal variation accounts for a major part of the variation in many quarterly, monthly, and daily time series.

However, some disagreement has existed on whether the seasonal components are very regular and constant over long periods of time or whether they are changing over the years.

Another controversy has concerned the interdependence between the seasonal components and other components of the time series such as the business cycle component.

These questions are both addressed in this monograph by Philip Hans Franses. The analysis shows that the seasonal variation is indeed changing and that there is information available from other components of the time series. Hence, seasonal adjustment is rejected as a mean of analysis.

The monograph covers most of the topics of interest within seasonal modelling. However, the parts containing the periodic models are especially interesting as they cover the important contribution made by Philip Hans Franses himself.

SVEND HYLLEBERG

University of Aarhus

Preface

The material in this book concerns the econometric analysis of seasonal time series in economics. Recent years have seen a revival of interest in the analysis of seasonality, both in theory and in practice. The intention of this book is to present an approach to modelling and forecasting seasonal economic time series that is based on periodic models with stochastic trends. Two key topics are periodic integration and periodic cointegration. Although the book surveys some concepts in time series, it is assumed that the reader is reasonably familiar with current techniques and approaches. The prime focus of this work is to show that an alternative approach to modelling seasonal time series can yield useful univariate and multivariate models for economic time series, which can have a reasonable interpretation in terms of economic behaviour and can result in, for example, improved forecasts. This monograph may prove useful for the applied econometrician and for graduate students also, since it surveys recent developments in seasonal time series analysis. Most theoretical material is illustrated using quarterly macroeconomic variables. It can be expected that the models can also be applied to time series in such fields as marketing, finance, and tourism. Of course, it is hoped that the material in this book will stimulate further empirical and theoretical research. The data used herein are available upon request.

The book was written mainly during my appointment as a research fellow of the Royal Netherlands Academy of Arts and Sciences, affiliated with the Econometric Institute at the Erasmus University Rotterdam. The financial support from the Academy throughout these years is gratefully acknowledged.

During the past few years I have benefited from the opportunity to work on the topic of periodicity in time series mainly with Peter Boswijk and Richard Paap. Parts of jointly published work appear in this book. This will be mentioned at the beginning of the relevant chapters and throughout the sections; full details are included in the References. The input of Peter Boswijk was especially valuable because of his impressive technical skills which facilitated progress in our joint research. Richard Paap was extremely helpful with the various simulation programs. Additionally, the detailed suggestions from Robert Engle, Clive Granger, David Hendry, Søren Johansen, Teun Kloek, Robert Kunst, Michael McAleer, Marius Ooms, and Denise Osborn are very much appreciated. Christiaan Heij and Marius Ooms read earlier versions of chapters of this book and supplied many useful suggestions. Of course, all errors are my responsibility. Furthermore, I wish to express my gratitude to Clive Granger, Graham Mizon, and Andrew Schuller for including this monograph in their series.

Last but not least, I thank Eric Ghysels and Denise Osborn for arousing my enthusiasm for the topic of modelling seasonality, and Svend Hylleberg for his continuing interest in my work and for his suggestion that I write this book.

PHILIP HANS FRANSES

Rotterdam, October 1995

Contents

List of abbreviations *xi*

1. Introduction 1

2. Concepts in time series analysis 4

2.1 Univariate time series models 4
2.2 Multivariate time series models 15
2.3 Unit roots in univariate time series 17
2.4 Unit roots in multivariate time series 23

3. An introduction to seasonal time series 32

3.1 Characteristics of seasonal time series 32
3.2 Univariate time series aspects 40

4. Seasonal adjustment 49

4.1 Seasonal adjustment methods 50
4.2 Evaluating seasonally adjusted data 54

5. Seasonal integration and cointegration 61

5.1 Seasonal integration 62
5.2 Seasonal cointegration 73
Appendix 78

6. Are seasons, trends, and cycles always independent? 79

6.1 Correlation between seasonal and nonseasonal components 80
6.2 Business cycles and changes in seasonal patterns 84
6.3 Seasonality in consumer confidence 88

7. Periodic autoregressive time series models 92

7.1 Notation and representation 93
7.2 Stationarity in periodic autoregression 98
7.3 Estimation and model selection 101

7.4 Effects of neglecting periodicity 117
7.5 Multivariate periodic autoregression 123

8. Periodic integration 126

8.1 Testing for unit roots in periodic autoregression 127
8.2 Periodic integration 150
8.3 Effects of nonperiodic analysis 166

9. Periodic cointegration 177

9.1 Representation of periodic VAR models 178
9.2 Common aspects in periodic integration 181
9.3 Periodic cointegration 193
9.4 Seasonal variation in business cycles 207

10. Conclusion 211

Data Appendix 214

References 215

Author index 225

Subject index 228

List of Abbreviations

ACF	autocorrelation function
ADF	augmented Dickey and Fuller statistic
ADL	autoregressive distributed lag
AIC	Akaike information criterion
AO	additive outlier
AR	autoregression, autoregressive
ARCH	autoregressive conditional heteroskedasticity
ARFIMA	autoregressive fractionally integrated moving average
ARIMA	autoregressive integrated moving average
ARMA	autoregressive moving average
BP	Box–Pierce test
BT	Box–Tiao method
CRDF	cointegrating regression Dickey–Fuller statistic
CRDW	cointegrating regression Durbin–Watson statistic
DF	Dickey–Fuller statistic
DGP	data-generating process
DS	difference-stationary
EG	Engle–Granger cointegration method
EGHL	Engle–Granger–Hylleberg–Lee method
ECM	error correction model/mechanism
FPE	final prediction error
GDP	gross domestic product
GG	Gonzalo–Granger method
GNP	gross national product
HEGY	Hylleberg–Engle–Granger–Yoo method
I(*d*)	integrated of order *d*
IMA	integrated moving average
IO	innovation outlier
IRF	impulse response function
JJ	Johansen–Juselius cointegration method
K	Kurtosis (test for excess)
LB	Ljung–Box test
LM	Lagrange multiplier
LR	likelihood ratio
LQAC	linear quadratic adjustment cost
MA	moving average
ML	maximum likelihood
MSE	mean squared error
MSEP	mean squared error of prediction
MAPE	mean absolute percentage error

NBER	National Bureau of Economic Research
NLS	nonlinear least squares
NSA	not seasonally adjusted
OCSB	Osborn–Chui–Smith–Birchenhall method
OLS	ordinary least squares
PACF	partial autocorrelation function
PADF	periodic ADF
PAR	periodic autoregression/autoregressive
PARI	periodic autoregression for integrated process
PARMA	periodic autoregressive moving average
PC	periodic cointegration
PCM	periodic cointegration model
PCSC	periodic cointegration model with seasonal cointegration
PECM	periodic error correction model/mechanism
PI	periodic integration
PIAR	periodically integrated autoregression
PMA	periodic moving average
PVAR	periodic vector autoregression
RMSE	root mean squared error
RSS	residual sum of squares
SA	seasonally adjusted/seasonal adjustment
SARIMA	seasonal ARIMA
SC	Schwarz criterion
SD	standard deviation
SE	standard error
SK	skewness (test for)
SPE	squared prediction error
STM	structural time series model
TS	trend-stationary
UC	unobserved components
VAR	vector autoregression/autoregressive
VARMA	vector autoregressive moving average
var	variance
VMA	vector moving average
VQ	vector of quarters

1

Introduction

This monograph deals with econometric models for seasonally observed economic time series. A typical approach to modelling seasonal time series is to remove the seasonal fluctuations using some seasonal adjustment method. Recently, however, there have appeared several econometric studies that indicate that it may be worthwhile to study seasonal fluctuations in their own right. The main reason for this is that such seasonal variation can also shed some light on the behaviour of economic agents. A further reason is that, in somehow detrended time series, the seasonal fluctuations dominate the remaining variation in the series to a large extent.

Recent empirical studies suggest that a straightforward incorporation of seasonal fluctuations in econometric models using simple deterministic terms does not seem feasible. This conjecture is based on the following two stylized facts (or empirical regularities). The first is that seasonal fluctuations in many quarterly or monthly observed macroeconomic time series do not appear to be constant over time. The second is that for several macroeconomic series it appears that the seasonal fluctuations and nonseasonal fluctuations are not independent, in the sense that one may observe different seasonal fluctuations in business cycle expansion periods from those in recession periods. Notice that, strictly speaking, the second empirical regularity violates the key assumption of seasonal adjustment methods, that is that one can identify independent seasonal and nonseasonal components.

In this monograph I focus on econometric time series models that can describe these two empirical regularities. In short, these models are periodic time series models with stochastic trends. For univariate time series I refer to these models as 'periodic integration', and for multivariate time series as 'periodic cointegration'. Although I do not explicitly derive these econometric models from economic theory, I argue that such periodic time series models can yield simple empirical representations of macroeconomic variables which can be given an economic interpretation. For completeness, I compare the proposed models with other approaches to modelling nonstationary seasonal time series using theoretical, simulation, and empirical arguments. A key feature of the model class is that the apparent correlation between seasonal and nonseasonal variation can be described using fairly simple linear time series models, which are reasonably parsimoniously parameterized. Of course, I also focus on out-of-sample forecasting.

Before I turn to a description of the content of the chapters in this monograph, some details are given on the general approach. A main characteristic of this book is its explicit focus on empirical modelling. In order to allow the reader to verify all empirical results, I give the source of the data in an appendix at the end of this book and I provide the data upon request. Of course, a selection is made from all available macroeconomic time series. The current selection is not guided by the intention to present results favourable only to my approach, but by the relative importance of these variables for, say, business cycle analysis and macroeconomic modelling in general. I therefore include such variables as US industrial production, UK total consumption, and Canadian unemployment. All data analysed are quarterly, and it should be stressed that my analysis naturally extends to monthly time series, although in the latter case one may face somewhat more effort needed to construct empirically adequate time series models.

All the models analysed in this book are in the time domain. An analysis of periodic time series using spectral techniques has (to my knowledge) not yet been developed; I adopt the classic statistical approach instead of a Bayesian approach. I have some experience with the Bayesian analysis of the proposed periodic models, and it appears that the empirical findings in this monograph seem robust to the adopted statistical approach. An implication of relying on the classic statistical approach is that all calculations in my empirical analysis can be performed using standard statistical packages. In fact, it suffices to use statistical programs such as MicroTSP (version 7.0) for the estimation of the suggested periodic models, while the Matlab and Gauss packages can be used for the simulation experiments. In only a few occasions, the Maple program was used. Furthermore, I do not provide technical details of useful theorems but refer the interested reader to the relevant studies.

The outline of this monograph is as follows. Roughly speaking, it can be divided into three parts. The first part concerns several preliminaries and comprises Chapter 2 for nonseasonal and Chapter 3 for seasonal time series. In these chapters I review several concepts in univariate and multivariate time series analysis that are relevant to the analysis in subsequent chapters; also, various notations and representations are introduced.

The second part concerns nonperiodic models for seasonal time series and consists of Chapters 4, 5, and 6. In Chapter 4 a few aspects of seasonal adjustment methods are discussed. In Chapter 5 I review a recently developed approach to analysing seasonal time series with stochastic trends at seasonal and nonseasonal frequencies in univariate and multivariate models (see Hylleberg *et al.* 1990). In Chapter 6 I question the assumption that seasons, trends, and cycles are always independent, i.e. the assumption made for the approaches in Chapters 3, 4, and 5.

The third and final part, which in fact subsumes the bulk of this monograph, concerns periodic time series models. Chapter 7 gives an introduction to periodic autoregression. A periodic autoregression extends the nonperiodic model by allowing the AR parameters to vary with the season. Chapters 8 and 9 cover the topic of unit roots in univariate and multivariate periodic autoregressive models. In Chapter 8 it is found that all considered example series are periodically integrated time series, i.e. that these series need a periodically varying differencing filter to remove the stochastic trend. These empirical results appear robust to sample variation, structural breaks, and an alternative data transformation. The various implications of periodic integration are discussed in detail. In Chapter 9 the focus is on variants of periodic cointegration models.

Chapter 10 concludes this monograph with a discussion of the main results and of the advantages and limitations of the various methods.

2

Concepts in Time Series Analysis

This chapter surveys some concepts for univariate and multivariate time series processes which are relevant for the material to be discussed in subsequent chapters of this book. For convenience, I postpone a discussion of seasonal time series to the next chapter. Examples of relevant concepts are stationarity, the autocorrelation function, unit roots, and several diagnostic measures to investigate the empirical adequacy of fitted models. The material does not provide a detailed analysis of all possible concepts, although I sometimes provide details of computational aspects. For detailed theoretical accounts, the reader is referred to standard textbooks in time series analysis such as T. W. Anderson (1971), Box and Jenkins (1970), Fuller (1976), Granger and Newbold (1986), and Hamilton (1994), among others.

The outline of this chapter is as follows. In Section 2.1, some concepts in stationary univariate time series are discussed. In Section 2.2, the focus is on multivariate models for stationary time series. Sections 2.3 and 2.4 deal with univariate and multivariate models for nonstationary time series, with an explicit focus on unit roots and cointegration.

2.1 Univariate Time Series Models

In addition to theoretical concepts in univariate time series, in this section I review several diagnostic measures and model selection criteria which I will use throughout this book.

A central concept in time series modeling is the white noise process ε_t, $t = 1, 2, \ldots, n$. The definition used is that ε_t is a sequence of uncorrelated random variables with zero mean and constant variance; or

$$\begin{aligned} E(\varepsilon_t) &= 0 \quad & t &= 1, 2, \ldots, n \\ E(\varepsilon_t^2) &= \sigma^2 \quad & t &= 1, 2, \ldots, n \\ E(\varepsilon_s \varepsilon_t) &= 0 \quad & s,t &= 1, 2, \ldots, n \text{ and } s \neq t, \end{aligned}$$

where E denotes the expectation operator.

Autoregressive moving average models

Consider the time series y_t, $t = 1,2,\ldots,n$, which can be described by the linear model

$$y_t - \phi_1 y_{t-1} - \ldots - \phi_p y_{t-p} = \varepsilon_t. \tag{2.1}$$

This model is called an autoregressive model of order p (AR(p)). It can be abbreviated as

$$\phi_p(B) y_t = \varepsilon_t, \tag{2.2}$$

with

$$\phi_p(B) = 1 - \phi_1 B - \ldots - \phi_p B^p, \tag{2.3}$$

where B is the backward shift operator defined by $B^k y_t = y_{t-k}$ for $k = 0, 1, 2, \ldots$ This AR(p) model can be represented by an infinite sum of weighted ε_t terms as

$$y_t = [\phi_p(B)]^{-1} \varepsilon_t = \sum_{i=0}^{\infty} \theta_i \varepsilon_{t-i}, \tag{2.4}$$

with parameters θ_i that converge to zero with increasing values of i, when the p solutions to

$$1 - \phi_1 z - \ldots - \phi_p z^p = 0 \tag{2.5}$$

lie outside the unit circle. For example, consider the AR(1) model

$$y_t = \phi_1 y_{t-1} + \varepsilon_t \tag{2.6}$$

for which (2.4) reads

$$y_t = \sum_{i=0}^{\infty} \phi_1^i \varepsilon_{t-i}. \tag{2.7}$$

Clearly, the parameters in (2.7) converge to zero with increasing i when $|\phi_1| < 1$. This can also be derived from (2.5), since the solution to $1 - \phi_1 z = 0$ is $z = 1/\phi_1$, which lies outside the unit circle when $|\phi_1| < 1$.

An alternative model for y_t, which can be viewed as an approximation to (2.4), is given by

$$y_t = \varepsilon_t + \theta_1 \varepsilon_{t-1} + \ldots + \theta_q \varepsilon_{t-q}, \tag{2.8}$$

or

$$y_t = \theta_q(B) \varepsilon_t, \tag{2.9}$$

with

$$\theta_q(B) = 1 + \theta_1 B + \ldots + \theta_q B^q, \tag{2.10}$$

which is called a moving average model of order q(MA(q)). The MA(q) model can be represented by $[\theta_q(B)]^{-1} y_t = \varepsilon_t$, which is $y_t = \Sigma_{i=1}^{\infty} \phi_i y_{t-i} + \varepsilon_t$, with parameters that converge to zero when the q solutions to $\theta_q(z) = 0$ lie outside the unit circle. For example, consider the MA(1) process,

$$y_t = \varepsilon_t + \theta_1 \varepsilon_{t-1}, \tag{2.11}$$

which can be rewritten as the AR process

$$y_t - \theta_1 y_{t-1} + \theta_1^2 y_{t-2} - \theta_1^3 y_{t-3} + \ldots = \varepsilon_t. \tag{2.12}$$

From this expression, it can be seen that for $|\theta_1| \geq 1$ the parameters on the left-hand side do not converge to zero. The MA(1) model is then said to be non-invertible. The MA(1) model is invertible when $|\theta_1| < 1$.

Finally, a third model for y_t combines (2.1) and (2.8) into

$$\phi_p(B)y_t = \theta_q(B)\varepsilon_t, \tag{2.13}$$

which is an autoregressive moving average model of order (p,q) (ARMA(p,q)).

Until now, it has been assumed that the mean μ of y_t is equal to 0. Where $\mu \neq 0$, the y_t in the above expressions can be replaced by $y_t - \mu$, which implies that its mean is subtracted before any modelling analysis. Notice that in practice one has to estimate this μ. Another possibility is to estimate a time series model with an intercept term, from which an estimate for μ can be calculated. For example, for an AR(p) model with a constant term δ,

$$\phi_p(B)y_t = \delta + \varepsilon_t, \tag{2.14}$$

the μ corresponds to $\delta/\phi_p(1)$. Henceforth, it will be assumed that in practical applications all estimated models include a constant term, but in this chapter continuing use will be made of zero mean variables for notational convenience.

A useful tool to obtain a first impression of the values of p and q in either (2.1), (2.8), or (2.13) may be the autocorrelation function (ACF) and the partial autocorrelation function (PACF). These measure the correlations between current and previous observations. To facilitate the interpretation of these functions, it is necessary to assume the stationarity of the process $\{y_t\}$ which generates the time series y_t, $t = 1,2, \ldots, n$. This process is defined to be covariance-stationary (or, briefly, stationary) when, for $t = 1,2, \ldots, n$ and $k = 0,1,2, \ldots$, it holds that

$$\text{(i)} \quad E(y_t) = \mu \qquad |\mu| < \infty \tag{2.15}$$

$$\text{(ii)} \quad E(y_t - \mu)^2 = \gamma_0 < \infty \tag{2.16}$$

$$\text{(iii)} \quad E(y_t - \mu)(y_{t-k} - \mu) = \gamma_k = \gamma_{-k} \qquad k = 0,1,2, \ldots \tag{2.17}$$

The γ_k in (iii) defines the autocovariance of order k. When requirement (ii) does not hold and the variance of the process is time-varying, it is clear that the estimation of the covariance between observations at times t and $t-1$ is difficult. The process y_t is called strictly stationary if the multivariate distribution of $(y_t, \ldots y_{t+i})$ is identical to that of the time shifted set $(y_{t+k}, \ldots, y_{t+k+i})$.

In practice, one often investigates the stationarity of a time series process by checking the properties of a time series model fitted to the time series realizations y_t. For example, suppose a useful model for y_t is $(1 - B)y_t = \varepsilon_t$, for $t = 1,2, \ldots, n$, and $y_0 = 0$; then y_t can be written as the MA model $y_t = \Sigma_{i=1}^{t}\varepsilon_i$. It is now easy to derive this, for the variance of y_t holds that $\gamma_0(t) = t\sigma^2$, where the argument t indicates that the variance γ_0 is time-dependent. Hence, when the model $(1 - B)y_t = \varepsilon_t$ is useful for y_t, one says that y_t is a nonstationary time series. In Section 2.3 I take a closer look at this type of nonstationary process.

Autocorrelation functions

The autocorrelation function is defined by

$$\rho_k = \gamma_k / \gamma_0, \tag{2.18}$$

with $\rho_0 = 1$, ρ_{-k}, and $-1 < \rho_k < 1$ where k takes discrete values other than zero. For example, for the white noise process ε_t it holds that $\rho_k = 0$ for $k \neq 0$. For the AR(1) model $y_t = \phi_1 y_{t-1} + \varepsilon_t$ with $|\phi_1| < 1$, it is easy to derive that $\rho_k = \phi_1^k$ for all $k < 0$. This derivation uses $E(\varepsilon_t y_{t-k}) = 0$ for all $k < 0$, which follows from (2.4). As an illustration of the determination of the autocorrelation coefficients for higher-order autoregressive models, consider the AR(2) model,

$$y_t - \phi_1 y_{t-1} - \phi_2 y_{t-2} = \varepsilon_t. \tag{2.19}$$

Multiplying both sides by y_{t-1}, taking expectations, and dividing them by γ_0 results in

$$\rho_1 - \phi_1 \rho_0 - \phi_2 \rho_1 = 0 \tag{2.20}$$

or

$$\rho_1 = \phi_1 / (1 - \phi_2).$$

To determine an expression for ρ_2, analogous operations are carried out to (2.19), yielding

$$\rho_2 - \phi_1 \rho_1 - \phi_2 \rho_0 = 0. \tag{2.21}$$

Substituting (2.20) into (2.21) gives

$$\rho_2 = \phi_1^2 / (1 - \phi_2) + \phi_2. \tag{2.22}$$

Analogously, it can be established that for larger values of k the ρ_k can be solved recursively using $\rho_k = \phi_1 \rho_{k-1} + \phi_2 \rho_{k-2}$. This expression for an AR(2) process indicates that this process can be characterized by an exponentially decaying pattern of the ACF. Hence, an AR model may be identified using its (estimated) ACF (see Box and Jenkins 1970). The ACF is however most useful to obtain a tentative estimate of the value of q in an MA(q) process. This is because for an MA(q) process it holds that

$$\begin{aligned} \gamma_k &= \sigma^2 \sum_{i=0}^{q-k} \theta_i \theta_{i+k} && \text{for } k = 0, 1, \ldots, q \\ &= 0 && \text{for } k > q, \end{aligned} \tag{2.23}$$

because of the uncorrelated ε_t variates, and where $\theta_0 = 1$. Hence, when $\rho_k = 0$ for $k \geq q + 1$, one may wish to consider an MA(q) model. Of course, if all the $\rho_j = 0, j = q - 1, \ldots, 1$, one may consider the MA(q) model $y_t = \varepsilon_t - \theta_q \varepsilon_{t-q}$. Such so-called subset MA processes may be useful for seasonal time series where q can equal 4 or 12 (see Chapter 3).

The partial autocorrelation function (PACF) can easily be illustrated by considering the AR(p) process, for which it can be shown that

$$\rho_k = \sum_{j=1}^{p} \phi_j \rho_{k-j}, \tag{2.24}$$

for $k > 0$. The partial autocorrelation of order K, denoted by ψ_{KK}, is given by the solution of the set of the linear equations

$$\rho_k = \sum_{j=1}^{K} \psi_{Kj}\rho_{k-j}, \tag{2.25}$$

for $k = 1, \ldots, K$. Clearly, $\rho_1 = \psi_{11}$ for all time series processes. The PACF is particularly useful to identify an AR(p) model for y_t, since then it holds for y_t that $\psi_{kk} = 0$ for $k > p$. Note that it may occur that only ψ_{pp} is not equal to zero while all other ψ_{ii}, $i = 1, \ldots, p-1$, are zero. In that case the AR(p) reduces to $y_t = \phi_p y_{t-p} + \varepsilon_t$, which is called a subset AR(p) process (see McClave 1975).

For stationary and invertible ARMA(p,q) processes, the patterns of the ACF and PACF do not cut off after a certain point. A simple example is the ARMA(1,1) model,

$$y_t - \phi_1 y_{t-1} = \varepsilon_t + \theta_1\varepsilon_{t-1}, \tag{2.26}$$

for which

$$\rho_k = \phi_1^{k-1}(1 + \phi_1\theta_1)(\phi_1 + \theta_1)/(1 + 2\phi_1\theta_1 + \theta_1^2) \tag{2.27}$$

for $k = 1,2, \ldots$. From this expression it can be seen that ρ_k can take a wide variety of values for distinct choices of ϕ_1 and θ_1. This suggests that the identification of an ARMA time series model from the patterns of the PACF alone may be rather difficult. One suggestion is then tentatively to specify several models, to estimate their parameters, and to use some model selection criterion to choose between them.

In practice, the correlation functions have to be estimated. The ACF can be estimated by means of

$$r_k = C_k / C_o, \tag{2.28}$$

where

$$C_k = \sum_{t=k+1}^{n} (y_t - \bar{y})(y_{t-k} - \bar{y}), \tag{2.29}$$

where $\bar{y}$ denotes the relevant sample mean of y_t. The sample equivalents of ψ_{KK}, to be denoted as a_{KK}, can be obtained by applying ordinary least squares (OLS) to

$$y_t - \bar{y} = a_{K1}(y_{t-1} - \bar{y}) + \ldots + a_{KK}(y_{t-K} - \bar{y}) + v_t. \tag{2.30}$$

This expression in (2.30) indicates that the parameters in AR(p) models can be estimated using OLS. The parameters in MA(q) and ARMA(p,q) models have to be estimated using iterative procedures. Intuitively, the main reason to use such iterative methods is that in an MA(q) model the unknown time series ε_t also serves as a regressor. For details on estimation methods, I refer the interested reader to Box and Jenkins (1970), T. W. Anderson (1971), and Granger and Newbold (1986).

I conclude this subsection on ARMA models, by mentioning that a different interpretation of these models can be given by considering the class of unobserved components (UC) models (see Harvey 1989). A simple example of a UC model is the so-called local-level model:

$$y_t = \mu_t + \varepsilon_t \tag{2.31}$$
$$\mu_t = \mu_{t-1} + \eta_t, \tag{2.32}$$

where ε_t and η_t are independent and uncorrelated white noise processes with variances σ^2 and ω^2. This local-level model can be written in ARMA format as

$$y_t - y_{t-1} = \zeta_t + \theta_1 \zeta_{t-1}, \tag{2.33}$$

with ζ_t a white noise process, and $\theta_1 = [(r^2+4r)^{1/2} - 2 - r]/2$, with $r = \omega^2/\sigma^2$ (see e.g. Harvey and Koopman 1992). Hence, when ω^2 is very close to zero, i.e. when the mean of y_t changes only very slowly over time, the θ parameter in (2.33) gets close to -1. We will return to such models in Chapter 4.

Diagnostic measures

Once a tentative model has been estimated, its empirical adequacy should be investigated. Much of this investigation concerns possibly undetected systematic patterns in the estimated residual process.

Testing for residual autocorrelation

An important check is whether the estimated residuals denoted as $\hat{\varepsilon}_t$ display any residual correlation. If so, the initial ARMA model may be modified to include more AR or MA terms. The ACF of the estimated residuals is given by

$$r_k(\hat{\varepsilon}) = \sum_{t=k+1}^{n} \hat{\varepsilon}_t \hat{\varepsilon}_{t-k} \Big/ \sum_{t=1}^{n} \hat{\varepsilon}_t^2. \tag{2.34}$$

It can be shown that the population equivalents of $r_k(\hat{\varepsilon})$ are asymptotically uncorrelated and have variances that can be approximated by $(n-k)/(n^2+2n) \cong n^{-1}$. Hence, under the additional assumption of normality, a rough check may be to test whether the estimated residual autocorrelations lie within the $\pm 2n^{-1/2}$ interval. Box and Pierce (1970) (BP) propose a joint test for the significance of the first m residual autocorrelations, which is given by

$$\mathrm{BP}(m) = n \sum_{k=1}^{m} r_k^2(\hat{\varepsilon}), \tag{2.35}$$

which asymptotically follows a $\chi^2(m-p-q)$ distribution under the hypothesis of no residual autocorrelation provided that m/n is small and m is moderately large. This type of test statistic, where the value of m is often set *a priori* and which should detect any kind of misspecification up to m, is usually called a portmanteau test. In small samples, it turns out that the BP test may not have an appropriate size. Therefore, Ljung and Box (1978) (LB) propose the statistic

$$\mathrm{LB}(m) = n(n+2) \sum_{k=1}^{m} (n-k)^{-1} r_k^2(\hat{\varepsilon}). \tag{2.36}$$

As expected, the drawback of a portmanteau test is that it may be able to detect whether the model is inadequate, but may not be useful in indicating how the model should be modified when necessary. Furthermore, if for example the m is selected too large, the BP and LB tests may lack power against low-order residual autocorrelation.

An alternative is to consider the nested hypotheses tests developed in, e.g., Godfrey (1979). With the Lagrange multiplier (LM) principle, these tests are relatively easy to calculate. For example, to test an AR(p) model against an AR($p + r$) or an ARMA(p,r) model (see also Poskitt and Tremayne 1980), the LM test is found by estimating

$$\hat{\varepsilon}_t = \alpha_1 y_{t-1} + \ldots + \alpha_p y_{t-p} + \alpha_{p+1}\hat{\varepsilon}_{t-1} + \ldots + \alpha_{p+r}\hat{\varepsilon}_{t-r} + \phi_t, \tag{2.37}$$

where $\hat{\varepsilon}_t$ are the estimated residuals of the AR(p) model. The test statistic is calculated as nR^2 where R^2 is the coefficient of determination from (2.37), and it is asymptotically $\chi^2(r)$ distributed under the null hypothesis that the AR(p) model is adequate. In the empirical sections of this book, the F-version of this LM test is used, denoted as $F_{\mathrm{AR},1-r}$. The simulation results in A. D. Hall and McAleer (1989) indicate that this LM test often has higher power than the BP and LB tests.

Normality and outliers

To facilitate the interpretation of, say, parameter estimates and t-ratios, the estimated residuals should be approximately normal. Under the assumption of no autocorrelation in the residuals, the test statistics for skewness (SK) and kurtosis (K) given in Lomnicki (1961) reduce to

$$\mathrm{SK} = (n/6)^{1/2}\,(\hat{m}_3^2/\hat{m}_2^3)^{1/2} \tag{2.38}$$

and

$$\mathrm{K} = (n/24)^{1/2}\,(\hat{m}_4/\hat{m}_2^2 - 3), \tag{2.39}$$

where

$$\hat{m}_j = n^{-1}\sum_{t=1}^{n}\hat{\varepsilon}_t^j. \tag{2.40}$$

Under the null hypothesis of normality, the SK and K test statistics have an asymptotic $N(0,1)$ distribution. The well-known Bera–Jarque (1982) (BJ—or commonly JB) test is

$$\mathrm{JB} = (\mathrm{SK}^2 + \mathrm{K}^2) \sim \chi^2(2). \tag{2.41}$$

Rejection of the normality null hypothesis may indicate that there are some outlying observations, or that the error process is not homoskedastic.

The time series literature contains many studies on detecting outlying observations (see e.g. Fox 1972; Abraham and Chuang 1989; and Chen and Liu 1993). Typically, one distinguishes two types of outlier model. When there is an additive outlier (AO) at time τ for an AR(p) process, this can be modelled as

$$z_t = y_t + \omega_1\delta_{t,\tau} = \phi_p(B)^{-1}\varepsilon_t + \omega_1\delta_{t,\tau} \tag{2.42}$$

or

$$\varepsilon_t = \phi_p(B)(z_t - \omega_1 \delta_{t,\tau}), \tag{2.43}$$

where $\delta_{t,\tau}$ equals 1 if $t = \tau$ and 0 elsewhere, ω_1 is the magnitude of the shift in the observations at $t = \tau$, and z_t is the observed time series. The innovation outlier (IO) model is given by

$$z_t = y_t + \phi_p(B)^{-1}\omega_2\delta_{t,\tau} = \phi_p(B)^{-1}(\varepsilon_t + \omega_2\delta_{t,\tau}) \tag{2.44}$$

or

$$\varepsilon_t = \phi_p(B)z_t - \omega_2\delta_{t,\tau}, \tag{2.45}$$

where ω_2 is the magnitude of the shift in the innovation at time τ. It is clear from (2.43) and (2.45) that in case of an AO the residuals in future times are affected, and that an IO affects the residual at time τ and p future observations are influenced.

In the empirical applications, I will abstain from explicitly testing for outlying observations using recently developed sophisticated techniques like those in Abraham and Yatawara (1988), Peña (1990), and Tsay (1988), among others. In fact, I will use the possibly crude method of inspecting the residual patterns with a focus on $\hat{\varepsilon}_t$ values which are much more than twice the estimated standard deviation. In case there is some suspicion that an observation is an outlier, this observation and several subsequent observations (depending on the model order) are deleted using single-dummy variables. Standard t-tests are used to check how many dummy variables are needed to remove the aberrant observations. Admittedly, such an approach may result in inappropriate outcomes because of the non-robustness of OLS to certain outliers, and it may be sensible (though more complicated) to use robust estimation methods such as advocated in, for example, Lucas (1995).

Autoregressive conditional heteroskedasticity

Another useful property of an estimated error process for time series models is that it is not conditionally heteroskedastic. An example of a process for which the assumption of homoskedasticity is not valid is the autoregressive conditional heteroskedasticity (ARCH) process (see Engle 1982; Bollerslev 1986). The simple ARCH(k) process is given by

$$E(\varepsilon_t^2 | I_{t-1}) = h_t = \alpha_0 + \sum_{i=1}^{k} \alpha_i \varepsilon_{t-i}^2, \tag{2.46}$$

where $\alpha_j \geq 0$ ($j = 0,1 \ldots, k$) and I_{t-1} denotes the information set at time t, where I_{t-1} may contain lagged y_t variables.

In this book the focus is on modelling the conditional mean of a time series y_t. Hence there will be no explicit discussion of ARCH type models and their use in forecasting error variances. However, the empirical models will be checked for the presence of ARCH in order to validate my approach to use tests for parameter significance that are based on the usual distributional assumptions. A simple LM test statistic for ARCH of order k is given by nR^2, where the R^2 is from the auxiliary regression

$$\hat{\varepsilon}_t^2 = \alpha_0 + \sum_{i=1}^{k} \alpha_i \hat{\varepsilon}_{t-i}^2 + u_t \qquad (2.47)$$

(see Engle 1982). Under the null hypothesis of no ARCH, this LM statistic asymptotically follows a $\chi^2(k)$ distribution. The *F*-version of this test will be denoted as $F_{\text{ARCH},1-k}$.

Parameter constancy and linearity

For several empirical models below, one of the so-called Chow (1960) tests will be calculated. The Chow test regularly used is

$$\text{Chow} = (RSS_n/RSS_m - 1)\,[(m-k)/(n-m)], \qquad (2.48)$$

where *RSS* denotes the residual sum of squares, *m* refers to the subset containing the first *m* observations, and *k* is the number of parameters. This statistic, which is particularly useful because it does not necessitate model estimation over $n - m$ observations, asymptotically follows an $F(n-m, m-k)$ distribution under the null hypothesis of parameter constancy and *a priori* known values of *m* and $n - m$. Rejection of the null hypothesis of parameter constancy may indicate the presence of a structural break. In many cases the timing of a likely structural break is unknown and is usually estimated from the data. Typically, one then considers the maximum Chow value. Asymptotic theory on the distribution of this *maxCHOW* test statistic is given in Andrews (1993).

In case parameter constancy is rejected, one may consider estimating ARMA models with time-varying parameters. This usually results in nonlinear time series models. (See Granger and Teräsvirta (1993) for a survey of such models.) In this book the focus will be on linear models.

Finally, there will be no detailed discussion on the selection of a possible transformation of the data prior to analysis. Typically, the y_t series can be x_t or $\log x_t$, where 'log' denotes the natural logarithm and x_t is some economic time series. The usual argument for the log transformation is that it can reduce the impact of outliers and heteroskedasticity. However, the transformation of the data can have large impact on the ACF and PACF and on possible nonlinear properties of the time series. (For an investigation of the impact of data transformations on, e.g., the stationarity property, see Granger and Hallman (1991) and Franses and McAleer (1995*a*).) In this book, y_t often denotes $\log x_t$, as is usual in applied econometric time series modeling. However, in some specific cases I investigate the robustness of the empirical findings to the data transformation.

Model selection

The identification and estimation stages may result in a set of tentative models, in the sense that these models cannot be rejected using some or all of the aforementioned diagnostic measures. One may now want to select between these remaining models. (A survey of model selection criteria is given in De Gooijer *et al.* (1985), among others.) One of the aspects of the model selection strategy regularly adopted in this book is that no model is considered to be true, or *a priori* preferable, and hence in a sense the models will be symmetrically treated. This corresponds with the views expressed in Granger *et al.* (1995). In general, this implies that the model is selected which minimizes the value of a certain criterion function, where this function combines the fit and the number of parameters. Of course, it may be possible to express these criterion functions in terms of test statistics (see Teräsvirta and Mellin 1986), but then one should anyhow have to fix a certain significance level.

The final prediction error (FPE) criterion developed in Akaike (1969) is often applied in practice. Consider an AR(p) model that is used to predict one step ahead at time t; then the theoretical forecast is

$$f_{t,1} = \phi_1 y_t + \ldots + \phi_p y_{t-p+1} \tag{2.49}$$

The estimated forecast $\hat{f}_{t,1}$ can be calculated by substituting the estimates of the parameters in (2.49). Note that the pre-sample data are set equal to zero. In Davisson (1965) it is shown that the asymptotic mean-square error of this one-step-ahead forecast is given by

$$E(y_{t+1} - \hat{f}_{t,1})^2 = \sigma^2(1 + p/n), \tag{2.50}$$

where σ^2 is the theoretical variance of the residual process. In case the AR(p) model is valid, an unbiased estimator of σ^2 is the OLS estimator $\hat{\sigma}^2_{\text{OLS}} = RSS/(n-p)$. The model selection rule based on these expressions is to choose the model that minimizes the FPE criterion

$$\text{FPE}(k) = \hat{\sigma}^2_{\text{OLS}}(1 + k/n), \tag{2.51}$$

where k is the order of the AR model, with $k \in [1, \ldots, P]$ and P is some value assumed *a priori*.

This FPE(k) criterion bears similarities with the often applied Akaike information criterion (AIC) proposed in Akaike (1974). In the case of AR models, it is given by,

$$\text{AIC}(k) = n\log\hat{\sigma}^2_{\text{ML}} + 2k, \tag{2.52}$$

where $\hat{\sigma}^2_{\text{ML}} = RSS/n$ and again 'log' denotes the natural logarithm. It is easy to show that the AIC and FPE criteria yield a similar choice between models.

The theoretical derivations of the above model selection criteria assume a true, but unknown, data-generating process. Alternatively, the criteria based on Bayesian methods, which consider model selection between pairs of models, do not need this assumption. For Bayesian inference it is necessary to have some *a priori* knowledge of the parameters in the model. Together with the information in the data, one may now calculate the posterior probabilities of the models. The well-known criterion developed in Schwarz (1978) (SC), however, ignores the part related to the prior density in the likelihood function. The SC criterion is given by

$$\mathrm{SC}(k) = n\log\hat{\sigma}^2_{\mathrm{ML}} + k\log n. \tag{2.53}$$

Note that the criterion proposed in Rissanen (1978) is the same although it is derived using information theory arguments. Finally, comparing the expressions in (2.52) and (2.53), it is clear that when $n \geq 8$ the SC criterion penalizes the inclusion of regressors more than the AIC does. This means that the model order selected with the SC criterion is usually smaller than the model order selected with the AIC.

Forecasting

An important use of ARMA models is in forecasting one or more steps ahead. Denote $f_{t,h}$ as the h-step forecast of y_{t+h}. First, consider an MA(q) model as in (2.9), for which this forecast equals

$$f_{t,h} = \sum_{i=0}^{q} \theta_{i+h}\varepsilon_{t-i}, \tag{2.54}$$

with $\theta_0 = 1$ (see e.g. Granger and Newbold 1986), with an h-step error,

$$e_{t,h} = y_{t+h} - f_{t,h} = \sum_{i=0}^{h-1} \theta_i\varepsilon_{t+h-i}. \tag{2.55}$$

Given the white noise assumption on ε_t, it follows that

$$E(e_{t,h}) = 0,$$

$$V(h) = E(e^2_{t,h}) = \sigma^2_{\varepsilon}\sum_{i=0}^{h-1}\theta^2_i. \tag{2.56}$$

From (2.23), we can see that for an MA(q) process $f_{t,h} = 0$ when $h > q$.

The forecasts from an ARMA(p,q) model are found in a similar way (see e.g. Granger and Newbold 1986: 133), by

$$f_{t,h} - \sum_{j=0}^{p}\phi_j f_{t,h-j} = \sum_{i=0}^{q}\theta_{i+h}\varepsilon_{t-i}, \tag{2.57}$$

where $f_{t,k} = y_{t+k}$ for $k \leq 0$. For an AR(p) the h-step forecasts can be derived from (2.57) by setting the θ_i equal to zero. The variance of these point forecasts is established analogous to (2.54). Denoting

$$\eta(B) = \theta_q(B)/\phi_p(B),$$

then

$$V(h) = \sigma^2_{\varepsilon}\sum_{i=0}^{h-1}\eta^2_i, \tag{2.58}$$

for the general ARMA(p,q) model.

Finally, if one has several one- or multi-step-ahead forecasts for two or more models made at time t, one may select between these rival models using the (root) mean squared forecast error (RMSE),

$$\mathrm{MSE} = (1/h)\left[\sum_{i=1}^{h}(f_{t,i} - y_{t+i})^2\right] \tag{2.59}$$

or the mean absolute percentage error (MAPE),

$$\text{MAPE} = (1/h)\left[\sum_{i=1}^{h} |(f_{t,i} - y_{t+i})/y_{t+i}|\right], \tag{2.60}$$

although other criteria are also available.

2.2 Multivariate Time Series Models

Consider an $(m \times 1)$ zero mean vector $\mathbf{Y}_t$ containing the observations at time t for m distinct time series, i.e. $\mathbf{Y}_t = (y_{1,t}, y_{2,t}, \ldots, y_{m,t})'$. A multivariate (or vector) autoregressive moving average model of order (p,q) (VARMA(p,q)) for this $\mathbf{Y}_t$ is written as

$$\mathbf{Y}_t - \Phi_1\mathbf{Y}_{t-1} - \ldots - \Phi_p\mathbf{Y}_{t-p} = \mathbf{e}_t + \Theta_1\mathbf{e}_{t-1} + \ldots + \Theta_q\mathbf{e}_{t-q}, \tag{2.61}$$

or, more compactly, as

$$\Phi_p(B)\mathbf{Y}_t = \Theta_q(B)\mathbf{e}_t, \tag{2.62}$$

where

$$\Phi_p(B) = \mathbf{I}_m - \Phi_1 B - \ldots - \Phi_p B^p, \tag{2.63}$$

and

$$\Theta_q(B) = \mathbf{I}_m + \Theta_1 B + \ldots + \Theta_q B^q, \tag{2.64}$$

with $\mathbf{I}_m$ the $(m \times m)$ identity matrix. The $\Phi_p(B)$ and $\Theta_q(B)$ in (2.62) are $(m \times m)$ matrices with elements that are polynomials in the backward shift operator B. Special cases of (2.61) are the vector autoregressive model of order p, a VAR(p), and the vector moving average model of order q, VMA(q). The $(m \times 1)$ vector process $\mathbf{e}_t = (\varepsilon_{1,t}, \ldots, \varepsilon_{m,t})'$ is defined to be a multivariate white noise process; i.e., for all t it holds that

$$\begin{aligned} E(\mathbf{e}_t) &= 0 \\ E(\mathbf{e}_t, \mathbf{e}_t') &= \Sigma \\ E(\mathbf{e}_u \mathbf{e}_t') &= 0, \quad u,t = 1,2,\ldots,n \text{ and } u \neq t, \end{aligned}$$

where Σ is an $(m \times m)$ matrix, which is not necessarily the identity matrix $\mathbf{I}_m$.

Similar to the univariate ARMA(p,q) model, a VARMA(p,q) model can be written as a meaningful VMA(∞) model $\mathbf{Y}_t = [\Phi_p(B)]^{-1}\Theta_q(B)\mathbf{e}_t$, i.e. a model where the effect of previous shocks disappears in the long run, when the solutions to the characteristic equation

$$|\Phi_p(z)| = 0 \tag{2.65}$$

are outside the unit circle. At most, (2.65) results in a polynomial of order pm. This expression is the multivariate equivalent of (2.5). Notice that the number of solutions to (2.65) depends on the values of p and m; this is in contrast to the univariate case, where only the value of p is relevant. If all solutions to (2.65) are outside the unit circle, the $\mathbf{Y}_t$ vector process is a stationary process.

The VARMA(p,q) model in (2.62) is said to be invertible, i.e. the VAR(∞) model $[\Theta_q(B)]^{-1}\Phi_p(B)\mathbf{Y}_t = \mathbf{e}_t$ is a meaningful model, when the solutions to

$$|\Theta_q(z)| = 0 \tag{2.66}$$

are outside the unit circle. It should be mentioned that in VARMA models one needs to impose several restrictions on the elements of the matrices $\Phi_p(B)$ and $\Theta_q(B)$ to ensure the uniqueness of these multivariate time series models. (For details see e.g. Hannan 1970, and Lütkepohl 1991.)

As a simple example of a VARMA model, consider the bivariate VARMA(1,1) model given by

$$\begin{bmatrix} 1-\phi_{11}B & -\phi_{12}B \\ -\phi_{21}B & 1-\phi_{22}B \end{bmatrix} \begin{bmatrix} y_{1,t} \\ y_{2,t} \end{bmatrix} = \begin{bmatrix} 1+\theta_{11}B & \theta_{12}B \\ \theta_{21}B & 1+\theta_{22}B \end{bmatrix} \begin{bmatrix} \varepsilon_{1,t} \\ \varepsilon_{2,t} \end{bmatrix} \tag{2.67}$$

of which the first equation can be written as

$$y_{1,t} - \phi_{11}y_{1,t-1} - \phi_{12}y_{2,t-1} = \varepsilon_{1,t} + \theta_{11}\varepsilon_{1,t-1} + \theta_{12}\varepsilon_{2,t-1}.$$

The vector process is stationary if the two solutions of

$$1 - (\phi_{11}+\phi_{22})z + (\phi_{11}\phi_{22} - \phi_{12}\phi_{21})z^2 = 0 \tag{2.68}$$

lie outside the unit circle. A similar expression to (2.68) can be given in order to investigate whether model (2.67) is invertible.

The identification of a stationary multivariate time series model may proceed along similar lines as for the univariate models. One can define a measure of multiple autocorrelations $E(\mathbf{Y}_t\mathbf{Y}_{t-k}')$, which can be shown to have similar characteristics to those of its univariate analogy. For example, for a VMA(q) model, $E(\mathbf{Y}_t\mathbf{Y}_{t-q-i}') = 0$ for $i > 0$. One can however imagine that in case of large m, i.e. when the dimension of $\mathbf{Y}_t$ is large, this strategy of identification may not be easy.

Typically, in practice one often confines the analysis to VAR(p) models. The parameters in such models can easily be estimated by the application of OLS when applied to each row of the VAR. (See e.g. Lütkepohl (1991) for more details on estimating parameters in general VARMA models.) When a tentative VAR(p) model is fitted, this model should be evaluated using diagnostic checks. First of all, one can check the properties of the residuals in each of the m equations using the univariate diagnostic checks discussed in Section 2.1. Furthermore, since such diagnostic measures do not consider any possible cross-equation relations, one may consider multivariate versions of the Box–Pierce test and of the test for normality of the estimated residuals. (See also Lütkepohl (1991) and Ooms (1994) for various diagnostic measures for VAR models and for suggestions for their useful practical application.)

Once one has obtained two or more VAR models that pass the diagnostic checks, one can select a final model using, for example, the multivariate extensions of the AIC and SC criteria in (2.52) and (2.53), respectively. These are given by

$$\text{AIC}(k) = n\log|\hat{\Sigma}| + 2km^2 \tag{2.69}$$

and

$$\text{SC}(k) = n\log|\hat{\Sigma}| + km^2\log n, \tag{2.70}$$

where $|\hat{\Sigma}|$ is the determinant of the estimated $(m \times m)$ residual covariance matrix of the $(m \times 1)$ residual process $\hat{e}_t$. Other model selection criteria that have similarities with the AIC and SC are discussed in Lütkepohl (1991).

Implied univariate models

An important aspect of a VARMA model is that it implies the maximum orders of univariate ARMA models for each of the y_{it}, $i = 1,2, \ldots ,m$, components of $\mathbf{Y}_t$. This result draws upon the fact that

$$\Phi_p(B)^{-1} = |\Phi_p(B)|^{-1}\Phi^*_p(B), \tag{2.71}$$

where $\Phi^*_p(B)$ is the adjoint matrix of $\Phi_p(B)$ containing the co-factors. When (2.62) is first multiplied by $\Phi_p(B)^{-1}$ and then both sides are multiplied by $|\Phi_p(B)|$, one obtains

$$|\Phi_p(B)|\mathbf{Y}_t = \Phi^*_p(B)\Theta_q(B)\mathbf{e}_t, \tag{2.72}$$

where $|\Phi_p(B)|$ is a scalar function of B. This expression implies that the m elements of $\mathbf{Y}_t$ can be described by ARMA models with maximum orders $(mp, (m-1)p + q)$. Furthermore, note that the AR parts of the m univariate ARMA models have the same parameters because each equation is multiplied by $|\Phi_p(B)|$. For example, for the bivariate model (2.66), the implied univariate models are ARMA(2,2), where the AR-polynomial reads $1 - (\phi_{11} + \phi_{22})B + (\phi_{11}\phi_{22} - \phi_{12}\phi_{21})B^2$ for both the $y_{1,t}$ and $y_{2,t}$ variables. Of course, it may occur that the AR and MA parts in these ARMA(2,2) models have common factors and hence that smaller models are sufficient.

Similar to (2.71), one may also want to consider the fact that

$$\Theta_q(B)^{-1} = |\Theta_q(B)|^{-1}\Theta^*_q(B), \tag{2.73}$$

where $\Theta^*_q(B)$ is the adjoint matrix of $\Theta_q(B)$, and hence that a VARMA model can be written as

$$\Theta^*_q(B)\Phi_p(B)\mathbf{Y}_T = |\Theta_q(B)|\mathbf{e}_t. \tag{2.74}$$

Each row of (2.74) represents an ARMA model for $y_{i,t}$ with lagged $y_{j,t}$ variables, $j = 1, \ldots ,i-1,i+1, \ldots ,m$. The models in (2.74) are called transfer-function models.

2.3 Unit Roots in Univariate Time Series

Two key properties of many economic time series is that they show trending behaviour and that shocks ε_t can have a permanent effect on the level or the trend of the time series. Examples of such shocks are oil crises or stock market crashes, which for some macroeconomic variables may result in a permanent shift in the mean or trend of their time series. In other words, such processes are then not mean-reverting. A useful class of models that can describe such permanent effects of shocks is a process that is integrated of order 1 or 2, to be denoted as I(1) or I(2). The term 'integrated' originates from the fact that, for example, an I(1) process needs to be differenced once to transform the time series such that shocks do not have a permanent effect. A simple example of an I(1) process is the random walk model

$$y_t = y_{t-1} + \varepsilon_t. \tag{2.75}$$

To understand the permanent effect of shocks, it is useful to rewrite (2.75) as

$$y_t = y_0 + \sum_{i=1}^{t} \varepsilon_i, \tag{2.76}$$

from which it is clear that the effect of ε_i does not die out and that its effect equals that of any $\varepsilon_j, j \neq i$. For (2.76) with $y_0 = 0$, it is easy to see that $E(y_t) = 0$ and

$$E(y_t^2) = t\sigma^2. \tag{2.77}$$

The random walk process is nonstationary since the variance of y_t depends on time. Furthermore, it can easily be derived that $E(y_t y_{t-k}) = (t-k)\sigma^2$ for $k > 0$, and hence that γ_k also depends on time. Notice that this means that the estimated ACF of a random walk process does not die out quickly, and hence that its interpretation is complicated. In terms of (2.7), there does not exist a useful moving average approximation to this random walk model.

With regard to the so-called random walk with drift model

$$y_t = y_{t-1} + \mu + \varepsilon_t, \tag{2.78}$$

it follows that this model can be written as

$$y_t = y_0 + \mu t + \sum_{i=1}^{t} \varepsilon_i. \tag{2.79}$$

The μt component in (2.79) is called the deterministic trend component, while $\Sigma_{i=1}^{t} \varepsilon_i$ is called the stochastic trend component. When this model is compared with the so-called trend-stationary (TS) model

$$y_t = \mu t + \varepsilon_t, \tag{2.80}$$

we observe that the main distinction between (2.79) and (2.80) is that the latter does not contain the stochastic trend component. This means that in the TS model shocks have only a transitory impact. Since processes such as (2.75) and (2.78) can be made stationary (conditional on the requirement that their residual processes are still invertible) by applying the differencing filter $\Delta_1 = (1 - B)$, where Δ_1 is defined by $\Delta_k y_t = y_t - y_{t-k}$ $(k = 1,2, \ldots)$, these processes are called difference-stationary (DS). Finally, notice that, when σ^2 of ε_t is small relative to the value of μ, it may be difficult to distinguish between TS and DS processes (see e.g. Hylleberg and Mizon 1989*a*).

Autoregressive integrated moving average processes

In general, a process y_t is defined to be integrated of order d [I(d)] when $(1 - B)^d y_t$ is a stationary and invertible process while $(1 - B)^{d-1} y_t$ is non-stationary and $(1 - B)^{d+1} y_t$ results in a noninvertible process. Typical values for d that are considered in practice are 0, 1, or 2. An I(d) series can be transformed to stationarity using the differencing filter Δ_1 d times. When $x_t = \Delta_1^d y_t$ is a stationary and invertible ARMA(p,q) process, then the y_t process is an ARIMA(p,d,q) process.

Another terminology for stochastic trend processes, and one that is often used in econometrics, is that an I(1) process has a unit root. The origin of this term is, for example for a random walk model, that the AR polynomial for y_t is $\phi_1(B) = 1 - B$. The solution to the corresponding characteristic polynomial $\phi_1(z) = 0$ is $z = 1$, and hence a random walk process is a unit root process. This unit root property extends to higher-order AR processes because

$$\phi_p(B) = \phi_p(1)B + \phi^*_{p-1}(B)(1 - B), \tag{2.81}$$

where

$$\phi_p(1) = 1 - \phi_1 - \phi_2 - \ldots - \phi_p. \tag{2.82}$$

Hence, when $\Sigma^p_{i=1}\phi_i = 1$, i.e. when the sum of the autoregressive parameters equals unity, $\phi_p(B)$ in (2.81) reduces to $\phi^*_{p-1}(B)(1 - B)$, for which at least one solution of the characteristic equation $\phi^*_{p-1}(z)(1 - z) = 0$ is $z = 1$. This implies that an AR(p) process with a single unit root can also be made stationary using the Δ_1 filter.

Testing for unit roots in the AR polynomial

To test for the presence of a unit root in an AR(p) model, one typically considers the so-called augmented Dickey–Fuller (ADF) regression (see Dickey and Fuller 1979), which is given by

$$\Delta_1 y_t = \mu + \tau t + \rho y_{t-1} + \phi^*_1 \Delta_1 y_{t-1} + \ldots + \phi^*_{p-1} \Delta_1 y_{t-(p-1)} + \varepsilon_t, \tag{2.83}$$

where t denotes a deterministic trend variable. Notice that the ρ parameter for y_{t-1} corresponds to $\phi_p(1)$ (see (2.82)). Under the null hypothesis of one unit root in y_t, the parameter ρ equals zero, and hence a natural test for such a unit root is the t-test for ρ, $t(\rho)$. Under the null hypothesis, the t-ratios for the lagged $\Delta_1 y_t$ variables are asymptotically distributed as standard normal, and hence can be used to select the appropriate order of the ADF model. The $t(\rho)$ test can be calculated for three important different cases, to be denoted as $t(\rho)$ where $\mu = \tau = 0$ in (2.83), $t_\mu(\rho)$ where only $\tau = 0$, and $t_\tau(\rho)$ where $\mu \neq 0$ and $\tau \neq 0$.

Since y_{t-1} in (2.83) is a stochastic trend process under the null hypothesis, the various $t(\rho)$ test statistics are not asymptotically distributed as standard normal. To provide some flavour of asymptotic results, consider the simple random walk process

$$y_t = y_{t-1} + \varepsilon_t,\ y_0 = 0, \tag{2.84}$$

which can be written as $y_t = \Sigma_{i=1}^{t}\varepsilon_i$, and the test regression

$$y_t = \phi y_{t-1} + \varepsilon_t, \tag{2.85}$$

for which one wants to investigate whether $\phi = 1$. Assuming normality of the ε_t process, the maximum likelihood (ML) estimator of ϕ based on n observations is

$$\hat{\phi}_n = \phi + \left(\sum_{t=1}^{n} y_{t-1}\varepsilon_t\right) \Big/ \left(\sum_{t=1}^{n} y_{t-1}^2\right). \tag{2.86}$$

Under the null hypothesis that $\phi = 1$, it is clear that the probability limits of $\Sigma_{t=1}^{n} y_{t-1}\varepsilon_t$ and $\Sigma_{t=1}^{n} y_t^2$ do not exist. In order to give the asymptotic distribution of functions of $\hat{\phi}_n$, one needs the so-called Brownian motion or Wiener process. For a Wiener process W, which is defined on the interval [0,1], it holds that $W(0) = 0$ and that $W(t) - W(s)$ follows a normal distribution with mean 0 and variance $t - s$ with $W(t) - W(s)$ independent of $W(r)$ for any $r \leq s$. In Phillips (1987) it is shown that

$$n^{-2}\sum_{t=1}^{n} y_{t-1}^2 \Rightarrow \int_0^1 W^2(r)\mathrm{d}r, \tag{2.87}$$

where '$\Rightarrow$' denotes convergence in distribution. Furthermore, Phillips (1987) shows that

$$n^{-1}\sum_{t=1}^{n} y_{t-1}\varepsilon_t \Rightarrow \tfrac{1}{2}[W(1)^2 - 1] = \int_0^1 W(r)\mathrm{d}W(r). \tag{2.88}$$

Combining the results in (2.87) and (2.88), it follows that

$$n(\hat{\phi}_n - 1) \Rightarrow \int_0^1 W(r)\mathrm{d}W(r) \Big/ \int_0^1 W^2(r)\mathrm{d}r; \tag{2.89}$$

and for the t-ratio for $\hat{\phi}_n - 1$, which corresponds to the $t(\hat{\rho})$ in (2.83) when there are no lags of $\Delta_1 y_t$, it follows that

$$t(\hat{\rho}) \Rightarrow \int_0^1 W(r)\mathrm{d}W(r) \Big/ \left[\int_0^1 W^2(r)\mathrm{d}r\right]^{1/2}. \tag{2.90}$$

Additional distributional results with the corresponding proofs are given in Phillips (1987), Banerjee *et al.* (1993), and Boswijk (1995), among others.

Some key fractiles of the distributions in (2.89) and (2.90) are tabulated in Fuller (1976). In Table 2.1, I give a subsample of these fractiles for the t-test statistics for the cases where the auxiliary regression (2.83) (with $p = 1$) contains (i) no constant and no trend, (ii) only a constant, and (iii) a constant and a linear deterministic trend. Notice that only the left-side fractiles of the distributions are given since the t-tests for ρ in (2.83) are one-sided t-tests; the alternative hypothesis is H_1: $\rho \leq 0$. Hence the explosive root case is excluded.

The critical values in Table 2.1 can also be used for the case where the auxiliary regression in (2.83) contains lags of $\Delta_1 y_t$, i.e. when $p > 1$ (see Dickey and Fuller 1979). Furthermore, Said and Dickey (1985) show that the same critical values can be used for ARMA processes, provided that the value of p is sufficiently large. The test statistic in (2.89) can be used in case of an AR(1) process. The simulation results in Dickey *et al.* (1986) indicate that the empirical power

TABLE 2.1. Some critical values for Dickey–Fuller test statistics

Test statistic	n	Significance level		
		0.01	0.05	0.10
Regression contains no constant and no trend				
$t(\rho)$	25	−2.66	−1.95	−1.60
	100	−2.62	−1.95	−1.61
	500	−2.58	−1.95	−1.62
Regression contains a constant				
$t_\mu(\rho)$	25	−3.75	−3.00	−2.63
	100	−3.51	−2.89	−2.58
	500	−3.44	−2.87	−2.57
Regression contains a constant and a trend				
$t_\tau(\rho)$	25	−4.38	−3.60	−3.24
	100	−4.04	−3.45	−3.15
	500	−3.98	−3.41	−3.13

Source: Fuller (1976: 371–3). The test regression is (2.83) with p set equal to 1.

of this statistic is higher than that of $t(\rho)$. For higher-order models, this test statistic has to be corrected for higher-order dynamics (see Said and Dickey 1984).

It is possible to design test statistics for the joint hypothesis that $\rho = 1$ and that the deterministic terms in (2.83) are insignificant. Consider the AR(1) model

$$y_t - \mu = \phi_1(y_{t-1} - \mu) + \varepsilon_t. \tag{2.91}$$

When $\phi_1 = 1$, the mean μ is not identified in (2.91) since then the model reads: $y_t = y_{t-1} + \varepsilon_t$. Hence, in the auxiliary regression model

$$y_t = \phi_1 y_{t-1} + \delta + \varepsilon_t, \tag{2.92}$$

with $\delta = \mu(1 - \phi_1)$, the $\delta = 0$ under H_0: $\phi_1 = 1$. A joint F-test can be constructed for H_0^*: $\phi_1 = 1$, $\delta = 0$ in (2.92). Asymptotic distributions and critical values are given in Dickey and Fuller (1981).

The fractiles in Table 2.1 are generated for the data-generating process (DGP) (2.84). Of course, in practice the regression models can be different from this DGP. In Hylleberg and Mizon (1989*a*) the critical values are given for the case where the DGP is a random walk with drift. In practice, one typically considers auxiliary models like (2.83) where both a constant and a trend are included. Simulation results in Campbell and Perron (1991), for example, indicate that inappropriately excluding a trend is worse for the empirical performance of the ADF test than inappropriately including it, in the sense that the loss of power can be much larger than the bias in size. Notice from the expression in (2.92) that a t-test for δ has a nonstandard distribution when $\phi_1 = 1$. Similarly, this applies to a parameter for a deterministic trend (see Nankervis and Savin 1987).

The ADF test for unit roots can be shown to be affected by outlying data points, by mean-shifts, and by breaking trends (see Perron 1989, and Lucas 1995). In Franses and Haldrup (1994) it is shown that neglecting additive outliers may bias the ADF test results towards incorrect rejection of the null hypothesis. This implies that one may too often find that time series are stationary when additive outliers are not taken care of in the auxiliary regression via the inclusion of dummy variables. In order to reduce the impact of aberrant data for testing for unit roots, Lucas (1995) proposes to use robust estimation methods to the auxiliary regression in (2.83).

A shift in the mean of a time series can be shown to bias the ADF test results towards nonrejection. For example, consider the model

$$y_t = \phi_1 y_{t-1} + \mu_1 I_{t \leq b} + \mu_2 I_{t>b} + \varepsilon_t, \tag{2.93}$$

where $I_{t \leq b}$ is an indicator function which takes a value 1 when $t \leq b$, and $I_{t>b}$ is a similar indicator function for $t > b$. In Perron (1989) it is shown that the OLS estimator for ϕ_1 in the AR(1) model, where the mean shift from $\mu_1 / (1 - \phi_1)$ to $\mu_2 / (1 - \phi_2)$ is neglected, is biased towards unity, and that part of this bias depends on the difference between μ_1 and μ_2. Hence, it can be of importance to allow for mean shifts when testing for unit roots (see Perron 1990). Similar results can be derived for the case where the trend parameter τ in (2.83) displays one or more breaks (see Perron 1989).

Integrated processes with d unequal to 1

The above discussion on tests for unit roots in the autoregressive polynomial assumes that y_t is at most I(1), and that the hypothesis of interest is whether y_t is either I(1) or I(0). It should be mentioned that, when this hypothesis is tested for the case when the y_t series is I(2), the asymptotic results in (2.89) and (2.90) do not hold (see e.g. Haldrup 1994*a*,*b*). The appropriate approach in the case where one suspects that y_t may be I(2) is to analyse the unit root properties of $\Delta_1 y_t$ first and, where this series is found to be stationary, to investigate y_t itself in a second step (see also Pantula 1989). Only when this sequence of tests is followed can it be shown that the asymptotic results in, e.g., (2.89) continue to hold.

There are economic time series of which the autocorrelation function suggests that the time series should be differenced, while the ACF of the $\Delta_1 y_t$ series indicates that $\Delta_1 y_t$ seems overdifferenced. One way of describing such time series is to use the so-called fractionally integrated ARIMA model (ARFIMA), where the differencing parameter d can take values other than 0, 1, and 2 (see Granger and Joyeux 1980, and Hosking 1981). In the ARFIMA model, the differencing operator $(1 - B)^d$ is defined by the expansion

$$(1-B)^d = 1 - dB - (1/2)d(1-d)B^2 - (1/6)d(1-d)(2-d)B^3 \ldots$$
$$\ldots - (1/j!)d(1-d)(2-d)\ldots((j-1)-d)B^j - \ldots, \tag{2.94}$$

where $j \to \infty$. When $0 < d < 0.5$, the ARFIMA process is long-memory stationary, and when $d \geq 0.5$, the process is nonstationary. (See Baillie (1996) for a recent survey of modelling and estimation aspects of ARFIMA models.)

Unit roots in the MA polynomial

Most attention in the econometrics and time series literature has been spent on testing for unit roots in the AR polynomial. Indeed, such a unit root implies that the y_t process contains a stochastic trend, which in a sense can be viewed as an unforecastable process: the one-step-ahead forecast $f_{t,1}$ from a random walk model $y_t = y_{t-1} + \varepsilon_t$ is y_t. Notice that the standard white noise process is unforecastable too. It can however be the case that a unit root test incorrectly suggests that the time series should be differenced. In that case, such an over-differenced time series will have a unit root in the MA polynomial. Hence, testing for such an MA unit root may then be considered as a check for over-differencing.

In the literature there are several approaches to test for an MA unit root. (See Breitung (1994) for a recent survey.) Testing for MA unit roots is somewhat complicated, since the estimator for the MA parameter in for example

$$y_t = \varepsilon_t - \theta_1 \varepsilon_{t-1} \tag{2.95}$$

is biased downwards when $\theta_1 = 1$ (see Plosser and Schwert 1977). One approach may then be to recognize the duality between MA and AR processes, and to base a test on (2.95) when considered in an AR framework (see Tsay 1993). An alternative method, which is followed in Franses (1995*a*), is to use the result that the autocorrelations in an MA time series with a unit root sum to $-\frac{1}{2}$. Suppose the autocovariances of the I(0) series y_t are denoted as g_k and those of $\Delta_1 y_t$ as γ_k; then

$$\gamma_k = E[(y_t - y_{t-1})(y_{t-k} - y_{t-k-1})] = 2g_k - g_{k-1} - g_{k+1} \tag{2.96}$$

Given this connection between γ_k and g_j, $j = k-1, k, k+1$, it follows that

$$\sum_{i=1}^{\infty} \rho_i = (g_1 - g_0)/(2g_0 - 2g_1) = -\tfrac{1}{2}, \tag{2.97}$$

where ρ_i is the ith autocorrelation of $\Delta_1 y_t$. The results in T. W. Anderson and Walker (1964) can then be used to construct test statistics for MA unit roots that follow standard normal distributions under the null hypothesis.

2.4 Unit Roots in Multivariate Time Series

When constructing multivariate time series models, it is important to take account of possible nonstationarity properties of univariate time series.

If univariate ARMA models for time series y_{it} have unit roots in the AR polynomials, and if a set of such series can be described using, say, a VAR model

$$\mathbf{Y}_t = \mu + \Phi_1 \mathbf{Y}_{t-1} + \ldots + \Phi_p \mathbf{Y}_{t-p} + \mathbf{e}_t, \tag{2.98}$$

where $\mathbf{Y}_t = (y_{1,t}, \ldots, y_{m,t})'$, then there is at least one $z = 1$ solution to

$$|\mathbf{I}_m - \Phi_1 z - \ldots - \Phi_p z^p| = 0. \tag{2.99}$$

An obvious and often used approach is to take first differences Δ_1 of each of the univariate time series and to consider a VAR model for only Δ_1 differenced time series. It may however occur that such a model is misspecified when there are linear relationships between the untransformed time series which convey information on long-run 'equilibrium' relationships. In that case we say that several time series are cointegrated (see Engle and Granger 1987, and Hylleberg and Mizon 1989*b*).

To formalize this discussion, consider the following simple bivariate model for $\{y_{1,t}, y_{2,t}\}$, i.e.

$$y_{1,t} + \beta y_{2,t} = v_t, \quad v_t = \rho_1 v_{t-1} + \varepsilon_{1,t} \quad 0 \leq \rho_1 \leq 1 \text{ and } \beta \neq 0 \tag{2.100}$$

$$y_{1,t} + \delta y_{2,t} = w_t, \quad w_t = \rho_2 w_{t-1} + \varepsilon_{2,t} \quad 0 \leq \rho_2 \leq 1 \text{ and } \delta \neq 0 \tag{2.101}$$

where it is assumed that $\beta \neq \delta$, and where $\varepsilon_{1,t}$ and $\varepsilon_{2,t}$ are standard white noise error processes. The bivariate model in (2.100) and (2.101) reflects that two linear combinations of $y_{1,t}$ and $y_{2,t}$ are AR(1) processes. Notice that these two equations can be written as a VAR(1) model for $(y_{1,t}, y_{2,t})'$; see also (2.102). An interpretation of these linear combinations depends on the values of ρ_1 and ρ_2. In this VAR(1) model, there are three possible situations with respect to these values and their impact on the interpretation of (2.100) and (2.101). The first is that $\rho_1 = \rho_2 = 1$, which implies that all linear combinations of $y_{1,t}$ and $y_{2,t}$ are random walk variables. In turn, this implies that $y_{1,t}$ and $y_{2,t}$ are I(1) variables and hence nonstationary themselves. This case is called the 'no cointegration' case. A second situation is that both ρ_1 and ρ_2 are unequal to 1, which implies that linear combinations of $y_{1,t}$ and $y_{2,t}$ are stationary AR(1) processes, and hence that $y_{1,t}$ and $y_{2,t}$ themselves are stationary variables. This is the case where both time series are stationary, and hence do not need to be differenced in order to obtain stationary time series. The final case occurs when either ρ_1 or ρ_2 is equal to 1 while the other is not. In that case there is a linear combination between $y_{1,t}$ and $y_{2,t}$ that is a stationary AR(1) process, while the other combination is a random walk. This implies that, although both $y_{1,t}$ and $y_{2,t}$ individually are I(1) time series, since any sum of an I(1) and an I(0) series yields an I(1) series there is one linear combination that is stationary. This third case is called the 'cointegration' case. From an economic point of view, this latter case is of interest if one wants to describe long-run equilibria between nonstationary variables.

Testing for cointegration

There are several methods to test for cointegration between a set of time series variables.

The Engle–Granger method

The often used Engle–Granger (1987) method amounts to estimating β or δ in (2.100) or (2.101), and then testing whether either ρ_1 or ρ_2 is equal to 1 using a test regression such as (2.83). This method is particularly useful when one analyses two time series; it may become less useful for more than two time series. This is because the number of possible cointegration relations increases with the number of time series, which implies an increasing ambiguity in determining the empirical validity of models such as (2.100) and (2.101). Asymptotic theory for the Engle–Granger method is given in Phillips and Ouliaris (1990).

In Table 2.2 some fractiles are displayed for the Engle–Granger (1987) method. From this table it can be seen that the ADF critical values in cases of more than one variable shift to the left, compared with those in Table 2.1. For small sample sizes one can use the critical values tabulated in MacKinnon (1991).

TABLE 2.2. Asymptotic critical values for the Engle–Granger (1987) method[a]

No. of variables m	Significance level		
	0.01	0.05	0.10
Regression contains no constant and no trend			
2	–3.39	–2.76	–2.45
3	–3.84	–3.27	–2.98
4	–4.30	–3.74	–3.44
Regressions contain a constant but no trend			
2	–3.96	–3.37	–3.07
3	–4.31	–3.77	–3.45
4	–4.73	–4.11	–3.83
Regressions contain a constant and a trend			
2	–4.36	–3.80	–3.52
3	–4.65	–4.16	–3.84
4	–5.04	–4.49	–4.20

[a] The test regression in (2.83) where the variable of interest is u_t, where u_t is the residual process from the regression of $y_{1,t}$ on $y_{2,t}$ to $y_{m,t}$, where both regressions may include a constant and/or a trend.

Source: Phillips and Ouliaris (1990).

Johansen cointegration testing approach

An alternative approach to investigate cointegration in a VAR model is the method developed in Johansen (1988, 1989, 1991; see also Johansen and Juselius 1990). To illustrate the core of the Johansen method, consider again the example in (2.100) and (2.101), which can be written as

$$\begin{bmatrix} 1 & \gamma \\ 1 & \delta \end{bmatrix} \begin{bmatrix} y_{1,t} \\ y_{2,t} \end{bmatrix} = \begin{bmatrix} \rho_1 & \beta\rho_1 \\ \rho_2 & \beta\rho_2 \end{bmatrix} \begin{bmatrix} y_{1,t-1} \\ y_{2,t-1} \end{bmatrix} + \begin{bmatrix} \varepsilon_{1,t} \\ \varepsilon_{2,t} \end{bmatrix}. \tag{2.102}$$

Simple manipulations enable one to rewrite (2.102) as

$$\begin{bmatrix} y_{1,t} - y_{1,t-1} \\ y_{2,t} - y_{2,t-1} \end{bmatrix} = \Pi \begin{bmatrix} y_{1,t-1} \\ y_{2,t-1} \end{bmatrix} + \begin{bmatrix} \varepsilon^*_{1,t} \\ \varepsilon^*_{2,t} \end{bmatrix}, \tag{2.103}$$

with

$$\Pi = \begin{bmatrix} (\delta\rho_1 - \beta\rho_2 - \delta + \beta)/(\delta - \beta) & \delta\beta(\rho_1 - \rho_2)/(\delta - \beta) \\ (\rho_2 - \rho_1)/(\delta - \beta) & (\delta\rho_2 - \beta\rho_1 - \delta + \beta)/(\delta - \beta) \end{bmatrix}, \tag{2.104}$$

where $\varepsilon^*_{1,t}$ and $\varepsilon^*_{2,t}$ are functions of the $\varepsilon_{1,t}$ and $\varepsilon_{2,t}$ processes in (2.102). When the $y_{1,t}$ and $y_{2,t}$ series are stationary, i.e. when both ρ_1 and ρ_2 are unequal to 1, the Π matrix in (2.104) has full rank 2. When $\rho_1 = \rho_2 = 1$, it can easily be observed that all elements of Π have value zero, and that the rank of Π is equal to 0. In the cointegration case, when, for example, $\rho_1 = 1$ and $\rho_2 \neq 1$, the Π matrix can be written as

$$\Pi = \begin{bmatrix} (-\beta\rho_2 + \beta)/(\delta - \beta) & \delta\beta(1 - \rho_2)/(\delta - \beta) \\ (\rho_2 - 1)/(\delta - \beta) & (\delta\rho_2 - \delta)/(\delta - \beta) \end{bmatrix}$$

$$= \begin{bmatrix} \beta(1 - \rho_2)/(\delta - \beta) \\ (\rho_2 - 1)/(\delta - \beta) \end{bmatrix} [1, \delta], \tag{2.105}$$

although this decomposition is not unique. It is clear that the Π matrix in (2.105) can be decomposed as $\gamma\alpha'$, where γ and α are (2×1) vectors, which means that the rank of Π is 1. The α vector contains the cointegration parameters, which are $(1, \delta)$ here, and the vector γ contains the so-called adjustment parameters. Model (2.103) with Π as in (2.105) is then an error correction model (ECM), since it becomes

$$\Delta_1 y_{1,t} = \gamma_1(y_{1,t-1} + \delta y_{2,t-1}) + \varepsilon^*_{1,t} \tag{2.106}$$

$$\Delta_1 y_{2,t} = \gamma_2(y_{1,t-1} + \delta y_{2,t-1}) + \varepsilon^*_{2,t}, \tag{2.107}$$

where γ_1 and γ_2 are the adjustment parameters.

In general, for $(m \times 1)$ vector time series there is cointegration when each univariate series is at most an I(1) series and the rank of the relevant Π matrix (as in (2.105)) has rank r with $0 < r < m$. Furthermore, when there is cointegration, the VAR model can be written in error correction form. (For more details on this so-called Granger representation theorem, see Johansen (1991).)

The Johansen maximum likelihood cointegration testing method amounts to testing the rank of the matrix Π in the rewritten version of (2.98), i.e.

$$\Delta_1\mathbf{Y}_t = \mu + \Gamma_1\Delta_1\mathbf{Y}_{t-1} + \ldots + \Gamma_{p-1}\Delta_1\mathbf{Y}_{t-p+1} + \Pi\mathbf{Y}_{t-p} + \mathbf{e}_t, \tag{2.108}$$

where

$$\Gamma_i = (\Phi_1 + \Phi_2 + \ldots + \Phi_i) - \mathbf{I}_m \text{ for } i = 1,2,\ldots,p-1,$$
$$\Pi = \Phi_1 + \Phi_2 + \ldots + \Phi_p - \mathbf{I}_m,$$

where Π is the matrix of interest since it corresponds to the level $\mathbf{Y}_t$ variables. Notice that (2.108) corresponds to the ADF regression in (2.83), and hence that the Johansen method can be seen as multivariate extension of the ADF approach. In fact, for the VAR(p) model with autoregressive polynomial $\Pi_p(B)$, one can write

$$\Pi_p(B) = \Pi(1)B^p - \Gamma_{p-1}(B)(1 - B), \tag{2.109}$$

where Π in (2.108) equals $-\Pi(1)$ and where $\Gamma_{p-1}(B)$ is a $(p - 1)$th order matrix polynomial as in (2.108). Obviously, in the vector AR model, (i) Π can be the zero-matrix; i.e., the rank of Π equals 0; (ii) Π can have full rank m; or (iii) $0 < \text{rank}\Pi < m$. Hence, the rank of Π determines the number of cointegrating relations; cf. (2.105). If there are r cointegration relations, it holds that $\Pi = \gamma\alpha'$, where γ and α are $(m \times r)$ matrices. Compare this with (2.105), where Π is a (2×2) matrix, $r = 1$, and γ and α are (2×1) matrices.

To investigate the rank of the Π matrix in (2.108), Johansen proposes to use the reduced rank regression technique based on canonical correlations. This amounts to the following computations, which I give in some detail here since this method will often be used in forthcoming chapters.

First, we regress $\Delta_1\mathbf{Y}_t$ and $\mathbf{Y}_{t-p}$ on a constant and the $\Delta_1\mathbf{Y}_{t-1}$ to $\Delta_1\mathbf{Y}_{t-p+1}$ variables, giving the $(m \times 1)$ vectors of residuals r_{0t} and r_{1t} and the residual product matrices

$$\mathbf{S}_{ij} = (1/n)\sum_{t=1}^{n}\mathbf{r}_{it}\mathbf{r}_{jt}', \text{ for } i,j = 0,1. \tag{2.110}$$

The next step is to solve the eigenvalue problem

$$|\lambda\mathbf{S}_{11} - \mathbf{S}_{10}\mathbf{S}_{00}^{-1}\mathbf{S}_{01}| = 0, \tag{2.111}$$

which gives the eigenvalues $\hat{\lambda}_1 \geq \ldots \geq \hat{\lambda}_m$ and the corresponding eigenvectors $\hat{\mathbf{v}}_1$ to $\hat{\mathbf{v}}_m$. In principle, a test for the rank of Π amounts to testing whether some or all λ_i are equal to unity. These eigenvalues can be interpreted as the partial correlations of $\mathbf{Y}_{t-p}$ and $\Delta_1\mathbf{Y}_t$ after the corrections for deterministic terms and lagged $\Delta_1\mathbf{Y}_t$ variables. In Johansen (1988, 1989) test statistics are given for the number of cointegration relations. The first is the likelihood ratio test statistic

$$\text{trace} = -n\sum_{i=r+1}^{m}\log(1 - \hat{\lambda}_i), \tag{2.112}$$

for the null hypothesis of at most r cointegration relations. Another test is given by testing the significance of the estimated eigenvalues themselves, or

$$\lambda_{\max} = -n\log(1 - \hat{\lambda}_r), \tag{2.113}$$

which can be used to test the null hypothesis of $r-1$ against r cointegration relations. Standard distributional results are not valid since the null hypothesis of r cointegrating relations involves nonstationary variables. Asymptotic distributions of the test statistics are derived in Johansen (1989, 1991). Critical values are given in Johansen and Juselius (1990) and in Osterwald–Lenum (1992). Some of these critical values are given in Table 2.3.

TABLE 2.3. Asymptotic critical values for the Johansen cointegration test statistics[a]

$l = m - r$[b]	Unrestricted constant[c]			Restricted constant[c]		
	0.20	0.10	0.05	0.20	0.10	0.05
λ_{max}						
1	4.82	6.50	8.18	5.91	7.52	9.24
2	10.77	12.91	14.90	11.54	13.75	15.67
3	16.51	18.90	21.07	17.40	19.77	22.00
4	22.16	24.78	27.14	22.95	25.56	28.14
Trace						
1	4.82	6.50	8.18	5.91	7.52	9.24
2	13.21	15.66	17.95	15.25	17.85	19.96
3	25.39	28.71	31.52	28.75	32.00	34.91
4	41.65	45.23	48.28	45.65	49.65	53.12

[a] The statistics are given in (2.112) and (2.113).

[b] The m is the number of elements in the vector $\mathbf{Y}_t$, r is the number of cointegration relationships.

[c] In the restricted constant case it is assumed that $\gamma'\mu = 0$, and hence that the constant term enters the cointegration relation. See Johansen (1991) for more details.

Source: Osterwald-Lenum (1992).

In case of r cointegrating relations, one may wish to test hypotheses on the cointegrating vectors. To test for linear restrictions on the cointegrating vectors in α, define the $(m \times q)$ matrix $\mathbf{H}$, where $r \leq q \leq m$, which reduces α to the $(q \times r)$ parameter matrix ϕ, or $\alpha = \mathbf{H}\phi$. For brevity, these restrictions are denoted by their matrix $\mathbf{H}$. Assuming the validity of the restrictions $\mathbf{H}$, one compares the estimated eigenvalues $\hat{\xi}_i$, $i = 1, \ldots, r$ from

$$|\xi \mathbf{H}'\mathbf{S}_{11}\mathbf{H} - \mathbf{H}'\mathbf{S}_{10}\mathbf{S}_{00}^{-1}\mathbf{S}_{01}\mathbf{H}| = 0, \tag{2.114}$$

with the $\hat{\lambda}_i$ from (2.111), $i = 1, \ldots, r$, via the test statistic

$$Q = n \sum_{i=1}^{r} \log\{ (1 - \hat{\xi}_i) / (1 - \hat{\lambda}_i) \}. \tag{2.115}$$

Under the null hypothesis, and conditional on the correct value of r, the test statistic Q asymptotically follows a $\chi^2(r(m-q))$ distribution.

The Boswijk method

A third method to test for cointegration, which is proposed in Boswijk (1992), is based on the recognition that cointegration implies error correction. Hence a test for the significance of error correction variables can indicate whether there is cointegration. Another specific aspect of this Boswijk method is the assumption that an error correction model for $y_{1,t}$ is of key interest, and that it can be assumed that the variables $y_{2,t}$ to $y_{m,t}$ are weakly exogenous for the cointegration parameters of interest. (See Engle *et al.* (1983) for a discussion of various exogeneity concepts.) Consider the bivariate AR case in (2.106) and (2.107) again; i.e.,

$$\Delta_1 y_{1,t} = \gamma_1(y_{1,t-1} - \delta y_{2,t-1}) + \varepsilon^*_{1,t}$$
$$\Delta_1 y_{2,t} = \gamma_2(y_{1,t-1} - \delta y_{2,t-1}) + \varepsilon^*_{2,t},$$

where δ is the cointegration parameter. In Johansen (1992*a*) it is shown that, if $\gamma_2 = 0$, the $y_{2,t}$ process is weakly exogenous for δ. If that is the case, one can perform cointegration analysis in a conditional ECM for $y_{1,t}$, which can be written as

$$\Delta_1 y_{1,t} = \gamma_1 y_{1,t-1} + \psi y_{2,t-1} + v\Delta_1 y_{2,t} + \varepsilon^{**}_{1,t}, \tag{2.116}$$

where $\psi = \delta\gamma_1$ (see Boswijk 1994 for details). This model can be estimated using OLS, and the significance of the error correction variable can be evaluated using a Wald test statistic for $\gamma_1 = 0$ and $\psi = 0$. Asymptotic results are given in Boswijk (1992, 1994). Since I use this method in Chapter 9, some critical values are displayed in Table 2.4.

TABLE 2.4. Asymptotic critical values for the Boswijk cointegration test

No. of variables m	Significance level		
	0.20	0.10	0.05
Regression contains no constant and no trend			
2	4.73	6.35	7.95
3	7.37	9.40	11.37
4	9.89	12.12	14.31
Regression contains a constant but no trend			
2	7.52	9.54	11.41
3	9.94	12.22	14.38
4	12.38	14.93	17.18
Regression contains a constant and a trend			
2	10.16	12.32	14.28
3	12.49	14.91	17.20
4	14.84	17.57	19.81

Source: Boswijk (1994).

In the case where the null hypothesis of no cointegration can be rejected, one can apply nonlinear squares (NLS) to estimate δ. A test for the weak exogeneity of $y_{2,t}$ can be performed by regressing $\Delta_1 y_{2,t}$ on the estimated error correction term, obtained in the first round, and lagged $\Delta_1 y_{2,t}$ and $\Delta_1 y_{1,t}$ variables, and testing the significance of this error correction variable (see Boswijk 1994). Conditional on cointegration, the relevant test statistic follows a standard χ^2 distribution. The same applies to tests on the values of γ_1 and δ in (2.116). I return to this method in more detail in Chapter 9.

Stochastic trends

Similar to the univariate case, a multivariate time series process with unit roots has stochastic trends. Formally, when there are r cointegration relations among m variables stacked in $\mathbf{Y}_t$, there are $m - r$ common stochastic trends in the system (see Stock and Watson 1989). For some purposes it is useful not only to obtain estimates of the cointegration relations, but also to obtain insight into the driving nonstationary forces, which are the stochastic trends.

There are several methods to estimate these stochastic trends, and two of them will briefly be discussed. The first method is proposed in Box and Tiao (1977), which in a sense searches for the most nonstationary linear combinations among the elements of $\mathbf{Y}_t$. For a VAR(1) process, the Box–Tiao method considers the eigenvalue problem (similar to (2.111))

$$|\lambda \mathbf{I}_m - (\mathbf{Y}_t'\mathbf{Y}_t)^{-1}\mathbf{Y}_t'\mathbf{Y}_{t-1}(\mathbf{Y}_{t-1}'\mathbf{Y}_{t-1})^{-1}\mathbf{Y}_{t-1}'\mathbf{Y}_t| = 0, \qquad (2.117)$$

where $\mathbf{Y}_t$ is corrected for its mean. (See Bewley *et al.* (1994) for some details.) The eigenvector that corresponds to the largest eigenvalue is the most nonstationary linear combination that can be constructed from the m $y_{i,t}$ series. This linear combination can be considered as the first stochastic trend. The other $m - r - 1$ stochastic trends can be found via the eigenvectors that correspond to the next $m - r - 1$ largest eigenvalues from (2.117).

In Gonzalo and Granger (1995) the common stochastic trends are found by solving the eigenvalue problem

$$|\lambda \mathbf{S}_{00} - \mathbf{S}_{01}\mathbf{S}_{11}^{-1}\mathbf{S}_{10}| = 0, \qquad (2.118)$$

which is the dual version of (2.111), and where the $\mathbf{S}_{ij}$ matrices are defined by (2.110). The solutions to (2.118) are the eigenvalues $\hat{\lambda}_1 \geq \ldots \geq \hat{\lambda}_m$ and eigenvectors $\hat{\mathbf{w}}_1, \ldots, \hat{\mathbf{w}}_m$. Gonzalo and Granger (1995) show that in the case of r cointegration relations the stochastic trend variables can be constructed as $\hat{\mathbf{w}}_{m-r+1}'\mathbf{Y}_t$ to $\hat{\mathbf{w}}_m'\mathbf{Y}_t$.

Some final practical issues

The above cointegration methods have been evaluated in many simulation studies that focus on the size and power properties of the methods where, for example, the lag order of the VAR has to be estimated from the data. For example, Boswijk and Franses (1992) document that too much cointegration can be found when the model order is too small, while the power of the tests decreases when there are too many lags in the VAR model. Boswijk and Franses (1992) recommend using the Schwarz model selection criterion in combination with diagnostic tests for residual autocorrelation in order to select the order of the VAR model.

In general, the power of univariate and multivariate unit root tests is typically found to be not very high. In practice, therefore, it is suggested that one not use the critical values in a rigid way, but for example also to allow for 20 per cent significance levels. Indeed, when spurious cointegration relationships are thereby obtained, this may be indicated by the values and relevance of adjustment parameters. Empirical examples in forthcoming chapters will illuminate these practical considerations.

Conclusion

In this chapter I have reviewed several concepts and some notational and computational issues that are useful for the content of subsequent chapters. The review was, of course, far from complete, but it should suffice for the present purposes. In the next chapter I will provide a brief introduction to seasonal time series, which constitute the prime focus of this book.

3
An Introduction to Seasonal Time Series

In this chapter I give an introduction to seasonal time series by discussing typical aspects of such time series. This discussion centres around graphs of some quarterly observed time series which serve as running examples throughout the chapter, around several auxiliary regressions, and around the estimated autocorrelation functions of (transformations) of the data. The sample series considered are the industrial production index for the USA, the number of unemployed in Canada, real GNP in Germany, and real total investment in the UK. To save space, I consider only quarterly observed time series. In forthcoming chapters I include additional time series; they display roughly similar characteristics to those considered here. The sources of the data used in this chapter are given in the Data Appendix. The data analysed here are all seasonally unadjusted. Furthermore, all are analysed after the log transformation has been applied, except for the unemployment in Canada series.

In Section 3.1 I first present a definition of seasonality as it is given in Hylleberg (1992). Next, I describe the notation that will be used throughout this book for seasonal time series. Finally, I present some graphs and the results of several auxiliary regressions. The regressions are used to indicate the relative importance of seasonal fluctuations in macroeconomic time series. In Section 3.2, the four quarterly sample series are analysed along the lines proposed in Box and Jenkins (1970). The seasonal ARIMA (SARIMA) model class is introduced, and four SARIMA models are fitted to the example series. Finally, I briefly discuss the class of seasonal unobserved component models.

3.1 Characteristics of Seasonal Time Series

A seasonally observed time series is a time series that is regularly observed during a time interval that is shorter than a year. It can relate to quarterly and monthly observations as well as hourly or daily observations. In this book the main focus is on quarterly data, although monthly data are sometimes discussed. The concept of a seasonal time series is often used for seasonally observed time series that display noteworthy seasonal fluctuations. However, not all seasonally observed time series necessarily display seasonality; for example, monthly observed nominal interest rates usually do not. From

Chapter 7 onwards the focus will be on periodic time series. For the moment, I define a periodic time series as a seasonal time series with the specific property that it can be described using different time series models for different seasons.

A definition of seasonality

The concept of a seasonal time series obviously depends on a definition of seasonality. However, similar to the definition of a trend, which is usually defined as a linear deterministic regressor or a smooth function of time in regression models, the definition of seasonality very much corresponds to the various approaches to modelling seasonality. Here I concur with the definition of seasonality as it is given in Hylleberg (1992) (see also Hylleberg 1986), which is that

> Seasonality is the systematic, although not necessarily regular, intra-year movement caused by changes of the weather, the calendar, and timing of decisions, directly or indirectly through the production and consumption decisions made by the agents of the economy. These decisions are influenced by the endowments, the expectations and the preferences of agents, and the production techniques available in the economy. (Hylleberg 1992: 4)

An important phrase in this definition of seasonality is that, although part of the seasonal fluctuations can be approximately deterministic because of, say, calendar and weather effects, some part of the seasonal fluctuations may be caused by the behaviour of economic agents. An example of such behaviour is that of producers of products that are harvested seasonally, such as coffee or rubber, who smooth their output using inventories, in turn generating a seasonal pattern in inventories. A possible result is that there may be almost no seasonal fluctuations in the corresponding commodity prices. Hence, economic agents may take seasonal fluctuations in some variables into account when making plans and forming expectations for other variables. These seasonal patterns may then change, not only because of changing weather conditions or the like, but also because of changing habits and changing utility functions of economic agents. For example, improved storage capacities and production in greenhouses allows one to buy many vegetables in all seasons. Another example is that nondurable consumption patterns may change when preferences for certain holiday seasons change. As final example, the start of the winter clearance sales, at least in countries where such sales occur, can depend on the current status of the economy. In recession periods typically the clearance sales start earlier, i.e. some weeks before Christmas instead of in the first weeks of January, to get rid of the remaining stock in order to be able to finance new stock. In sum, for some economic time series it may be the case that the seasonal fluctuations are not constant over time.

Notation

Before turning to the analysis of graphs and auxiliary regressions, additional notation has to be introduced. In general, I use y_t for a quarterly or monthly observed time series, where t runs from 1 to n. In specific empirical examples, the y_t will always denote the seasonal time series that is then studied; i.e., I do not introduce different notation for different variables unless strictly necessary. The y_t series is assumed to be observed for S seasons per year, where $S = 4$ or 12, for example. The seasons are denoted as season s, where $s = 1,2,\ldots,S$. For convenience, I often assume that y_t is observed for N complete years, i.e. $n/S = N$. The annual index is T with $T = 1,2,\ldots,N$.

Since the discussion focuses on seasonal time series, where the data may have different characteristics across seasons, it is convenient to consider the notation $Y_{s,T}$, which denotes the observation on y_t in season s in year T. When there are n observations covering N years, i.e. $t = 1,2,\ldots,n = SN$, the annually observed $Y_{s,T}$ series are defined as

$$\mathbf{Y}_1' = \begin{bmatrix} y_1 \\ y_{S+1} \\ y_{2S+1} \\ \vdots \\ y_{(N-1)S+1} \end{bmatrix}, \quad \mathbf{Y}_2' = \begin{bmatrix} y_2 \\ y_{S+2} \\ y_{2S+2} \\ \vdots \\ y_{(N-1)S+1} \end{bmatrix}, \quad \ldots, \quad \mathbf{Y}_S' = \begin{bmatrix} y_S \\ y_{2S} \\ y_{3S} \\ \vdots \\ y_{NS} \end{bmatrix} \tag{3.1}$$

In some occasions (see Chapters 7, 8, and 9) I will analyse multiple time series models for the $(S \times 1)$ vector process $\mathbf{Y}_T$, which is defined as $\mathbf{Y}_T = (Y_{1,T}, Y_{2,T}, \ldots, Y_{S,T})'$. Similarly, one can denote the error process ε_t as $\varepsilon_T = (\varepsilon_{1,T}, \varepsilon_{2,T}, \ldots, \varepsilon_{S,T})'$, where the $\varepsilon_{s,T}$ components are defined similar to (3.1). A seasonally varying constant term is denoted as the $(S \times 1)$ vector $\mu = (\mu_1, \mu_2, \ldots, \mu_S)'$. Notice from (3.1) that, when $y_t = \mu_s + \varepsilon_t$ is an adequate model for y_t, the μ_s can be estimated by $(1/N)\Sigma_{T=1}^{N} Y_{s,T}$ $(s = 1,2,\ldots,S)$. An alternative regression-based notation considers the use of S seasonal dummy variables $D_{s,t}$. For quarterly time series these seasonal dummy variables are defined by the sth row of the $(4 \times n)$ matrix,

$$\mathbf{D} = \begin{bmatrix} \mathbf{D}_1' \\ \mathbf{D}_2' \\ \mathbf{D}_3' \\ \mathbf{D}_4' \end{bmatrix} = \begin{bmatrix} 1\,0\,0\,0\,1\,0\,0\,0\,1\,0\,0\,0 \ldots 1\,0\,0\,0 \\ 0\,1\,0\,0\,0\,1\,0\,0\,0\,1\,0\,0 \ldots 0\,1\,0\,0 \\ 0\,0\,1\,0\,0\,0\,1\,0\,0\,0\,1\,0 \ldots 0\,0\,1\,0 \\ 0\,0\,0\,1\,0\,0\,0\,1\,0\,0\,0\,1 \ldots 0\,0\,0\,1 \end{bmatrix} \tag{3.2}$$

The regression model to estimate the seasonal constants μ_s in $y_t = \mu_s + \varepsilon_t$ can now be represented by

$$y_t = \sum_{s=1}^{S} \mu_s D_{s,t} + \varepsilon_t \tag{3.3}$$

Throughout this book I will often use the notation $y_t = \mu_s + \varepsilon_t$ to indicate a regression model such as (3.3).

A final remark on notation concerns the backward shift operator B. In the previous chapter it was defined as $B^k y_t = y_{t-k}$. In case of annually observed time series such as $Y_{s,T}$, the backward shift operator is similarly defined as $B^k Y_{s,T} = Y_{s,T-k}$; i.e., it operates on the time series in the same season. In order to avoid confusion, I do not introduce an alternative operator for the time series in the same year. For example, the $Y_{s,T}$ one season lagged will be denoted as $Y_{s-1,T}$ for $s = 1, \ldots, S$, where $Y_{0,T}$ is defined as $Y_{S,T-1}$. In general, I denote

$$Y_{-j,T} = Y_{S-j,T-1} \quad \text{for } j = 0,1,2,\ldots,S-1,\ T = 2,\ldots,N. \tag{3.4}$$

Since differencing filters like Δ_1 and Δ_S involve the backward shift operator, the above notation carries over to differencing filters. The Δ_S filter defined by $\Delta_S y_t = y_t - y_{t-S}$ corresponds to the Δ filter for annually observed time series, where Δ is defined by

$$\Delta Y_{s,T} = Y_{s,T} - Y_{s,T-1}, \quad s = 1,2,\ldots,S \text{ and } T = 2,\ldots,N. \tag{3.5}$$

Again to avoid confusion, I do not introduce a new differencing filter for the $Y_{s,T}$ series which is an equivalent to the Δ_1 filter for the y_t series. Instead, I define that

$$\Delta_1 y_t \Leftrightarrow \begin{array}{ll} Y_{s,T} - Y_{s-1,T} & \text{for } s = 2,3,\ldots,S \\ Y_{s,T} - Y_{S,T-1} & \text{for } s = 1. \end{array} \tag{3.6}$$

Along similar lines, one can find expressions for any differencing filter $(1 - B^k)$ for some k.

Some graphs

A first and natural impression of the properties of seasonal time series can be obtained by depicting their time series plots (see also Hylleberg 1994). The first time series to be considered here is the log of the quarterly industrial production index for the USA, observed from 1960.1–1991.4. Denoting this time series as y_t, the graph of y_t is given in Fig. 3.1, while the four graphs of $Y_{s,T} - Y_{s-1,T}$ ($s = 2,3,4$) and $Y_{1,T} - Y_{4,T-1}$ are given in Fig. 3.2. Hence the graphs in Fig. 3.2 concern the approximate quarterly growth rates of y_t, i.e. $y_t - y_{t-1}$, in each of the four quarters.

The key feature of the y_t series in Fig. 3.1 is its upward trend, while any seasonal fluctuations are not very pronounced. It may now be more useful to consider graphs as in Fig. 3.2. From these it can be observed that the growth rate in the second quarter seems highest and that there seems to be an upward-moving trend in the third-quarter growth rate. Note that this trending pattern suggests that the seasonal pattern in the growth rates seems not constant over time. Whether this may be due to the (possibly incorrectly) assumed Δ_1 filter for this industrial production series will be discussed in later chapters.

Fig. 3.3, presents the quarterly observed number of unemployed in Canada for the sample 1960.1–1987.4. Here no logs are taken in order to reduce the

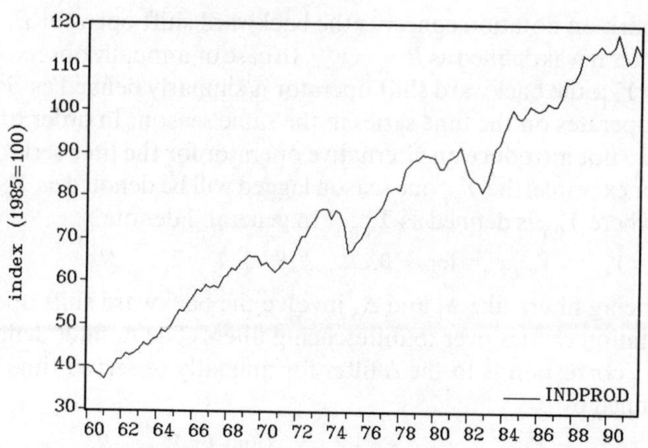

FIG. 3.1 *Industrial production, USA, 1960.1–1991.4*

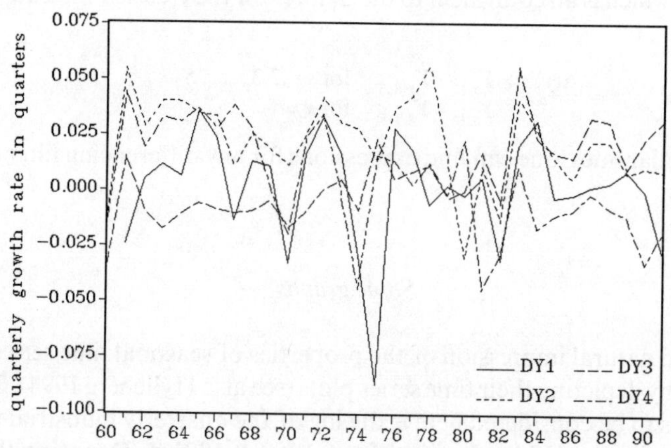

FIG. 3.2 *Quarterly growth rate of industrial production, USA, in each of the quarters for 1960–1991*

impact of the observations in the first few years. The graph indicates a marked seasonal pattern in this series, as well as shifts in mean around 1975 and 1982. Clearly, if there is a trend in the unemployment series, it seems to be a stochastic trend more than a deterministic trend. In Fig. 3.4 the first differences in each of the seasons are displayed. Evidently, the change in unemployment is largest in the first quarter, while the differences between the changes in other seasons seem to become less pronounced towards the end of the sample.

In Figs. 3.5 and 3.6, the graphs for y_t and $\Delta_1 y_t$ are given for the German real GNP series, which is observed for the sample 1960.1–1990.4. In addition to a marked trending pattern, the series in Fig. 3.5 shows significant seasonal fluctuations. These are even more clear from the four graphs of the growth rates in

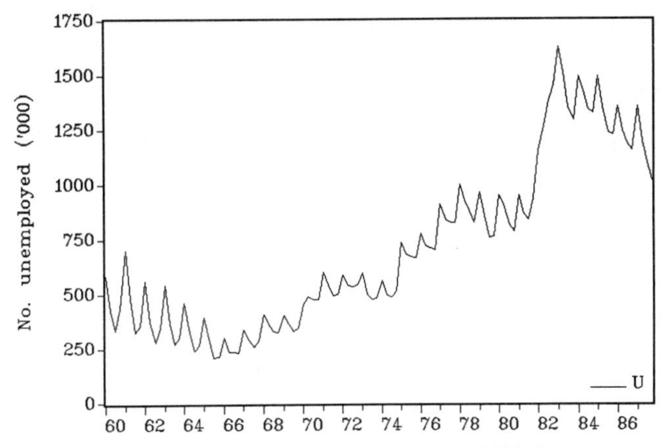

FIG. 3.3 *Unemployment in Canada, 1960.1–1987.4*

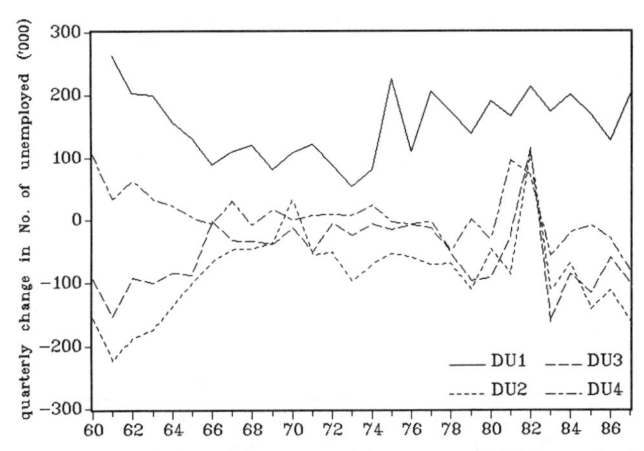

FIG. 3.4 *First differences of unemployment, Canada, in each of the quarters for 1960–1987*

Fig. 3.6. From this figure it can be seen that the growth rate in the first quarter seems roughly constant, while the growth rates in the other quarters seem to converge towards the same value.

Finally, in Fig. 3.7 and 3.8 the graphs are displayed for real total investment in the UK for the sample 1955.1–1988.4. From Fig. 3.7 it can be observed that there is a marked seasonal pattern in investment, and that the time series shows an upward trend and cyclical behaviour. The seasonal movements seem to become more volatile towards the end of the sample. This seems confirmed by the four growth rate graphs in Fig. 3.8, where the growth rate time series tend to diverge. Also, for this series the growth rates do not seem to be stationary, as each of them displays some trending behaviour.

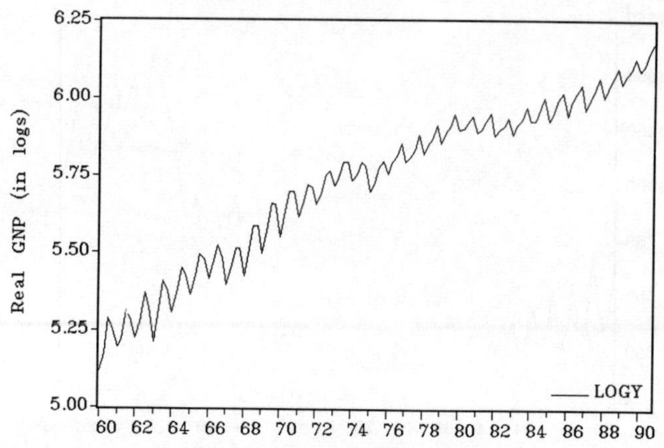

FIG. 3.5 *Real GNP, Germany, 1960.1–1990.4*

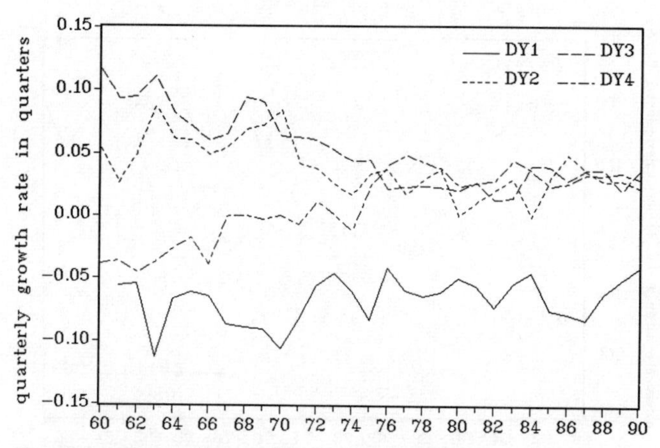

FIG. 3.6 *Quarterly growth rate of real GNP, Germany, in each of the quarters for 1960–1990*

Auxiliary regressions

To obtain a first and tentative impression of the amount of seasonal variation using a method that is somewhat more formal than the visual evidence obtained from the graphs above, I use the approach advocated in Miron (1994). This method amounts to the regression of the first-order differenced variables on four seasonal dummies,

$$\Delta_1 y_t = \delta_1 D_{1,t} + \delta_2 D_{2,t} + \delta_3 D_{3,t} + \delta_4 D_{4,t} + u_t \tag{3.7}$$

where u_t is some error process (see also e.g. Barsky and Miron 1989 and Beaulieu *et al.* 1992). In these studies it is usually assumed that u_t is a stationary and invertible ARMA process, and hence that the Δ_1 filter is sufficient to

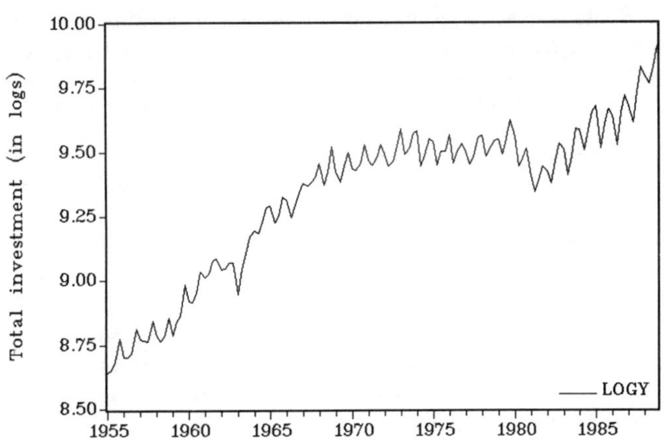

FIG. 3.7 *Total investment, UK, 1955.1–1988.4*

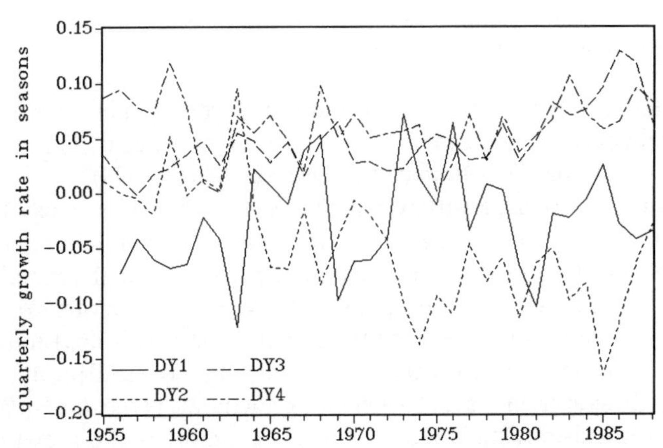

FIG. 3.8 *Quarterly growth rate of total investment, UK, in each of the quarters for 1955–1988*

remove the stochastic trend from the time series y_t. It appears that this assumption may be debatable (see e.g. Hylleberg *et al.* 1993). Therefore, I use (3.7) as a tentative model framework that can give some indication of the amount of seasonal variation in a quarterly time series. Since the error process u_t is unlikely to be a white noise process, and also since I do not aim formally to test hypotheses on the δ_s, the t-ratios for $\hat{\delta}_s$, $s = 1,2,3,4$, are not considered.

Table 3.1 reports the R^2 of (3.7) and the estimates for δ_s ($s = 1,2,3,4$). These values are given when the model is considered for the entire sample as well as for two subsamples, each of which contains roughly the same number of observations. When it is assumed that the Δ_1 filter is sufficient to remove the trend from the time series, and that the seasonal dummies are sufficient to describe seasonality (i.e. u_t does not depend on, say, u_{t-4}), the estimated R^2 values in

TABLE 3.1. Some estimation results from the auxiliary regression $\Delta_1 y_t = \delta_1 D_{1,t} + \delta_2 D_{2,t} + \delta_3 D_{3,t} + \delta_4 D_{4,t} + u_t$, where $D_{s,t}$ $(S = 1,2,3,4)$ are seasonal dummy variables

Variable	Sample[a]	R^2	$\hat{\delta}_1$	$\hat{\delta}_2$	$\hat{\delta}_3$	$\hat{\delta}_4$
Industrial production, USA[b]	1960.2–91.4	0.12	0.11	2.19	0.54	0.42
	1960.2–75.4	0.27	–0.10	2.65	–0.63	1.58
	1976.1–91.4	0.25	0.31	1.73	1.71	–0.74
Unemployment, Canada	1960.2–87.4	0.73	152.52	–85.32	–54.21	7.93
	1960.2–74.4	0.73	129.43	–94.13	–54.27	22.87
	1975.1–87.4	0.77	177.38	–75.15	–54.15	–9.31
Real GNP, Germany[b]	1960.2–90.4	0.77	–6.73	3.86	5.18	0.89
	1960.2–75.4	0.89	–7.46	5.14	7.53	–1.43
	1976.1–90.4	0.93	–6.00	2.49	2.68	3.37
Total investment, UK[b]	1955.2–88.4	0.56	–2.47	–4.73	4.68	6.18
	1955.2–73.4	0.49	–3.14	–1.60	3.19	6.28
	1974.1–88.4	0.78	–1.66	–8.69	6.58	6.06

[a] The first observation is discarded since the regression involves first differences of y_t.
[b] The parameter estimates for the δ_s are multiplied with 100.

Table 3.1 tentatively indicate that the amount of variation in y_t accounted for by seasonality can range from about 25 to about 90 per cent. The latter value seems typical for macroeconomic time series (cf. Miron 1994).

A further tentative observation from the results in Table 3.1 is that the estimates for δ_s $(s = 1,2,3,4)$ often do not seem to be constant over time. For example, for the US industrial production, the $\hat{\delta}_1$ in the first part of the sample equals –0.10, while in the second part it equals 0.31. Even more pronounced is the difference, for the same series, between the $\hat{\delta}_3$ in the first part (–0.63) and its value of 1.71 in the second part. Note that these results for industrial production correspond to the graphs in Fig. 3.2, where the growth rate in the third quarter displays an upward-moving trend. A further example is the growth rate in German GNP, for which the estimates $\hat{\delta}_2$, $\hat{\delta}_3$, and $\hat{\delta}_4$ are equal to 5.14, 7.53, and –1.43 in the first half of the sample, while they become about equal in the second part, i.e. 2.49, 2.68, and 3.37, respectively. Notice that for all four sample series the sample is split around 1973–1975, which is the period that corresponds to the first major oil crisis. Whether this major economic event may have caused the seasonal fluctuations to change will be discussed later.

To summarize this section, macroeconomic time series seem able to display significant seasonal fluctuations, which may not be constant over time.

3.2 Univariate Time Series Aspects

The univariate ARIMA time series models reviewed in Chapter 2 can be applied to economic variables that are observed with any frequency, such as daily, monthly, or annual time series. Sometimes it may then be convenient to restrict the model parameterization by explicitly taking into account any recurrent patterns arising from significant seasonality.

Seasonal ARIMA models

For the purpose of modelling seasonal time series, Box and Jenkins (1970) propose the class of so-called seasonal ARIMA (SARIMA) models. A SARIMA model consists of two parts, one part for the seasonal fluctuations and one for the nonseasonal fluctuations. The seasonal model for a time series y_t reads

$$\phi_{P,S}(B^S)(1-B^S)^D Y_t = \theta_{Q,S}(B^S)v_t, \tag{3.8}$$

where

$$\phi_{P,S}(B^S) = 1 - \phi_{1,S}B^S - \phi_{2,S}B^{2S} - \ldots - \phi_{P,S}B^{PS} \tag{3.9}$$
$$\theta_{Q,S}(B^S) = 1 + \theta_{1,S}B^S + \theta_{2,S}B^{2S} + \ldots + \phi_{Q,S}B^{QS}. \tag{3.10}$$

Note that $(1 - B^S)$ corresponds to the application of the seasonal differencing filter Δ_S. In the case where $D = 1$, it is said that y_t is transformed to annual growth rates (when $y_t = \log x_t$). Obviously, the annual growth rate $\Delta_S y_t$ equals the moving average sum of S seasonal growth rates $\Delta_1 y_t$ because $(1 + B + B^2 + \ldots + B^{S-1})\Delta_1 = \Delta_S$ (see also Chapter 5). For the remaining nonseasonal process v_t in (3.8), it is assumed that it can be described by the standard ARIMA model, i.e.

$$\phi_p(B)(1-B)^d v_t = \theta_q(B)\varepsilon_t, \tag{3.11}$$

where ε_t is white noise. Substituting (3.11) in (3.8) yields the so-called multiplicative SARIMA model,

$$\phi_p(B)\phi_{P,S}(B^S)(1-B)^d(1-B^S)^D y_t = \theta_q(B)\theta_{Q,S}(B^S)\varepsilon_t, \tag{3.12}$$

which can be abbreviated as the SARIMA $(P,D,Q)_S \times (p,d,q)$ model. Note again that this SARIMA model considers two distinct time series models, i.e. one for the seasonal and one for the nonseasonal fluctuations.

Similar to the nonseasonal case discussed in Chapter 2, it is possible to derive the ACF and PACF for various SARIMA models in order to facilitate the identification of an empirical SARIMA model using the estimated ACF and PACF. However, it is clear from the expression in (3.12) that a sheer endless list of 'typical' ACFs and PACFs can be constructed (see Hylleberg 1986, app. C for some examples), and hence that identification in practice does not seem easy. Furthermore, an additional decision in the case of SARIMA models involves the value of D, which is how often the seasonal differencing filter should be applied.

For seasonally observed economic time series, one may often find that the observations in season s $(s = 1,2, \ldots, S)$ are highly dependent on the values of those in the same season in previous years. In the notation of (3.1), this implies that $Y_{s,T}$ is highly correlated with $Y_{s,T-i}$ $(i = 1,2,\ldots)$. Furthermore, it is also likely that the nonseasonal time series v_t in (3.11) needs to be first-order differenced because the values of, e.g., v_t and v_{t-1} can be highly correlated. Taking these two suggestions together, a first tentative model for y_t may be based on the $(1-B)(1-B^S)$ differenced time series, i.e. on $(1-B)(1-B^S)y_t$. In words, when $S=4$, this time series is the quarterly change in the annual growth rate. This double differencing transformation of the y_t series is regularly applied in Box–Jenkins (1970) type identification procedures. Furthermore, in practical occasions it often appears that a useful model for y_t is the so-called 'airline model', which is given by

$$\Delta_1\Delta_S y_t = (1-B)(1-B^S)y_t = (1+\theta_1 B)(1+\theta_{1,S}B^S)\varepsilon_t. \tag{3.13}$$

Notice that this SARIMA$(0,1,1)_S \times (0,1,1)$ model contains only two parameters to be estimated. This airline model has been applied and evaluated in a host of empirical studies and it seems to be useful for many economic time series (see Nelson 1973: ch. 7; Nerlove *et al.* 1979: ch. 10; Abraham and Ledolter 1983: ch. 6; and Granger and Newbold 1986: ch. 3, *inter alia*). Notice that, when an airline model describes y_t, the only nonzero autocorrelations of $\Delta_1\Delta_S y_t$ are those at lags 1, $S-1$, S, and $S+1$, which suggests that the airline model should be easy to identify.

Despite its apparent empirical success in fitting economic time series, a possible theoretical drawback of the airline model (3.13) is that the $(1-B^S)$ component assumes that the seasonal fluctuations are generated by a seasonal stochastic trend if $\theta_{1,S} \neq -1$. For example, $(1-B^S)y_t = \varepsilon_t$ with $y_{-j}=0$ for $j=0,1,\ldots,S$ equals

$$y_t = \varepsilon_t + \varepsilon_{t-S} + \varepsilon_{t-2S} + \ldots \tag{3.14}$$

Notice that (3.14) implies that certain shocks, i.e. certain values of ε_t, can permanently change the seasonal pattern. A possibly useful alternative to the airline model is the model where the seasonal fluctuations are assumed to be deterministic instead of generated by the seasonal stochastic trend, i.e.

$$\phi_p(B)(1-B)^d y_t = \sum_{s=1}^{S} \delta_s D_{s,t} + \theta_q(B)\varepsilon_t, \tag{3.15}$$

where the $D_{s,t}$ are again the seasonal dummies as in (3.2). Of course, the AR and MA polynomials in (3.15) may contain elements as B^S, such that part of the seasonal fluctuations is stochastic. Given the possibility that (3.15) can provide a reasonable description of the data, it seems useful at the identification stage to analyse the ACF of $(1-B)y_t$ after a regression of this variable on S seasonal dummies. Below I will denote this residual time series as $\Delta_1 y_t - \hat{\delta}_s$, where $\hat{\delta}_s$ reflects the estimated mean of the $\Delta_1 y_t$ series in season s. Note that, given the visual evidence in Fig. 3.2, 3.4, 3.6, and 3.8, the $\hat{\delta}_s$ may not be constant over time.

Estimated SARIMA models for example series

In order to illustrate the Box–Jenkins (1970) approach and also to study possibly typical patterns in the ACF of seasonal time series, consider the estimated

TABLE 3.2. Estimated autocorrelation function: industrial production, USA[a]

Lags	y_t	$\Delta_1 y_t$	$\Delta_1 y_t - \hat{\delta}_s$	$\Delta_4 y_t$	$\Delta_1\Delta_4 y_t$
1	0.975**	0.162	0.242**	0.851**	0.535**
2	0.947**	0.140	0.196**	0.586**	0.162
3	0.918**	–0.110	–0.061	0.295**	–0.051
4	0.888**	0.300**	0.205**	0.036	–0.328**
5	0.853**	–0.268**	–0.264**	–0.126	–0.296**
6	0.821**	–0.046	–0.032	–0.220**	–0.190**
7	0.789**	–0.249**	–0.224**	–0.274**	–0.165
8	0.761**	0.120	0.008	–0.296**	–0.204**
9	0.732**	–0.257**	–0.253**	–0.262**	–0.066
10	0.705**	0.015	0.044	–0.207**	0.080
11	0.676**	–0.198**	–0.165	–0.172	0.025
12	0.649**	0.199**	0.099	–0.138	0.018
13	0.621**	–0.177**	–0.167	–0.111	–0.027
14	0.595**	–0.024	–0.003	–0.076	–0.095
15	0.569**	–0.158	–0.115	–0.014	–0.013
16	0.546**	0.245**	0.158	0.051	0.040

[a] The y_t series is in logs. The autocorrelations are calculated for y_t, $\Delta_1 y_t = y_t - y_{t-1}$, the residuals of the regression of $\Delta_1 y_t$ on four seasonal dummy variables, $\Delta_4 y_t = y_t - y_{t-4}$, and $\Delta_1\Delta_4 y_t = y_t - y_{t-1} - y_{t-4} + y_{t-5}$.

** Significant at the 5 per cent level, when the 95 per cent confidence interval for the estimated autocorrelations is assumed to be $\pm 2n^{-1/2}$ ($\approx$ + 0.178, here).

ACFs (up to 16 lags) for the four empirical sample series in Tables 3.2–3.5. The ACFs are estimated for y_t, $\Delta_1 y_t$, $\Delta_1 y_t - \hat{\delta}_s$, $\Delta_4 y_t$, and $\Delta_1\Delta_4 y_t$. In Table 3.2 the ACFs are given for US industrial production. It appears that the ACF of y_t dies out only very slowly, while the ACF of $\Delta_1 y_t$ dies out slowly only at the seasonal lags (4, 8, 12, and 16). In contrast, the $\Delta_1 y_t - \hat{\delta}_s$ series shows an ACF that dies out quickly, although the pattern does not correspond to the theoretical ACF pattern of a simple ARMA process. Hence, the regression on seasonal dummies seems to remove part of the seemingly nonstationary seasonal movements. The ACF of $\Delta_4 y_t$ shows a cyclical pattern which eventually appears to die out, while the ACF of $\Delta_1\Delta_4 y_t$ seems to suggest a reasonable SARIMA model for y_t. When I estimate the parameters in a model for $\Delta_1\Delta_4 y_t$ in unrestricted form, i.e. when the MA polynomial is $1 + \theta_1 B + \theta_4 B^4 + \theta_5 B^5$ instead of $(1 + \theta_1 B)(1 + \theta_{1,4} B^4)$, I obtain the following estimation results (for 122 effective observations):

$$\underset{}{\Delta_1\Delta_4 y_t} = \underset{(0.002)}{-0.0004} + \underset{(0.073)}{0.386}\Delta_1\Delta_4 y_{t-1} + \hat{\varepsilon}_t + \underset{(0.099)}{0.139}\hat{\varepsilon}_{t-1} - \underset{(0.055)}{0.764}\hat{\varepsilon}_{t-4} - \underset{(0.094)}{0.290}\hat{\varepsilon}_{t-5}, \qquad (3.16)$$

(estimated standard errors in parentheses), with diagnostic measures

$$F_{\text{AR},\,1\text{–}1} = 0.698, \quad JB = 52.962^{**}, \quad F_{\text{ARCH},\,1\text{–}1} = 10.535^{**},$$
$$F_{\text{AR},\,1\text{–}4} = 1.570, \qquad F_{\text{ARCH},\,1\text{–}4} = 3.122^{**},$$

where ** denotes significance at the 5 per cent level. The parameters are estimated using the MicroTSP (version 7.0) program. The diagnostic results indicate that the model suffers from some outliers and from heteroskedasticity, and hence that the estimated standard errors should be treated with care.

The identification of a simple SARIMA model for the number of unemployed in Canada seems to be slightly hampered by some outlying observations. Again, the ACF of y_t dies out only very slowly, while similar results apply for $\Delta_1 y_t$ at the seasonal lags; for example, $r_{16} = 0.557$ for this $\Delta_1 y_t$ series. Similar to the US industrial production series, the regression of $\Delta_1 y_t$ on seasonal dummies reduces this seemingly nonstationary seasonal pattern; for example, r_{16} for $\Delta_1 y_t - \hat{\delta}_s$ is now equal to 0.084. The ACF of $\Delta_4 y_t$ suggests an AR model, while the ACF of $\Delta_1\Delta_4 y_t$ seems to suggest that an airline model for y_t may be more appropriate. However, several autocorrelations of $\Delta_1\Delta_4 y_t$ at lags other than 1, 3, 4, and 5 are also significant (see r_{12} to r_{14}), although this may be due to some outlying observations. The estimation results for an airline model for Canadian unemployment are (for 107 effective observations)

$$\underset{}{\Delta_1\Delta_4 y_t} = \underset{(4.612)}{-0.704} + \hat{\varepsilon}_t + \underset{(0.085)}{0.466}\hat{\varepsilon}_{t-1} - \underset{(0.082)}{0.512}\hat{\varepsilon}_{t-4} - \underset{(0.090)}{0.436}\hat{\varepsilon}_{t-5}, \tag{3.17}$$

TABLE 3.3. Estimated autocorrelation function: unemployment, Canada[a]

Lags	y_t	$\Delta_1 y_t$	$\Delta_1 y_t - \hat{\delta}_s$	$\Delta_4 y_t$	$\Delta_1\Delta_4 y_t$
1	0.958**	−0.071	−0.276**	0.860**	0.502**
2	0.919**	−0.423**	−0.137	0.600**	0.140
3	0.909**	−0.176	0.017	0.312**	−0.134
4	0.910**	0.754**	0.304**	0.068	−0.501**
5	0.859**	−0.147	−0.120	−0.052	−0.410**
6	0.817**	−0.412**	−0.245**	−0.066	−0.218**
7	0.804**	−0.173	0.006	−0.034	−0.037
8	0.803**	0.706**	0.296**	0.002	0.181
9	0.750**	−0.110	−0.021	−0.003	0.199**
10	0.705**	−0.350**	−0.120	−0.063	0.179
11	0.683**	−0.181	−0.023	−0.164	0.011
12	0.672**	0.590**	0.053	−0.259**	−0.188**
13	0.621**	−0.130	−0.167	−0.299**	−0.204**
14	0.578**	−0.349**	−0.223**	−0.293**	−0.266**
15	0.559**	−0.190**	−0.046	−0.228**	−0.154
16	0.553**	0.557**	0.084	−0.128	0.139

[a] The data on y_t are untransformed, i.e. not in logs. The autocorrelations are calculated for y_t, $\Delta_1 y_t = y_t - y_{t-1}$, the residuals of the regression of $\Delta_1 y_t$ on four seasonal dummy variables, $\Delta_4 y_t = y_t - y_{t-4}$, and $\Delta_1\Delta_4 y_t = y_t - y_{t-1} - y_{t-4} + y_{t-5}$.

** Significant at the 5 per cent level, when the 95 per cent confidence interval for the estimated autocorrelations is assumed to be $\pm 2n^{-1/2}$ ($\approx \pm 0.188$, here).

with diagnostic measures

$$F_{\text{AR},1\text{–}1} = 0.343, \quad JB = 31.166^{**}, \quad F_{\text{ARCH},1\text{–}1} = 1.491,$$
$$F_{\text{AR},1\text{–}4} = 0.420, \quad F_{\text{ARCH},1\text{–}4} = 1.679.$$

These diagnostics indicate that (3.17) seems an adequate model, although the estimated residuals do not display normality.

The estimation autocorrelations in Table 3.4 for real GNP in Germany come close to the 'typical textbook case'. The ACF of y_t dies out slowly, and the ACF of $\Delta_1 y_t$ does not show a rapidly decaying pattern at seasonal lags. Even the regression on seasonal dummies does not establish that the ACF of $\Delta_1 y_t - \hat{\delta}_s$ dies out quickly at seasonal lags. On the other hand, the ACF of $\Delta_4 y_t$ corresponds to an AR(1) model, while that of $\Delta_1\Delta_4 y_t$ suggests an even simpler model. For the parameters in a model for $\Delta_1\Delta_4 y_t$, one obtains (for 115 effective observations)

$$\Delta_1\Delta_4 y_t = \underset{(0.0015)}{0.00003} + \underset{(0.091)}{0.494}\Delta_1\Delta_4 y_{t-4} + \hat{\varepsilon}_t - \underset{(0.058)}{0.799}\hat{\varepsilon}_{t-4}, \tag{3.18}$$

for which the diagnostic results are

$$F_{\text{AR},1\text{–}1} = 0.584, \quad JB = 0.263, \quad F_{\text{ARCH},1\text{–}1} = 4.001^{**}$$
$$F_{\text{AR},1\text{–}4} = 1.515, \quad F_{\text{ARCH},1\text{–}4} = 3.310^{**}.$$

TABLE 3.4. Estimated autocorrelation function: real GNP, Germany[a]

Lags	y_t	$\Delta_1 y_t$	$\Delta_1 y_t - \hat{\delta}_s$	$\Delta_4 y_t$	$\Delta_1\Delta_4 y_t$
1	0.948**	–0.111	–0.150	0.740**	–0.056
2	0.904**	–0.683**	–0.468**	0.523**	–0.088
3	0.895**	–0.124	–0.079	0.367**	0.084
4	0.889**	0.910**	0.710**	0.167	–0.198**
5	0.841**	–0.106	–0.152	0.065	–0.142
6	0.799**	–0.654**	–0.424**	0.018	0.053
7	0.790**	–0.128	–0.052	–0.060	0.054
8	0.783**	0.854**	0.540**	–0.158	–0.247**
9	0.736**	–0.077	–0.084	–0.098	–0.005
10	0.697**	–0.625**	–0.402**	–0.057	0.058
11	0.689**	–0.146	–0.062	–0.055	–0.159
12	0.685**	0.804**	0.490**	0.017	0.023
13	0.637**	–0.046	–0.054	0.054	0.067
14	0.599**	–0.572**	–0.338**	0.076	–0.063
15	0.591**	–0.146	–0.009	0.148	0.152
16	0.587**	0.747**	0.415**	0.136	0.058

[a] The y_t series is in logs. The autocorrelations are calculated for y_t, $\Delta_1 y_t = y_t - y_{t-1}$, the residuals of the regression of $\Delta_1 y_t$ on four seasonal dummy variables, $\Delta_4 y_t = y_t - y_{t-4}$, and $\Delta_1\Delta_4 y_t = y_t - y_t - y_{t-1} - y_{t-4} + y_{t-5}$.

** Significant at the 5 per cent level, when the 95 per cent confidence interval for the estimated autocorrelations is assumed to be $\pm\, 2n^{-1/2}$ ($\approx \pm 0.180$, here).

A closer look at the residuals indicates that the significance of the ARCH test statistics is due mainly to the first few observations.

Finally, in Table 3.5 I present the estimated ACFs for total investment in the UK. The patterns of the ACFs are very similar to those for the first three sample series; i.e., again the ACF of $\Delta_1\Delta_4 y_t$ seems easily interpretable. Estimation of a simple MA process for $\Delta_1\Delta_4 y_t$ yields (for 131 observations)

$$\Delta_1\Delta_4 y_t = \underset{(0.0035)}{-0.0002} + \hat{\varepsilon}_t - \underset{(0.062)}{0.200}\hat{\varepsilon}_{t-1} - \underset{(0.063)}{0.724}\hat{\varepsilon}_{t-4}, \tag{3.19}$$

with diagnostics

$$F_{AR,1-1} = 1.307,\ JB = 1.801,\ F_{ARCH,\,1-1} = 1.715,$$
$$F_{AR,1-4} = 1.849,\qquad\qquad F_{ARCH,\,1-4} = 1.558.$$

Notice that model (3.19) appears adequately to describe 131 observations with only two parameters. In fact, this is one of the main advantages of the Box–Jenkins method to specify time series models for seasonal data.

To summarize the results so far, simple models for $\Delta_1\Delta_4$ transformed series often seem adequate to describe the data. Furthermore, the auxiliary regression of $\Delta_1 y_t$ on seasonal dummies does not always remove the, in a sense, long-memory properties of the time series; i.e., the ACF may still show only slowly decaying patterns at seasonal lags.

TABLE 3.5. Estimated autocorrelation function: total investment, UK[a]

Lags	y_t	$\Delta_1 y_t$	$\Delta_1 y_t - \hat{\delta}_s$	$\Delta_4 y_t$	$\Delta_1\Delta_4 y_t$
1	0.945**	–0.137	–0.320**	0.711**	–0.189**
2	0.900**	–0.530**	0.051	0.528**	–0.011
3	0.881**	–0.063	–0.141	0.348**	0.158
4	0.867**	0.733**	0.433**	0.080	–0.335**
5	0.816**	–0.119	–0.286**	0.002	–0.062
6	0.774**	–0.506**	0.051	–0.037	–0.017
7	0.759**	–0.055	–0.127	–0.067	–0.038
8	0.748**	0.622**	0.239**	–0.068	–0.137
9	0.706**	–0.084	–0.185**	0.009	0.068
10	0.669**	–0.474**	0.073	0.050	0.032
11	0.657**	–0.027	–0.081	0.074	–0.015
12	0.645**	0.579**	0.201**	0.113	0.065
13	0.603**	–0.075	–0.155	0.109	–0.087
14	0.565**	–0.448**	0.071	0.149	0.033
15	0.551**	–0.003	–0.018	0.171**	–0.010
16	0.533**	0.508**	0.104	0.187**	–0.073

[a] The y_t series is in logs. The autocorrelations are calculated for y_t, $\Delta_1 y_t = y_t - y_{t-1}$, the residuals of the regression of $\Delta_1 y_t$ on four seasonal dummy variables, $\Delta_4 y_t = y_t - y_{t-4}$, and $\Delta_1\Delta_4 y_t = y_t - y_t - y_{t-1} - y_{t-4} + y_{t-5}$.

** Significant at the 5 per cent level, when the 95 per cent confidence interval for the estimated autocorrelations is assumed to be $\pm 2n^{-1/2}$ ($\approx \pm 0.172$, here).

Moving average unit roots

A typical aspect of the estimated models in (3.16)–(3.19) is that the solutions of the characteristic equation for the MA polynomials all seem very close to the unit circle. In fact, for (3.16)–(3.19) these solutions are

$$z^5 + 0.139z^4 - 0.764z - 0.290 = 0 \Rightarrow z \in \{0.982, 0.055 \pm 0.951i, -0.841, -0.388\},$$

$$z^5 + 0.466z^4 - 0.512z - 0.436 = 0 \Rightarrow z \in \{0.900, 0.059 \pm 0.890i, -0.742 \pm 0.244i\},$$

$$z^4 - 0.799 = 0 \Rightarrow z \in \{\pm 0.954, \pm 0.954i\},$$

$$z^4 - 0.200z^3 - 0.734 = 0 \Rightarrow z \in \{0.977, 0.050 \pm 0.918i, -0.876\},$$

respectively. It is clear that several of these solutions are close to the unit circle, and that several solutions are quite close to either ± 1 or $\pm i$. Note that, in the case of the four solutions ± 1 and $\pm i$, the MA polynomial contains the component $(1 - B)(1 + B)(1 - iB)(1 + iB) = 1 - B^4$. (This refers to the case where the Δ_4 part of the $\Delta_1\Delta_4$ filter for the y_t series seems not necessary, and hence only the Δ_1 filter may be sufficient to transform the data to stationarity.)

For SARIMA models it may occur that the $(1 - B^4)$ filter results in a non-invertible MA process. One way to decide on the appropriate differencing filter for y_t is a test for unit roots in the MA polynomial after the $(1 - B^4)$ filter has been applied. Bell (1987) shows that, when $\theta_{1,S} = -1$ in the model $(1 - B^S)y_t = (1 + \theta_{1,S}B^S)\varepsilon_t$, and given certain assumptions on the starting values, the model $y_t = \Sigma_{s=1}^{S}\delta_s D_{s,t} + \varepsilon_t$ emerges. In Franses (1991*a*) a simple procedure is proposed to test whether θ_1 and/or $\theta_{1,S}$ in the airline model (3.13) are equal to -1. The results in Anderson and Walker (1964) provide the basis for this procedure, which can be used as a quick check for the possibility of noninvertibility of the MA polynomial. The three test statistics proposed in Franses (1991*a*) are

$$T(\theta_1 = -1 | \theta_{1,S} = -1) = (2n)^{1/2}(r_1 + 0.5) \sim N(0,1) \qquad (3.20)$$

$$T(\theta_{1,S} = -1 | \theta_1 = -1) = (4n/3)^{1/2}(r_S + 0.5) \sim N(0,1) \qquad (3.21)$$

$$T(\theta_1 = -1 \text{ and } \theta_{1,S} = -1) = (2n/3)^{1/2}(r_1 + r_S + 1) \sim N(0,1), \qquad (3.22)$$

where r_1 and r_S are the estimated autocorrelations at lags 1 and S of the $\Delta_1\Delta_S y_t$ series. The unconditional tests for $\theta_1 = -1$ and $\theta_{1,S} = -1$ involve more complicated expressions. As an example, for Canadian unemployment I obtain the values 14.660**, –0.012, and 5.978**, respectively. These results suggest that the Δ_4 part in the $\Delta_1\Delta_4$ filter for Canadian unemployment is superfluous. For UK total investment the values 5.034**, 2.181**, and 4.448** are obtained, which suggests that the $\Delta_1\Delta_4$ filter for this series may be appropriate. The practical application of the test statistics in (3.20)–(3.22) is however limited, since they can be used only where the airline model can be fitted to the data. In Chapter 5 more useful tests will be described to select the appropriate differencing filter for a quarterly time series.

Seasonal unobserved components models

Finally, an alternative framework for modelling seasonal time series which is proposed in Harvey (1984) amounts to an extension of the unobserved components models (see Section 2.1) to include a seasonal component. One version of this so-called structural model for a quarterly time series is written as

$$y_t = \mu_t + \gamma_t + \varepsilon_t \tag{3.23}$$
$$\mu_t = \mu_{t-1} + \beta_{t-1} + \eta_t \tag{3.24}$$
$$\beta_t = \beta_{t-1} + \zeta_t \tag{3.25}$$
$$\gamma_t = \omega_t \,/\,(1 + B + B^2 + B^3), \tag{3.26}$$

where γ_t reflects the seasonal component, which can evolve smoothly over time when ω_t is a random error process. Similar to the nonseasonal UC models, the model in (3.23)–(3.26) can be written as a SARIMA model (see Harvey 1984). The SARIMA model for (3.23)–(3.26) contains a $(1 - B^4)$ part and a seasonal MA part with roots very close to unity. Note that this compares to the empirical findings above that the MA part of the model seems close to noninvertibility. Hence, the $(1 - B^4)$ polynomial may seem to cancel out from both sides of the model because of slowly changing seasonal patterns. See Harvey (1984) for more details of this approach.

Conclusion

In this chapter I have discussed various features of seasonal time series. It appears that seasonal variation can take account of a large part of the total variation in an approximately detrended time series. Furthermore, the Box–Jenkins type SARIMA models seem useful to describe the four quarterly macroeconomic time series which served as running examples in this chapter. The estimation results for these SARIMA models however tentatively suggest that the regularly applied double differencing filter $\Delta_1\Delta_4$ may be superfluous in some occasions. In Chapter 5, I consider formal testing methods to investigate the adequacy of this and other filters. Before I turn to that, however, I first discuss some aspects of seasonal adjustment methods in the next chapter.

4

Seasonal Adjustment

In this chapter some aspects of seasonally adjusting economic time series are discussed. The basic assumption underlying seasonal adjustment methods is that seasonality is some form of data contamination that can be removed from the data in order to facilitate the analysis of, for example, trends and business cycles, and hence that a seasonal time series can be decomposed into various separate unobserved components which can be estimated from the data. The assumption that seasonality may be removed from a time series has led to three approaches, two of which will briefly be discussed in Section 4.1. The first and simplest approach amounts to a regression of the time series on functions of time. The second approach is that the seasonal fluctuations can be filtered out using a sequence of moving average filters. This approach is followed in the regularly applied Census X-11 method. The third and so-called model-dependent approach considers the construction of ARIMA type time series models for the various unobserved components including the seasonal component. Such an ARIMA model can then be used to indicate how to adjust the time series seasonally. In Section 4.1 these last two approaches are discussed. In Section 4.2 I describe some techniques to evaluate the quality of seasonally adjusted (SA) time series. One property of SA time series is that ARIMA models for such data can contain roots on the unit circle in the MA polynomial. This property is highlighted by an empirical analysis of the official seasonally adjusted real GNP series for Germany. The chapter is concluded with a discussion of some of the advantages and disadvantages of the practical application of SA time series.

It should be noted that the discussion in this chapter only touches upon the issue of seasonal adjustment and hence is not intended to give a detailed and concise account of seasonal adjustment methods. The main purposes of the chapter are to provide some intuitive arguments on how statistical agencies may apply seasonal adjustment methods and to mention its effects on testing for unit roots and the invertibility of MA polynomials. The latter purpose is motivated since the topic of stochastic trends in seasonal time series is one of the key features in this book. Furthermore, this chapter aims to survey the concepts that will be relevant in forthcoming chapters. For detailed discussions on seasonal adjustment, the interested reader is referred to Bell and Hillmer (1984), Hylleberg (1986), Maravall (1985), and several papers in Hylleberg (1992), among others.

4.1 Seasonal Adjustment Methods

The main underlying idea of seasonal adjustment is that a seasonally observed time series y_t can be decomposed into two unobserved components,

$$y_t = y_t^{ns} + y_t^s, \tag{4.1}$$

with y_t^{ns} denoting the nonseasonal component containing the trend, cycles, and irregular components, and y_t^s denoting the seasonal component. Since the two components have to be estimated from the data, I will use the notation $\hat{y}_t^{ns}$ and $\hat{y}_t^s$ for SA data obtained using some SA method. In the text I sometimes refer to y_t^{ns} as the SA series, while y_t is the NSA (nonseasonally adjusted) series. Furthermore, for later purposes I also define the seasonal and nonseasonal components for the annual observations $Y_{s,\mathrm{T}}$, with $s = 1,2,3,4$, $T = 1, \ldots, N$, i.e.

$$\mathbf{Y}_\mathrm{T} = \mathbf{Y}_\mathrm{T}^{ns} + \mathbf{Y}_\mathrm{T}^s, \tag{4.2}$$

where $\mathbf{Y}_\mathrm{T}$, $\mathbf{Y}_\mathrm{T}^{ns}$ and $\mathbf{Y}_\mathrm{T}^s$ are (4×1) vector processes with elements

$$Y_{s,\mathrm{T}} = Y_{s,\mathrm{T}}^{ns} + Y_{s,\mathrm{T}}^s, \quad s = 1,2,3,4, \tag{4.3}$$

where $Y_{s,\mathrm{T}}^{ns}$ is the seasonally adjusted observation in season s in year T, and $Y_{s,\mathrm{T}}^s$ is the seasonal component in season s in year T. Of course, in the case where the seasonal fluctuations are multiplicative, that is when the seasonal variation in y_t increases with the trend in the nonseasonal component, one may wish to consider (4.1) in multiplicative form, i.e. $y_t = y_t^{ns}\, y_t^s$. In this case one usually log-transforms the data prior to the application of certain moving average filters.

Where the trend and seasonal fluctuations in a time series can be assumed to be deterministic, one may construct an expression for y_t^{ns} as a polynomial function of time t and an expression for y_t^s as a function of sine and cosine functions. In the case of a stochastic trend and deterministic seasonals (as in $y_t = y_{t-1} + \mu_s + \varepsilon_t$, which is (3.7)), the trend and seasonal fluctuations can be removed by first-differencing the data and then subtracting estimated μ_s values; i.e., $\Delta_1 y_t - \hat{\mu}_s$ reflects a trend and seasonal free time series. However, for many macroeconomic time series it is often observed (as in the previous chapter) that the seasonal pattern is not constant over time. This empirical regularity leads to the consideration of certain moving average filters to characterize a changing trend and changing seasonality. In many practical occasions, such moving average filters are linear, symmetric, and centred around the current observation. When I use the forward shift operator F, which is defined by $F = B^{-1}$ and hence $F^k y_t = y_{t+k}$ for $k = 0,1,2, \ldots$, such a linear moving average filter is given by

$$C_m(B,F) = c_0 + \sum_{i=1}^{m} c_i(B^i + F^i), \tag{4.4}$$

where $c_0, c_i, \ldots, c_m$ are the moving average weights. A simple example of (4.4) is the $C_1(B,F)$ filter with $c_0 = ½$ and $c_1 = -(¼)$, which equals $-¼(B^2 - 2B + 1)F$

since $FB = 1$. This simple filter concerns two unit roots because $(B^2 - 2B + 1) = (1 - B)^2$. In general, it is the case that $C_m(1,1) = c_0 + 2\Sigma^m_{i=1}c_i = 0$ when one wants to remove a changing or stochastic trend. Notice that the simple $(1 - B)$ filter, which is often applied to remove a stochastic trend (see Section 2.3), corresponds to the asymmetric filter $C_1(B,0)$ with weights $c_0 = 1$ and $c_1 = -1$ and where the forward shift operator is excluded from (4.4).

In the case of a simple seasonal time series generated by $y_t = -y_{t-1} + \varepsilon_t$ (where there are only two seasons), one may use the $C_1(B,F)$ filter with $c_0 = ½$ and $c_1 = ¼$ since in that case $C_1(B,F)$ corresponds to $(¼)(B^2 + 2B + 1) = (¼)(1 + B)^2$. In general, it is the case that $c_0 + 2\Sigma^m_{i=1}c_i = 1$ when one wants to remove a changing seasonal pattern. Additional details of the use of linear moving average filters are given in Maravall (1995) and Grether and Nerlove (1970), where in the latter study it is shown that filters like (4.4) result in certain optimal properties of the estimates of seasonal and nonseasonal components.

Census X-11

The Census X-11 seasonal adjustment method is often applied in practice. Early references for this approach include Shiskin and Eisenpress (1957) and Shiskin *et al* (1967). A recent extensive documentation of this method is given in Hylleberg (1986). Apart from the treatment of holiday, trading-day, and calendar effects, the additive version of the Census X-11 method involves two main actions. The first is the sequential application of several linear moving average filters as in (4.4) to characterize the seasonal and trend fluctuations. The filters have to be selected by the practitioner. The second action is the removal of outlying observations in about each round of moving average filtering and the replacement of such influential observations by weighted data points. Again, this involves decisions that should be made by the practitioner and can vary with the time series at hand. The outlier weighting part makes the overall procedure an intrinsically nonlinear method in the sense that for example, such data weights depend on the choice of moving average filters. A recent discussion of nonlinearity aspects of Census X-11 is given in Ghysels *et al.* (1995).

Despite the nonlinear outlier removal part of the official Census X-11 method, it is possible to give a linear symmetric moving average approximation to an often applied sequence of moving average filters in the Census X-11 program. For quarterly time series, the weights in this $C_{28}(B,F)$ filter are given in Laroque (1977); they are also given in Table 4.1 for completeness. In the Chapters 7 and 8, this approximation to the Census X-11 filter will be used to evaluate the impact of seasonal adjustment on specific properties of certain seasonal time series processes. Of course, Table 4.1 indicates that some c_j in the $C_{28}(B,F)$ filter are negligibly small. An approximate version of the $C_{28}(B,F)$ filter is given in Ghysels and Perron (1993).

TABLE 4.1. The weights c_j, in the linear moving average filter ($j = 0,1,\ldots,m$) $C_m(B,F) = c_0 + \Sigma_{i=1}^{m} c_i(B^i + F^i)$, which is a close approximation to the official Census X–11 seasonal adjustment filter

j	c_j	j	c_j	j	c_j
0	8.55948077E–01	10	2.50000000E–02	20	–2.97403828E–03
1	5.13376778E–02	11	1.17071519E–02	21	3.64145664E–04
2	4.08588430E–02	12	–5.27873744E–02	22	1.59377573E–03
3	4.99324371E–02	13	2.08468094E–02	23	–4.32672757E–04
4	–1.40109868E–01	14	1.58162521E–02	24	–3.43123088E–04
5	5.52281637E–02	15	–5.37050420E–03	25	8.88671669E–05
6	3.41436181E–02	16	–9.54700875E–03	26	4.01298070E–05
7	2.91552844E–02	17	5.00235635E–04	27	–2.09372906E–06
8	–9.72115645E–02	18	7.54738126E–03	28	–1.06140432E–06
9	3.82928481E–02	19	–1.70182345E–03		

Source: Laroque (1977) and Ooms (personal communication, 1995)

In Laroque (1977: Table 3) it is shown that the linear $C_{28}(B,F)$ filter in Table 4.1 approximately contains the component

$$(1 + B + B^2 + B^3)^2 = (1 + B)^2(1 - iB)^2(1 + iB)^2. \tag{4.5}$$

This implies that the $C_{28}(B,F)$ filter given in Table 4.1 corresponds to six roots on the unit circle, i.e. two times –1, two times $-i$ and two times $+i$. These unit roots are so-called seasonal unit roots (see Hylleberg *et al.* 1990). The unit root 1, which corresponds to the first-differencing filter $(1 - B)$, is usually called the nonseasonal unit root. In the next chapter I will discuss several methods of testing for the presence of such seasonal and nonseasonal unit roots. However, anticipating somewhat the content of Chapter 5, it should be mentioned that for many economic time series the maximum number of seasonal unit roots is three; i.e., the filter $(1 + B + B^2 + B^3)$ is usually more than sufficient to remove changing seasonal patterns from the time series. Indeed, in the previous chapter we have seen that the double-differencing filter $\Delta_1\Delta_4 = (1 - B)^2(1 + B + B^2 + B^3)$ often yields a time series for which low-order ARMA models can be estimated. In other words, the application of linear moving average filters as in (4.5) may yield noninvertible MA processes for the SA time series. Since the Census X-11 filter can be reasonably approximated by the linear filter $C_{28}(B,F)$ with weights as in Table 4.1, one may expect that similar results hold for Census X-11 corrected time series. (See Maravall (1995) for a recent discussion of the noninvertibility of SA time series.) The next section deals with the empirical content of this result for one sample SA series.

Finally, in order to adjust observations seasonally at a certain time t, e.g. with the $C_{28}(B,F)$ filter, one needs the observations over the sample $y_{t-28}, \ldots, y_{t+28}$. Since such observations are not available at the beginning and at the end of a sample, one needs to obtain backcasts and forecasts in order seasonally to adjust the y_t series for all observations. A simple approach is now to estimate SARIMA models as in Chapter 3 for the y_t series, and to generate $\hat{y}_{-27},\ldots,\hat{y}_0, \hat{y}_{n+1}, \ldots, \hat{y}_{n+28}$, when the sample contains n observations. A recent empirical evaluation of this procedure is given in Ooms (1994).

Model-based methods

Although the $C_{28}(B,F)$ filter in Table 4.1 can be approximated by a SARIMA type of polynomial (see Cleveland and Tiao 1976), the outlier removal and the sequence of moving average filters in the Census X-11 method are usually made dependent on the properties of a specific time series. An obvious drawback is that one cannot reconstruct the NSA series from SA data. Furthermore, the statistical properties of Census X-11 SA time series are hard to assess. This has led many researchers to recommend statistical agencies to publish both the NSA and SA series in their official publications.

In response to the possible ambiguities involved in the application and evaluation of the Census X-11 procedure, Hillmer and Tiao (1982) propose the so-called ARIMA-model-based approach to seasonal adjustment. The assumption is that a time series y_t can be decomposed as

$$y_t = y_t^{\mathrm{T}} + y_t^s + u_t, \tag{4.6}$$

where y_t^{T} is the trend component and u_t is the noise component, and where $y_t^{\mathrm{T}} + u_t$ equals the nonseasonal component y_t^{ns} as in (4.1). The model-based approach assumes that each component in (4.6) can be described using a certain ARIMA model. Typical examples are the ARIMA $(0,d,d-1)$ model,

$$(1-B)^d y_t^{\mathrm{T}} = \phi_{d-1}(B)\zeta_t, \tag{4.7}$$

for the trend component and the ARIMA (3,0,3) model for a quarterly series,

$$(1 + B + B^2 + B^3)y_t^s = \psi_{S-1}(B)\phi_t, \tag{4.8}$$

for the seasonal component, where $\phi_{d-1}(B)$ and $\psi_3(B)$ are polynomials in B of maximum order $d-1$ and 3, respectively. When an ARIMA model for the noise process u_t is assumed too, it can be seen that y_t can be described by a certain SARIMA model. Hillmer and Tiao (1982) show for some sample models that such a SARIMA model can yield estimates for the variances of the residual processes ζ_t and ϕ_t in (4.7) and (4.8) and of the parameters in, say, the $\psi_3(B)$ polynomial in (4.8). Given such estimates, it is possible to estimate the seasonal component y_t^s. As with the Census X-11 method, the latter estimate in principle involves the application of a two-sided linear symmetric filter which is based on the estimation results for (4.8). In Bell (1984) it is shown that, if the variance of the ϕ_t process in (4.8) exceeds zero, the application of such a filter results in the minimum mean squared error estimator of the trend component. Hence, notice that for the model-based method similar results to those above apply; i.e., the SA filter corresponds to twice the number of seasonal unit roots as given in (4.8). A lucid exposition of the model-based method is given in Maravall and Pierce (1987).

An alternative model-based seasonal adjustment method is proposed in Harvey (1984). It amounts to estimating a basic structural time series (STM) model such as

$$\begin{aligned} y_t &= \mu_t + \gamma_t + \varepsilon_t, \\ \mu_t &= \mu_{t-1} + \beta_{t-1} + \eta_t, \\ \beta_t &= \beta_{t-1} + \zeta_t, \end{aligned}$$

with the seasonal component

$$(1 + B + B^2 + B^3)\gamma_t = \omega_t. \tag{4.9}$$

Kalman-filtering-based estimation methods yield the estimates of the various parameters in the STM. Similarly, one may now obtain an estimate of $\hat{\gamma}_t$, i.e. the seasonal component. Given the relationships between unobserved component models and ARIMA models, as mentioned in Chapter 2, it is clear that there are several relationships between this and the above model-based approach (see Harvey 1989).

4.2 Evaluating Seasonally Adjusted Data

There are several criteria that can be used to evaluate the quality of the SA data obtained from one of the above procedures. An extensive discussion of these criteria is given in Hylleberg (1986: ch. 3). In this section I will make some general remarks. Part of the discussion concerns recent studies which analyse the unit root properties of SA data.

Evaluation criteria

Obviously, important requirements for seasonal adjustment procedures are that $(y_t + z_t)^{ns} = y_t^{ns} + z_t^{ns}$ for the additive model, that $(y_t z_t)^{ns} = y_t^{ns} z_t^{ns}$ for the multiplicative model, and that $(ay_t)^{ns} = ay_t^{ns}$. Since the estimates of the seasonal and nonseasonal components at the end of the sample are based on multi-step-ahead out-of-sample forecasts, an indication of the robustness of a seasonal adjustment procedure is that no major revisions are needed should new data become available. Pierce (1980) provides some theoretical results on data revision in the context of moving average filters, and Burridge and Wallis (1985) consider SA data.

Another important property of seasonal adjustment methods is that the key assumption of orthogonality of the seasonal and nonseasonal components as in (4.1) is not systematically violated. In practice these two components have to be estimated, and there may always be some correlation between the two estimated components. However, it may be viewed as inconvenient that, for example, the $\hat{y}_t^{ns}$ time series can be used to predict the $\hat{y}_t^s$ series, either in- or out-of-sample. If so, there would be some seasonality left in the $\hat{y}_t^{ns}$ series. Shiskin and Eisenpress (1957) consider the simple regression of $\hat{y}_t^s$ on $\hat{y}_t^{ns}$ for each season to investigate the orthogonality property. Note that the significance of the parameter for $\hat{y}_t^{ns}$ can also indicate that one should have considered the multiplicative version of (4.1) instead of the additive version.

Given earlier discussions on stochastic trends, and also given the tentatively identified models in Chapter 3, it is unlikely that $\hat{y}_t^{ns}$, for example, is a stationary time series and that the relevant t-ratio in the 'orthogonality-regression' of $\hat{y}_t^s$ on $\hat{y}_t^{ns}$ is approximately distributed as standard normal. Hence, it seems more useful to consider, say, a regression of $\Delta_4 \hat{y}_t^s$ on $\Delta_4\hat{y}_t^{ns}$(see also Franses and Ooms 1995). To illustrate the practical application of this regression, consider the officially (i.e. published) seasonally adjusted time series $\hat{y}_t^{ns}$ for the US industrial production, Canadian unemployment, and real GNP in Germany. (The relevant sources of the data are given in the Data Appendix.) These data correspond to the unadjusted time series used in Chapter 3. The latter unadjusted data are used to construct the estimated seasonal component as $\hat{y}_t^s = y_t - \hat{y}_t^{ns}$, where these time series are the log-transformed time series for the US industrial production and German GNP series and the untransformed series for Canadian unemployment. Figs. 4.1–4.3 depict this estimated seasonal component. The graphs strongly suggest that for each of these series the seasonal pattern has changed over time. The $\hat{y}_t^s$ series for US industrial production shows a structural break around 1969, which is possibly due to data revisions. The variation in the estimated seasonal component for unemployment in Canada in Fig. 4.2 decreases quite rapidly until 1967, is roughly constant until 1975, and increases after 1975. Finally (Fig. 4.3), the seasonal component for German GNP becomes less important after 1972. Notice the similarities of the graph in this figure with the graphs in Fig. 3.6; there, the quarterly growth rates

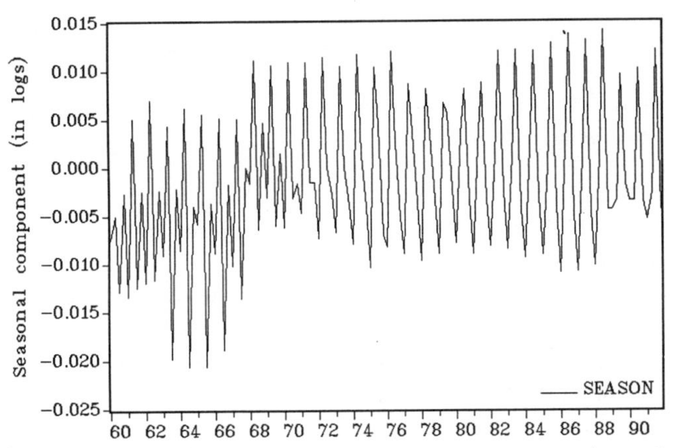

FIG. 4.1 *The official estimated seasonal component: industrial production in the USA, 1960.1–1991.4*

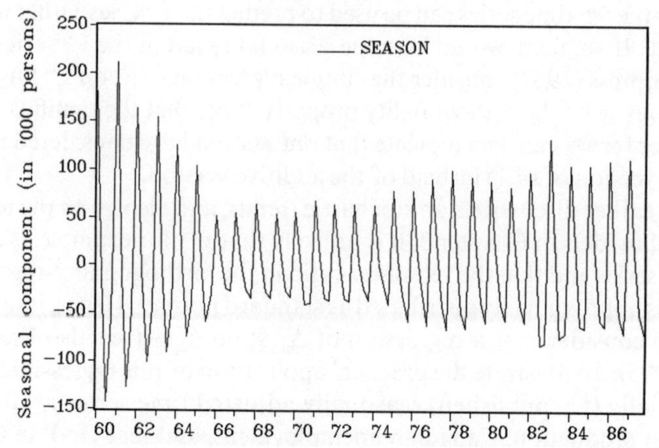

FIG. 4.2 *The official estimated seasonal component: unemployment in Canada, 1960.1–1987.4*

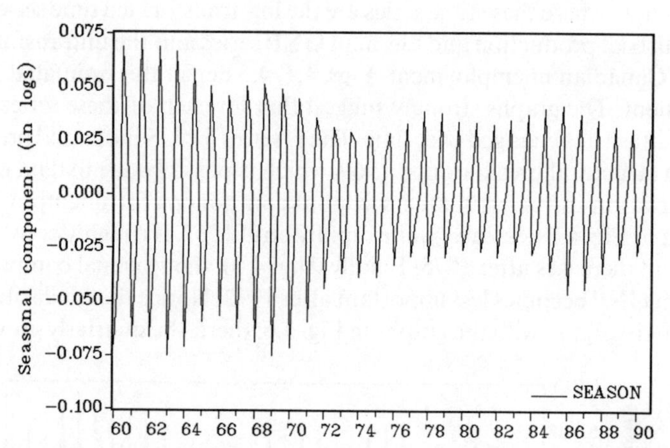

FIG. 4.3 *The official estimated seasonal component: real GNP in Germany, 1960.1– 1990.4*

of the NSA real GNP series indicate that the differences between these growth rates seem to disappear over time.

Figs. 4.4–4.6 depict the regression lines in the orthogonality regressions based on annual changes. It is clear, at least when averaged over all quarters, that the explanatory power of $\Delta_4\hat{y}_t^{ns}$ for $\Delta_4\hat{y}_t^s$ seems very small, if not equal to zero. At first sight therefore it seems that the orthogonality assumption for the three seasonally adjusted sample series is not violated. In Chapter 6 I return to the empirical assessment of the orthogonality assumption using a modified version of this orthogonality regression.

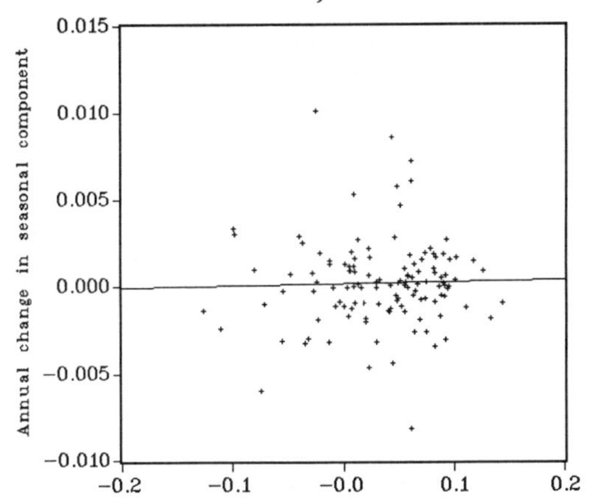

FIG. 4.4 *Regression of annual changes in the estimated seasonal component $\Delta_4\hat{y}_t^s$ on annual changes in the estimated nonseasonal component $\Delta_4\hat{y}_t^{ns}$: industrial production in the USA*

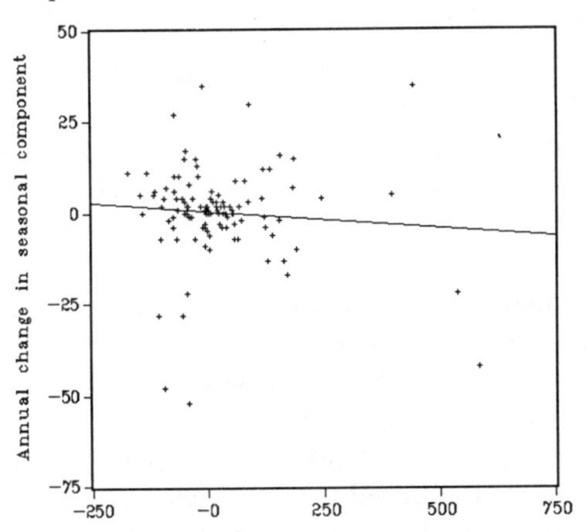

FIG. 4.5 *Regression of annual changes in the estimated seasonal component $\Delta_4\hat{y}_t^s$ on annual changes in the estimated nonseasonal component $\Delta_4\hat{y}_t^{ns}$: unemployment in Canada*

Additional properties of SA time series

An additional requirement for SA data can be that the nonseasonal unit root and/or long-memory properties in the NSA time series are somehow preserved in the SA series. This seems particularly relevant in the case where one wants to study trend and business cycle fluctuations. Recent research, however, suggests that this may not be the case. For example, in Bhargava (1990), Ghysels (1990), and Jaeger and Kunst (1990), it is shown that shocks to seasonally adjusted

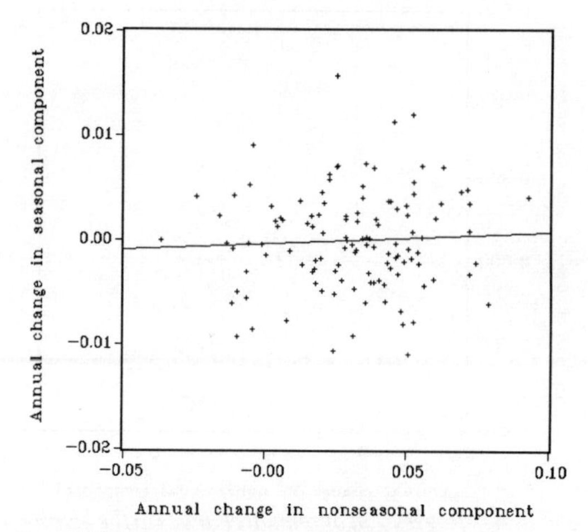

FIG. 4.6 *Regression of annual changes in the estimated seasonal component $\Delta_4\hat{y}_t^s$ on annual changes in the estimated nonseasonal component $\Delta_4\hat{y}_t^{ns}$: real GNP in Germany*

time series are more persistent. Consider again the three SA sample series and their estimated ACFs which are given in Table 4.2. Comparing these ACFs with those for the NSA time series reported in Tables 3.2, 3.3, and 3.4, it can be observed that the estimated autocorrelations of the SA series are either equal to or slightly higher than those of the NSA series. It is however likely that the SA series all have a nonseasonal unit root. Note that Ghysels and Perron (1993) show that for a stationary time series a linear two-sided moving average filter as in (4.4) biases unit root tests towards nonrejection. Table 4.2 also reports the estimated ACF for the $\Delta_1\hat{y}_t^{ns}$ time series. Theoretically, these ACFs should resemble the ACFs of the $\Delta_4 y_t$ series in Tables 3.2–3.4.

To illustrate the presence of MA unit roots in an estimated SA series, consider again the seasonally adjusted German real GNP series. The estimated ACF in Table 4.2 indicates that $\hat{y}_t^{ns}$ is nonstationary and that this ACF for $\Delta_1\hat{y}_t^{ns}$ is more easily interpretable. Furthermore, note that several autocorrelations take reasonably large values, although these are not all significant at the 5 per cent level. The (unreported) estimated PACF shows several significant partial autocorrelations. The latter is reflected by the tentatively adequate AR(9) model that is fitted to 114 effective observations:

$$\begin{aligned}\Delta_1\hat{y}_t^{ns} = \underset{(0.0015)}{0.0041} + \underset{(0.091)}{0.414}\Delta_1\hat{y}_{t-1}^{ns} + \underset{(0.081)}{0.021}\Delta_1\hat{y}_{t-2}^{ns} + \underset{(0.081)}{0.323}\Delta_1\hat{y}_{t-3}^{ns} - \underset{(0.087)}{0.559}\Delta_1\hat{y}_{t-4}^{ns}\\ + \underset{(0.086)}{0.417}\Delta_1\hat{y}_{t-5}^{ns} - \underset{(0.079)}{0.098}\Delta_1\hat{y}_{t-6}^{ns} + \underset{(0.076)}{0.085}\Delta_1\hat{y}_{t-7}^{ns} - \underset{(0.075)}{0.403}\Delta_1\hat{y}_{t-8}^{ns} + \underset{(0.080)}{0.267}\Delta_1\hat{y}_{t-9}^{ns} + \hat{\varepsilon}_t.\end{aligned}$$

The (unreported) diagnostic test results indicate that this model cannot be rejected by the data. From these estimation results, where the parameters at lags

TABLE 4.2. Estimated autocorrelation functions of three seasonally adjusted time series.

Lags	Industrial production, USA[a]		Unemployment, Canada[b]		Real GNP, Germany[c]	
	y_t^{ns}	$\Delta_1 y_t^{ns}$	y_t^{ns}	$\Delta_1 y_t^{ns}$	y_t^{ns}	$\Delta_1 y_t^{ns}$
1	0.976**	0.538**	0.986**	0.553**	0.971**	0.154
2	0.948**	0.173	0.965**	0.265**	0.943**	0.091
3	0.919**	0.074	0.939**	0.148	0.915**	0.178
4	0.887**	–0.012	0.911**	–0.096	0.887**	–0.141
5	0.854**	–0.152	0.884**	–0.157	0.862**	0.130
6	0.821**	–0.111	0.858**	–0.036	0.835**	0.085
7	0.790**	–0.150	0.831**	0.015	0.807**	–0.047
8	0.761**	–0.243**	0.803**	0.006	0.780**	–0.187**
9	0.733**	–0.168	0.773**	0.096	0.754**	0.009
10	0.705**	–0.019	0.742**	0.106	0.729**	0.033
11	0.676**	–0.063	0.707**	–0.125	0.704**	–0.145
12	0.648**	–0.101	0.673**	–0.231**	0.679**	0.115
13	0.621**	–0.080	0.641**	–0.166	0.653**	0.019
14	0.595**	–0.063	0.610**	–0.218**	0.628**	–0.082
15	0.570**	–0.026	0.581**	–0.185	0.604**	0.223**
16	0.545**	0.021	0.553**	0.009	0.581**	0.055

[a] The estimated standard error is approximately 0.088.
[b] The estimated standard error is approximately 0.094.
[c] The estimated standard error is approximately 0.090.
** Significant at the 5 per cent level.

2, 6, and 7 appear to be insignificant, it seems that the AR(9) polynomial may be approximated by an ARMA polynomial. For the same effective 114 observations (i.e. the estimation sample runs from 1962.3–1990.4), an alternative model is

$$\Delta_1 \hat{y}_t^{ns} = \underset{(0.0013)}{0.0001} + \underset{(0.082)}{0.346}\Delta_1 \hat{y}_{t-1}^{ns} + \underset{(0.081)}{0.359}\Delta_1 \hat{y}_{t-3}^{ns} + \underset{(0.077)}{0.149}\Delta_1 \hat{y}_{t-5}^{ns} + \hat{\varepsilon}_t - \underset{(0.066)}{0.668}\, \hat{\varepsilon}_{t-4},$$

for which the (unreported) diagnostic measures again suggest that the model cannot be rejected by the data. The AR polynomial in this ARMA model can be decomposed as $(1 - 0.941B)(1 - 0.876B + 0.465B^2)(1 + 0.280B + 0.341B^2)$, while the MA polynomial can be decomposed as $(1 - 0.904B)(1 + 0.904B)$ $(1 - 0.904iB)\,(1 + 0.904iB)$. It is clear that the roots of the MA polynomial are close to the unit circle, and that the first components of the AR and MA polynomials are very similar. When it is assumed that $(1 - 0.941B)$ and $(1 - 0.904B)$ cancel out, the resulting MA part in the ARMA model for $\Delta_1 \hat{y}_t^{ns}$ contains three approximate seasonal unit roots. Given the earlier discussion, this is exactly what one would expect when seasonal adjustment methods are used to a time series with only three seasonal unit roots. See Maravall (1995) for similar examples.

Conclusion

In this chapter I have discussed some aspects of seasonal adjustment which can be relevant for the material in the next chapters. The main empirical argument for analysing seasonally adjusted time series is that practitioners may find it hard to assign an economic interpretation to seasonal fluctuations (which are not weather and holiday effects). Hence, they may be interested only in the underlying trend or cyclical pattern. Indeed, in static regression models the use of SA time series instead of NSA time series does not seem to affect the estimation of key parameters when all variables are adjusted using the same filter (see Sims 1974). However, as Wallis (1974) shows, in the case of dynamic regression models such as autoregressive distributed lag (ADL) models, seasonal adjustment may result in an estimation bias. Additionally, the recent discussion on SA and NSA data in Ericsson *et al.* (1994) shows that SA data can affect exogeneity properties. The results concerning the impact of seasonal adjustment on long-memory properties of economic time series suggests that the estimation of the stochastic trend may be affected by the SA filter. Furthermore, the presence of implied MA unit roots may also blur the inference on trends and cycles. In addition to these results, Ghysels *et al.* (1995) document that the Census X-11 method may result in SA time series that have nonlinear properties while the NSA time series are linear. All in all, these results may provide a motivation to consider econometric time series models that explicitly incorporate a description of seasonal variation.

A final remark concerns the crucial assumption of the orthogonality of seasonal and nonseasonal fluctuations. Although this assumption serves mainly technical purposes, since it allows the identification of the unobserved components, there are no important economic arguments why the seasonal and nonseasonal fluctuations should be strictly independent. One can think of several examples where the seasonal movements can depend on the business cycle and/or the trend, other than pure multiplicative seasonality. For example, when clearance sales start earlier because of the current poor state of the economy, one may observe changing seasonality because of business cycle fluctuations. In Chapter 6 I will therefore return to this assumption of orthogonality in more detail.

5

Seasonal Integration and Cointegration

When the estimated (partial) autocorrelation functions are used to identify a tentative time series model, as is done in the standard Box–Jenkins (1970) approach discussed in Chapter 3, one often obtains the impression that the double-differencing filter $\Delta_1\Delta_4 = (1 - B)(1 - B^4)$ is useful to remove unit roots from a seasonal time series. For quarterly logarithmic transformed data, the application of this filter results in a time series of quarterly changes in the annual growth rates. This chapter reviews recent developments in formally testing whether this double filter or the Δ_4 filter are useful transformations in practice. If a Δ_4 differencing filter is required, a time series is said to be seasonally integrated. Since differencing filters assume the presence of one or more seasonal or nonseasonal unit roots, most methods to test for an appropriate differencing filter are based on statistical tests for such unit roots. These tests are all extensions of the univariate Dickey–Fuller tests discussed in Chapter 2.

In the first section of this chapter I discuss testing for unit roots in univariate seasonal time series with an explicit focus on seasonal unit roots. This is because such roots are typically found for time series with changing seasonal patterns. The tests will be illustrated for the four sample time series that were discussed in Chapter 3, as well as for seven new sample series. These new series all concern data for the UK: GDP, total consumption, consumption of nondurables, exports, imports, public investment and the work-force, where the first six variables are all in real terms. Including the total investment series analysed in Chapter 3, eight UK macroeconomic time series are analysed in this chapter. These eight (and other) time series have been previously analysed in Osborn (1990) with respect to seasonal unit roots. The data source is given in the Data Appendix. All data are analysed in logs. In addition to the presentation of unit root test results in Section 5.1, I also focus on the robustness of the obtained results on seasonal integration to variation in the sample size and to the possible presence of deterministic mean shifts. Such sensitivity analysis will also be performed for the estimated models in later chapters. In Section 5.2 the focus is on multivariate models for seasonal time series with seasonal and nonseasonal unit roots, i.e. the so-called seasonal cointegration models.

5.1 Seasonal Integration

In Chapter 3 it was documented that, when a quarterly observed time series is differenced using the differencing filter $\Delta_1\Delta_4 = (1 - B)(1 - B^4)$, the resulting ACF could often be easily interpreted. In practice, it regularly appears that, when one follows the Box–Jenkins (1970) identification strategy, the $\Delta_1\Delta_4$ filter emerges as the most useful differencing filter. It may however be sensible to test formally for the adequacy of such a filter. Since

$$\begin{aligned}\Delta_1\Delta_4 &= (1-B)(1-B^4)\\ &= (1-B)(1-B)(1+B)(1-iB)(1+iB)\\ &= (1-B)^2(1+B)(1+B^2)\\ &= (1-B)^2(1+B+B^2+B^3),\end{aligned} \tag{5.1}$$

it is clear that this double filter assumes the presence of five unit roots in the autoregressive polynomial of an ARMA model for a quarterly series y_t. Two of the unit roots in the $\Delta_1\Delta_4$ filter correspond to the nonseasonal frequency, i.e. $(1 - B)^2$. The $(1 + B)(1 - iB)(1 + iB)$ part corresponds to the three seasonal unit roots -1 and $\pm i$. Hence, the $\Delta_1\Delta_4$ filter in fact assumes that the time series is an I(2) series.

There are several methods by which to investigate whether the $\Delta_1\Delta_4$ filter is appropriate to remove stochastic trends from a time series y_t. Here two such methods are considered. The first is developed in Hasza and Fuller (1982), and it concerns the model

$$y_t = \phi_1 y_{t-1} + \phi_4 y_{t-4} + \phi_5 y_{t-5} + u_t, \tag{5.2}$$

where u_t is some stationary and invertible ARMA process. The null hypothesis of interest is that jointly $\phi_1 = \phi_4 = 1$ and $\phi_5 = -1$. Hasza and Fuller (1982) develop the relevant test statistics and their asymptotic distributions. The second approach, which also allows for intermediate outcomes such as Δ_1 and Δ_4, is developed in Osborn *et al.* (1988)(OCSB). This method amounts to estimating the parameters in the auxiliary regression

$$\phi_k(B)\Delta_1\Delta_4 y_t = \sum_{s=1}^{4}\delta_s D_{s,t} + \beta_1\Delta_4 y_{t-1} + \beta_2\Delta_1 y_{t-4} + \varepsilon_t, \tag{5.3}$$

where $\phi_k(B)$ is some polynomial in B of order k and the $D_{s,\,t}$ are seasonal dummy variables. If needed, this regression can be extended with the inclusion of a deterministic trend or with seasonal deterministic trends. The latter regression model can be used to investigate series with increasing seasonal variation (see Bowerman *et al.* 1990, and Franses and Koehler 1994). For (5.3) it holds that when $\beta_1 = 0$ and $\beta_2 < 0$ the Δ_1 filter is appropriate, when $\beta_1 < 0$ and $\beta_2 = 0$ the Δ_4 filter is appropriate, when $\beta_1 = \beta_2 = 0$ the $\Delta_1\Delta_4$ filter is required, and when both β_1 and β_2 are smaller than 0 no differencing filter is needed. Critical values of the various relevant t- and F-type test statistics are given in Osborn (1990) for model (5.3) and in Franses and Koehler (1994) for some of its extended versions.

The application in Osborn (1990) of the auxiliary regression (5.3) to the eight UK example series results in a clear rejection of the empirical adequacy of the $\Delta_1\Delta_4$ filter for each of the UK variables. My (unreported) findings are that also for the non-UK sample time series, i.e. US industrial production, Canadian unemployment, and real GNP in Germany, the $\Delta_1\Delta_4$ filter does not seem appropriate. Osborn (1990) additionally analyses some 20 other UK series, and again no clear evidence is obtained in favour of the double-differencing filter. This result can be considered as indicating that in practice often at most the Δ_4 filter may be useful.

One can now argue that these empirical findings for the double filter are caused by the fact that models like (5.3) are autoregressive regression models, and that finite-order AR models are not sufficiently approximating ARMA models with near unit roots in the MA polynomial. Indeed, when the airline model described in Chapter 3, i.e. $\Delta_1\Delta_4 y_t = (1 - \theta_1 B)(1 - \theta_{1,4} B^4)\varepsilon_t$, is the DGP, the empirical size of t-tests for β_1 and β_2 and the joint F-test for β_1 and β_2 in the regression (5.3) can exceed the nominal size to a large extent when the θ_1 and $\theta_{1,4}$ parameters get close to unity (see e.g. the simulation results in Franses and Koehler 1994). Agiakloglou and Newbold (1992), among others, discuss a similar bias in Dickey–Fuller type tests for nonseasonal time series. It is my view that this does not imply that Dickey–Fuller type tests are useless in practice, but rather that residual correlation diagnostics for auxiliary regressions like (5.3) should indicate whether there is a (near) unit root in the MA polynomial. If so, the estimated residuals of (5.3) should be correlated even when many lags are included. When such model diagnostics indicate that there is no obvious misspecification when only a few lags of $\Delta_1\Delta_4 y_t$ in (5.3) are included, i.e. when k is reasonably small, one may assume that an AR model yields an approximately adequate description of the data and that unit root tests can be based on this AR model.

Testing for seasonal unit roots

The empirical results obtained using the auxiliary regression (5.3), that the Δ_4 filter may be sufficient to remove one or more stochastic trends from y_t, naturally lead to a closer investigation of the empirical adequacy of the Δ_4 filter. If a Δ_4 filter is needed, a time series is called seasonally integrated. Similar to testing for the adequacy of the $\Delta_1\Delta_4$ filter, one may also test whether this Δ_4 filter amounts to an adequate data transformation. Dickey *et al.* (1984) propose to analyse

$$y_t = \phi_4 y_{t-4} + u_t, \tag{5.4}$$

where u_t is some stationary and invertible ARMA process, and to test the null hypothesis that $\phi_4 = 1$. The alternative hypothesis is that $\phi_4 < 1$. Given that $(1 - B^4)$ can be decomposed into four components (see (5.1)), it is clear that the

Dickey *et al.* (1984) method considers the null hypothesis of four unit roots, i.e. one nonseasonal unit root and three seasonal unit roots, versus the alternative hypothesis of no unit roots. Transforming a time series with the $(1 - B^4)$ filter yields an overdifferenced series when only one unit root is present. This is the case, for example, when the $(1 - B)$ filter already yields a stationary time series and when seasonality can be handled by seasonal dummies. Overdifferencing can complicate the construction of univariate time series models because, for example, the PACF pattern may become hard to interpret. Furthermore, one may expect estimation problems because of the introduction of moving average polynomials with roots on the unit circle. On the other hand, underdifferenced series may yield unit roots in their autoregressive parts, and then arguments such as those in Granger and Newbold (1974) for time series containing neglected unit roots apply; see e.g. Abeysinghe (1994), where it is shown that spurious relationships between two univariate time series may appear when seasonal unit roots are neglected.

Hylleberg *et al.* (1990)(HEGY) propose a simple procedure to test for seasonal and nonseasonal unit roots in a quarterly time series, and hence to test the $(1 - B^4)$ filter versus its nested variants like $(1 - B)$ or $(1 + B)$. Beaulieu and Miron (1993) and Franses (1991*b*) present extensions of the HEGY method to monthly time series. The HEGY approach is based on a proposition on roots of polynomials, which is reproduced in the Appendix to this chapter. An application of the expression in equation (5.A5) to an AR polynomial $\phi(B)$ for a quarterly time series results in

$$\begin{aligned}\phi(B) = {} & \lambda_1 B\phi_1(B) + \lambda_2(-B)\phi_2(B) + \lambda_3(-i - B)B\phi_3(B) \\ & + \lambda_4(i - B)B\phi_3(B) + \phi^*(B)\phi_4(B),\end{aligned} \tag{5.5}$$

where the $\phi_i(B)$ polynomials ($i = 1,2,3,4$) are defined by

$$\begin{aligned}\phi_1(B) &= (1 + B + B^2 + B^3) \\ \phi_2(B) &= (1 - B)(1 + B^2) = (1 - B + B^2 - B^3) \\ \phi_3(B) &= (1 - B^2) \\ \phi_4(B) &= (1 - B^4).\end{aligned} \tag{5.6}$$

When the λ_i ($i = 1,2,3,4$) in (5.5) are redefined as

$$\begin{aligned}\lambda_1 &= -\pi_1 \\ \lambda_2 &= -\pi_2 \\ \lambda_3 &= (-\pi_3 + i\pi_4)/2 \\ \lambda_4 &= (-\pi_3 - i\pi_4)/2,\end{aligned} \tag{5.7}$$

the expression in (5.5) can be written as

$$\phi(B) = -\pi_1 B\phi_1(B) + \pi_2 B\phi_2(B) + (\pi_3 B + \pi_4)B\phi_3(B) + \phi^*(B)\phi_4(B). \tag{5.8}$$

When it is assumed that y_t can be described by a certain AR process, i.e.

$$\phi(B)y_t = \mu_t + \varepsilon_t, \tag{5.9}$$

where

$$\mu_t = \delta_0 + \sum_{s=1}^{3} \delta_s D_{s,t} + \beta t, \tag{5.10}$$

with t denoting a deterministic trend variable, then, given (5.8), the HEGY auxiliary regression to test for the presence of nonseasonal and seasonal unit roots becomes

$$\phi^*(B) y_{4,t} = \mu_t + \pi_1 y_{1,t-1} + \pi_2 y_{2,t-1} + \pi_3 y_{3,t-2} + \pi_4 y_{3,t-1} + \varepsilon_{\cdot t}, \tag{5.11}$$

where

$$\begin{aligned} y_{1,t} &= \phi_1(B) y_t \\ y_{2,t} &= -\phi_2(B) y_t \\ y_{3,t} &= -\phi_3(B) y_t \\ y_{4,t} &= \phi_4(B) y_t, \end{aligned}$$

with the $\phi_i(B)$ polynomials defined as in (5.6).

The application of ordinary least squares to (5.11) gives estimates of the π_i parameters ($i = 1,2,3,4$), where the order of the AR polynomial $\phi^*(B)$ is selected such that the estimated residuals seem approximately white noise. Because the π_i parameters are zero in the case where the corresponding roots of the AR polynomial are on the unit circle (see (5.7) and (5.A5)), testing the significance of the estimated π_i parameters implies testing for unit roots. There will be no seasonal unit roots if π_2 and π_3 (or π_4) are different from zero. If $\pi_1 = 0$, then the presence of the nonseasonal unit root 1 cannot be rejected.

The t-tests for π_1 and π_2 are denoted as $t(\pi_1)$ and $t(\pi_2)$. The alternative hypotheses for the unit roots 1 and -1 are that the roots are smaller than unity in an absolute sense. This is the case when the relevance of π_1 and π_2 can be tested using one-sided $t(\pi_1)$- and $t(\pi_2)$-tests. A test strategy for π_3 and π_4 may be to test π_4 in a two-sided procedure and, when the insignificance of π_4 cannot be rejected, to check the significance of π_3 with a one-sided test. To understand why π_4 requires a two-sided test, consider for example the AR polynomial

$$\phi(B) = (1 + \alpha B^2)\,[(1 - B^4)/(1 + B^2)] = (1 + \alpha B^2)\,(1 - B^2), \tag{5.12}$$

where $|\alpha| < 1$ in case of stationarity. Using (5.A1) and (5.A2), it can be shown that for this $\phi(B)$ it holds that

$$\phi(-i) = \phi(i) = 2(1 - \alpha),$$

and

$$\delta_1(-i)\,\delta_2(-i)\,\delta_4(-i) = \delta_1(i)\,\delta_2(i)\,\delta_3(i) = 4,$$

where these ϕ and δ functions are given in Appendix 5.1. Furthermore, given (5.A2), $\lambda_3 = \lambda_4 = (1 - \alpha)/2$, such that both λ_3 and $\lambda_4 > 0$ under the stationary alternative. The transformations in (5.7) give that

$$\begin{aligned} \pi_3 &= -\lambda_3 - \lambda_4 \\ \pi_4 &= i(\lambda_4 - \lambda_3). \end{aligned}$$

These expressions for π_3 and π_4 show that, when $\lambda_3, \lambda_4 > 0$, the tests for π_3 and π_4 are one- and two-sided, respectively. In many practical occasions one usually considers the joint F-test for π_3 and π_4 to be denoted as $F(\pi_3,\pi_4)$. In addition to the t-tests for π_1 and π_2 and the F-test for π_3 and π_4, one may also consider the F-tests for π_2, π_3, and π_4 ($F(\pi_2, \ldots, \pi_4)$) and for $\pi_1 - \pi_4$ ($F(\pi_1, \ldots, \pi_4)$). The latter two tests are proposed in Ghysels *et al.* (1994*b*). Note that the $F(\pi_1, \ldots, \pi_4)$ test can be compared with the Dickey *et al.* (1984) test in (5.4).

The null hypothesis in the HEGY test procedure is that the $(1-B^4)$ filter is the appropriate filter to remove unit roots. This implies that the asymptotics for the various t- and F-tests are nonstandard. Discussions of the relevant asymptotic distributions are given in Engle *et al.* (1993), Hylleberg *et al.* (1990) and Ghysels *et al.* (1994*b*). Like the standard Dickey–Fuller type tests discussed in Section 2.3, these asymptotic distributions depend on the deterministic terms that are included in the auxiliary regression (5.11). Tables with critical values are displayed in HEGY for the $t(\pi_1)$, $t(\pi_2)$, and $F(\pi_3,\pi_4)$ test statistics and in Ghysels *et al.* (1994*b*) for the other two F-statistics. For completeness, I give some critical values in Table 5.1 for 10, 20, 30, and 40 years of quarterly observations and in the cases where (5.11) includes a constant and three seasonal dummies and possibly a linear deterministic trend. For many practical purposes, the latter two choices for the μ_t in (5.10) are the most relevant. Additional tables can be found in the aforementioned references and in Franses and Hobijn (1995).

The HEGY test approach has been evaluated in the simulation exercises in Hylleberg (1995) and in Ghysels *et al.* (1994*b*). The results in the latter study indicate that the size of the tests deteriorates when the DGP is a seasonal MA process with a parameter close to the unit circle. This again corresponds to the findings in Agiakloglou and Newbold (1992). For practical applications, therefore, it seems necessary to check thoroughly the ACF of the estimated residuals. If this estimated ACF does not die out at seasonal lags, one may be very cautious with the interpretation of HEGY test outcomes. Furthermore, and similar to the standard Dickey–Fuller tests, the power of the HEGY tests is not high when the DGP is a model like $y_t = 0.95y_{t-4} + \varepsilon_t$, which is very close to the null hypothesis. Hence, in practice one may too often find seasonal and nonseasonal unit roots. In sum, one should exercise some care with the interpretation of HEGY test outcomes. Somewhat more confidence in the empirical results may however be obtained when the test results appear to be robust to different lag structures and different samples.

Empirical results

The empirical application of the HEGY test procedure is illustrated by investigating the presence of seasonal and nonseasonal unit roots in the 11 sample time series. The results are summarized in Table 5.2. For this illustration, the auxiliary regression (5.11) contains four seasonal dummies and a trend. Similar results are however obtained when the trend is deleted. The lag

TABLE 5.1. Critical values of test statistics for seasonal and nonseasonal unit roots in quarterly time series[a]

Test	Years	Regression I				Regression II			
		0.01	0.025	0.05	0.10	0.01	0.025	0.05	0.10
$t(\pi_1)$	10	–3.42	–3.06	–2.77	–2.44	–4.02	–3.64	–3.34	–3.02
	20	–3.43	–3.09	–2.81	–2.51	–3.97	–3.66	–3.37	–3.06
	30	–3.43	–3.10	–2.83	–2.53	–3.96	–3.65	–3.40	–3.09
	40	–3.41	–3.11	–2.84	–2.54	–3.96	–3.65	–3.39	–3.10
$t(\pi_2)$	10	–3.40	–3.07	–2.77	–2.45	–3.40	–3.06	–2.77	–2.45
	20	–3.40	–3.07	–2.80	–2.51	–3.41	–3.08	–2.81	–2.51
	30	–3.41	–3.10	–2.82	–2.53	–3.40	–3.10	–2.83	–2.53
	40	–3.41	–3.09	–2.83	–2.53	–3.41	–3.10	–2.82	–2.53
$F(\pi_3,\pi_4)$	10	9.32	7.80	6.63	5.44	9.30	7.77	6.56	5.38
	20	8.94	7.65	6.62	5.47	8.86	7.58	6.57	5.44
	30	8.97	7.72	6.70	5.62	8.91	7.67	6.66	5.59
	40	8.79	7.57	6.57	5.52	8.79	7.54	6.55	5.48
$F(\pi_2,\ldots,\pi_4)$	10	8.49	7.18	6.15	5.21	8.41	7.10	6.12	5.15
	20	7.93	6.88	6.04	5.12	7.86	6.84	6.03	5.10
	30	7.91	6.84	6.02	5.18	7.85	6.82	6.01	5.16
	40	7.63	6.73	5.95	5.09	7.62	6.71	5.93	5.09
$F(\pi_1,\ldots,\pi_4)$	10	8.03	6.90	6.00	5.06	9.18	7.83	6.92	5.84
	20	7.42	6.42	5.70	4.94	8.26	7.26	6.47	5.64
	30	7.25	6.35	5.67	4.93	8.12	7.16	6.41	5.65
	40	7.07	6.23	5.56	4.86	7.93	7.01	6.31	5.55

[a] The DGP is $(1-B^4)y_t = \varepsilon_t$, with $\varepsilon_t \sim N(0,1)$, and the auxiliary regression is

$$\text{(I)}\; y_{4,t} = \Sigma^4_{s=1}\delta_s D_{s,t} + \pi_1 y_{1,t-1} + \pi_2 y_{2,t-1} + \pi_3 y_{3,t-2} + \pi_4 y_{3,t-1} + u_t, \text{ or}$$

$$\text{(II)}\; y_{4,t} = \Sigma^4_{s=1}\delta_s D_{s,t} + \beta t + \pi_1 y_{1,t-1} + \pi_2 y_{2,t-1} + \pi_3 y_{3,t-2} + \pi_4 y_{3,t-1} + u_t,$$

where the $y_{s,t}$ variables are defined in (5.11). Based on 25,000 Monte Carlo replications.

Source: Franses and Hobijn (1995); see also Hylleberg *et al.* (1990) and Ghysels *et al.* (1994 *b*) for critical values for some of these and other sample sizes.

structure is determined by first setting the order equal to five, then checking the autocorrelation properties of the residuals using the $F_{\text{AR},\,1\text{–}k}$ tests; when these tests do not indicate misspecification, the model is then simplified by deleting insignificant lags of $\Delta_4 y_t$. Of course, the residuals of the final model should be uncorrelated. From the bottom panel of Table 5.2, it can be observed that sometimes the residuals do not possess white noise properties; i.e. , the null hypotheses of no ARCH and/or normality are sometimes rejected at the 5 per cent level. An inspection of the various error processes reveals that this is due mainly to only a few outlying observations in each of the series. Removal of these outliers using single observation dummy variables does not result in very different conclusions for the HEGY test statistics as reported in the first panel of Table 5.2. Notice that some results in this table differ from those reported in Osborn (1990) because of a different lag structure.

TABLE 5.2. Testing for seasonal and nonseasonal unit roots in 11 quarterly sample series[a]

Variable	Test statistics					
	n^*	$t(\pi_1)$	$t(\pi_2)$	$F(\pi_3,\pi_4)$	$F(\pi_2,..,\pi_4)$	$F(\pi_1,..,\pi_4)$
Industrial production, USA	120	–2.329	–1.533	9.388**	10.352**	13.241**
Unemployment, Canada	103	–3.145	–2.028	7.822**	6.473**	7.543**
Real GNP, Germany	119	–1.845	–3.035**	3.391	5.430	5.091
UK GDP	131	–1.772	–3.113**	9.820**	11.585**	12.792**
UK total consumption	127	–2.724	–1.603	4.373	3.066	3.400
UK consumer nondurables	127	–2.792	–2.429	4.462	2.930	2.050
UK exports	132	–1.589	–5.449**	29.678**	38.970**	33.477**
UK imports	131	–2.747	–4.571**	19.698**	21.014**	20.796**
UK total investment	131	–1.957	–3.124**	11.058**	13.169**	15.132**
UK public investment	100	–1.910	–0.696	2.321	1.772	2.489
UK workforce	130	–2.106	–5.046**	17.109**	20.464**	18.380**
	Diagnostics					
	Lags	$F_{AR,1-1}$	$F_{AR,1-4}$	$F_{ARCH,1-1}$	$F_{ARCH,1-4}$	JB
Industrial production, USA	1–4	1.200	1.058	0.363	0.837	10.186**
Unemployment, Canada	1,2,4,5	0.777	0.223	2.839	0.276	31.132**
Real GNP, Germany	1	0.868	1.553	1.411	1.618	5.278
UK GDP	1	0.466	0.193	0.001	2.212	6.321**
UK total consumption	1–5	1.195	2.332	0.107	3.485**	7.862**
UK consumer nondurables	1,2,4,5	3.004	2.072	0.032	1.231	4.133
UK exports	0	0.435	0.373	16.727**	5.571**	3.503
UK imports	1	2.180	0.574	3.419	1.082	0.445
UK total investment	1	0.396	0.529	6.076**	1.730	6.812**
UK public investment	1,4	1.562	0.453	0.047	0.072	21.156**
UK workforce	1,2	2.251	1.757	0.001	2.097	1.274

[a] The auxiliary regression contains four seasonal dummies and a trend. n^* denotes number of effective observations. Lags indicates which lags of $\Delta_4 y_t$ are included, e.g. 1–4 means 1,2,3,4 and 1,4 means 1 and 4. The model for UK workforce includes a dummy for 1959.2, which is an extreme outlier.

** Significant at the 5 per cent level.

The HEGY test results in Table 5.2 indicate that the nonseasonal unit root appears to be present in all considered time series. Furthermore, three of the sample series have all four unit roots; i.e. , the $(1 - B^4)$ differencing filter seems necessary for UK total and nondurables consumption and for UK public investment. The seasonal unit root –1 is present for US industrial production and for Canadian unemployment. The German real GNP series appears to require the $(1 - B)(1 + B^2)$ differencing filter since it has the unit roots 1 and $\pm i$. The other five series can be differenced using the $(1 - B)$ filter. Broadly speaking, the results in Table 5.2 suggest that the root 1 is often found and that a few series have one or more seasonal unit roots. This corresponds to the findings in Osborn (1990), Beaulieu and Miron (1993), Otto and Wirjanto (1990), and Hylleberg *et al.* (1993).

Robustness of empirical results

As mentioned, additional confidence in the empirical results obtained using the HEGY auxiliary regression in (5.11) is gained when these results are robust to small changes in the lag structure, to the inclusion of single observation dummy variables for outliers, and to variation in the sample size. To investigate the robustness to the selected sample, one can estimate (5.11) recursively in order to detect possible parameter changes. As an illustration, such a recursive analysis of the $F(\pi_3, \pi_4)$ test statistic for the German real GNP series is given in Table 5.3. This table reports the values of this $F(\pi_3, \pi_4)$ test when (5.11) is estimated for the subsamples 1962.1–1970.4 to 1962.1–1990.4 (which can be called forward recursive samples) and for 1965.1–1990.4 to 1983.1–1990.4 (which can be called backward recursive samples). Recall from Table 5.2 that for the sample 1961.2–1990.4 this $F(\pi_3,\pi_4)$ test statistic is insignificant, and hence that there seem to be two seasonal unit roots, $\pm i$. The forward recursive $F(\pi_3,\pi_4)$-test results in Table 5.3 indicate that these seasonal unit roots seem present in all forward subsamples. However, the backward recursive F-test results in the last column indicate a rejection of the null hypothesis when the estimation sample starts in 1970.1–1980.1. Combining the results in these two columns suggests that the empirical finding of seasonal unit roots $\pm i$ may be due to observations in the first part of the sample.

One possibility that may explain the observed finding of only temporarily present seasonal unit roots is that there has been a shift in one or more of the seasonal means. In fact, the results in Perron (1990) show that neglecting deterministic shifts in the mean biases unit root test statistics towards nonrejection. Hence, one may too often find unit roots when a structural shift in a stationary time series is neglected. Similar arguments can be given for seasonal unit roots and seasonal mean shifts. (see also Ghysels 1994). If one suspects that seasonal mean shifts have occurred at time T_b, where for convenience it is assumed that T_b corresponds to an observation in the fourth quarter, it may be useful to consider a modification of the HEGY regression in (5.11). This modification is

TABLE 5.3. Recursive testing for the seasonal unit roots $\pm i$ for real GNP Germany[a]

Sample	$F(\pi_3,\pi_4)$	Sample	$F(\pi_3,\pi_4)$
1962.1–1970.4	3.448	1965.1–1990.4	3.882
1962.1–1971.4	3.886	1966.1–1990.4	3.582
1962.1–1972.4	2.614	1967.1–1990.4	4.286
1962.1–1973.4	1.794	1968.1–1990.4	3.979
1962.1–1974.4	1.824	1969.1–1990.4	6.437
1962.1–1975.4	2.090	1970.1–1990.4	8.688**
1962.1–1976.4	0.698	1971.1–1990.4	11.803**
1962.1–1977.4	0.826	1972.1–1990.4	11.550**
1962.1–1978.4	1.195	1973.1–1990.4	9.922**
1962.1–1979.4	1.323	1974.1–1990.4	9.547**
1962.1–1980.4	1.158	1975.1–1990.4	12.064**
1962.1–1981.4	1.461	1976.1–1990.4	14.901**
1962.1–1982.4	1.650	1977.1–1990.4	9.615**
1962.1–1983.4	1.463	1978.1–1990.4	9.392**
1962.1–1984.4	1.804	1979.1–1990.4	8.047**
1962.1–1985.4	2.740	1980.1–1990.4	6.814**
1962.1–1986.4	2.923	1981.1–1990.4	5.222
1962.1–1987.4	3.000	1982.1–1990.4	4.952
1962.1–1988.4	3.002	1983.1–1990.4	4.135
1962.1–1989.4	2.765	1984.1–1990.4	2.772
1962.1–1990.4	3.039	1985.1–1990.4	2.091

[a] $F(\pi_3,\pi_4)$ denotes the F-test for the significance of π_3 and π_4 in the auxiliary regression $y_{4,t} = \Sigma_{s=1}^{4}\delta_s D_{s,t} + \beta t + \pi_1 y_{1,t-1} + \pi_2 y_{2,t-1} + \pi_3 y_{3,t-2} + \pi_4 y_{3,t-1} + u_t$, where the $y_{s,t}$ variables are defined in (5.11). The samples denote the effective samples for which (5.11) is estimated. Each estimation sample contains complete years; i.e., the number of effective observations is a multiple of 4.

** Significant at the 5 per cent level.

$$\phi^*(B)y_{4,t} = \pi_1 y_{1,t-1} + \pi_2 y_{2,t-1} + \pi_3 y_{3,t-2} + \pi_4 y_{3,t-1} + \beta t + \sum_{s=1}^{4}\delta_s D_{s,t} + \sum_{s=1}^{4}\mu_s DU_{s,t} + \sum_{s=1}^{4}\theta_s D(T_b)_{s,t} + \varepsilon_t \tag{5.13}$$

where the $DU_{s,t}$ are seasonal dummy variables with $DU_{s,t} = 1$ when $t > T_b$ for season $s = 1,2,3,4$ and $DU_{s,t} = 0$ elsewhere, and where the $D(T_b)_{s,t}$ are single observation dummy variables with $D(T_b)_{s,t} = 1$ when $t = T_b + s$, for $s = 1,2,3,4$ and $D(T_b)_{s,t} = 0$ elsewhere. This model in (5.13) is called the innovative outlier model (see Perron 1990, and Franses and Vogelsang 1995). Depending on $\phi^*(B)$, model (5.13) can be enlarged by including lags of $D(T_b)_{s,t}$. Obviously, the critical values of the t-tests for π_1 and π_2, the F-test for (π_3, π_4), and the other joint F-tests for this modified regression differ from those in Table 5.1. In Table 5.4, I give the critical values for a sample of 20 years in the case where $T_b = n/2$, i.e. when the break is known to occur halfway the sample. When the

TABLE 5.4. Critical values for test statistics for seasonal and nonseasonal unit roots in quarterly time series when allowing for structural shifts in the mean to occur halfway the sample[a]

Test	Regression I				Regression II			
	0.01	0.025	0.05	0.10	0.01	0.025	0.05	0.10
$t(\pi_1)$	−3.90	−3.54	−3.26	−2.96	−4.28	−3.92	−3.64	−3.32
$t(\pi_2)$	−3.85	−3.54	−3.27	−2.96	−3.85	−3.52	−3.25	−2.95
$F(\pi_3,\pi_4)$	11.91	10.34	9.09	7.69	11.76	10.30	9.01	7.63
$F(\pi_2,\ldots,\pi_4)$	10.88	9.47	8.45	7.27	10.78	9.42	8.37	7.22
$F(\pi_1,\ldots,\pi_4)$	10.22	9.05	8.15	7.12	10.96	9.68	8.66	7.61

[a] The DGP is $(1-B^4)y_t = \varepsilon_t$, with $\varepsilon_t \sim N(0,1)$, and the auxiliary regression is (I) regression model (5.13) with $\beta = 0$ (no trend), or (II) regression model (5.13) with $\beta \neq 0$ (with trend). Based on 25,000 Monte Carlo replications. Effective sample size is 20 years.

Source: Franses and Hobijn (1995).

critical values in this table are compared with those in Table 5.1, it can be observed that the distribution of the t-test statistics shifts to the left, while the distributions of the F-test statistics shift to the right. The asymptotic distributions are given in Franses and Vogelsang (1995).

To illustrate the empirical application of the auxiliary regression in (5.13), I consider six of the 11 time series in Table 5.2 for which there seem to be one or more seasonal unit roots. The results are presented in Table 5.5, where it is assumed from the outset (and just for illustrative purposes) that there is a break at $T_b = n/2$. The second panel of this table presents the diagnostic results. It appears that for some series the number of lags of $\Delta_4 y_t$ to be included in (5.13) is smaller than in (5.11). The results in the first panel of the table clearly indicate that the previous inference on seasonal unit roots may not be robust to structural mean shifts. Only for the UK public investment series is the $(1 - B^4)$ filter still appropriate. For Canadian unemployment, (5.13) results in two more seasonal unit roots than (5.11) does. For the other four series, the earlier evidence for seasonal unit roots disappears completely. For example, the $F(\pi_3, \pi_4)$-test value for UK total consumption increases from the insignificant value of 4.373 for (5.11) to the highly significant value of 23.144 for (5.13). Similar results apply to other sample series. Notice however that the finding of a unit root at the nonseasonal frequency is robust to shifts in the mean; i.e. , for all six series in Table 5.5, I still obtain evidence that the $(1 - B)$ filter is useful.

The results in Table 5.5 merely serve illustrative purposes in order to indicate that some seasonal unit root inference can be fragile. Of course, in practice one usually does not know the value of T_b. Franses and Vogelsang (1995) propose to select T_b, for example, when the F-test statistic for the joint significance of the μ_s parameters in (5.13) obtains its largest value. These authors also consider the additive outlier model, but that extension is not considered here.

TABLE 5.5. Testing for seasonal and nonseasonal unit roots in six quarterly series when allowing for a structural shift in the seasonal means halfway the sample[a]

Variable	Test statistics					
	T_b	$t(\pi_1)$	$t(\pi_2)$	$F(\pi_3,\pi_4)$	$F(\pi_2,.,\pi_4)$	$F(\pi_1,.,\pi_4)$
Industrial production, USA	1976.4	–2.359	–3.715**	14.962**	14.300**	22.908**
Unemployment Canada	1974.4	–3.302	–2.633	7.346	7.096	8.305
Real GNP, Germany	1974.4	–1.635	–3.632**	8.908**	10.192**	8.552**
UK total consumption	1973.4	–2.459	–3.751**	23.144**	19.341**	16.427**
UK consumer nondurables	1973.4	–2.626	–5.803**	23.367**	41.523**	34.915**
UK public investment	1975.4	–2.192	–1.272	3.969	3.052	3.718
	n^*	Lags	$F_{AR,1-1}$	$F_{AR,1-4}$	$F_{ARCH,1-1}$	$F_{ARCH,1-4}$	JB
Industrial production, USA	122	2	2.355	2.162	3.866	1.352	28.809**
Unemployment, Canada	103	1,2,4,5	1.305	0.366	4.933**	1.244	73.038**
Real GNP, Germany	119	1	1.602	1.531	4.166**	1.568	5.587
UK total consumption	131	1	1.046	1.397	3.625	2.117	10.716**
UK consumer nondurables	127	1,2,4,5	3.004	2.072	0.032	1.231	4.133
UK public investment	100	1,4	1.135	0.450	0.146	0.189	36.664**

[a] The auxiliary regression is (5.13). n^*denotes number of effective observations. Lags indicates which lags have been included; e.g., 1–4 means 1,2,3,4, while 1,4 means 1 and 4. When the auxiliary regression for Canadian unemployment is enlarged with a dummy for 1982.2, the JB test does not indicate misspecification, while the test results for nonseasonal and seasonal unit roots remain virtually the same.

** Significant at the 5 per cent level.

Some remarks

The presence of seasonal unit roots in an autoregressive polynomial for y_t implies that y_t contains so-called seasonal stochastic trends, which in turn implies that the seasonal fluctuations in y_t can change over time because of certain large values of the shocks ε_t. For example, consider the seasonal unit root process

$$y_t = -y_{t-1} + \varepsilon_t, \tag{5.14}$$

which (with $y_0 = 0$) can be written as

$$y_t = \sum_{i=0}^{t-1} (-1)^i \varepsilon_{t-i}. \tag{5.15}$$

This expression indicates that some shocks to y_t can have a permanent effect on the seasonal pattern of y_t, where the seasonal pattern is loosely defined as $y_t - y_{t-1}$. Indeed, if we suppose that (5.15) is the DGP, and consider the regression model as in (3.7) in Chapter 3, i.e. with the $\Delta_1 y_t$ variable regressed on four seasonal dummy variables, we can expect that the estimated seasonal dummy parameters are not constant over time. Using formal asymptotic arguments, Franses *et al.* (1995) show that, for example in the case where $y_t = y_{t-2} + \varepsilon_t$ is the DGP process, and the regression model (3.7) is used to investigate the seasonal patterns, the asymptotic distribution of the R^2 and the parameters for the seasonal dummies are functions of Brownian motions. In other words, we may obtain spuriously large R^2 values and spuriously significant dummy parameters in (3.7) if we neglect seasonal unit roots. Therefore, Franses *et al.* (1995) suggest caution in interpreting such empirical results as in Table 3.1 as suggesting that seasonality is predominantly deterministic.

With respect to the practical application of the test equation (5.11), it should be mentioned that this regression is useful in the case where the AR order for y_t equals or exceeds 4. It can also be used for smaller models, but then one should consider an additional step to the standard HEGY procedure. This is easily demonstrated by the AR(1) process $y_t = \phi_1 y_{t-1} + \varepsilon_t$, for which the ACF is ϕ_1^k for $k = 0,1,2, \ldots$ Suppose one considers the process $x_t = (1 + B)y_t$, which resembles the $y_{1,t}$ variable in the HEGY regression (5.11). It is easy to see that the first autocorrelation of this x_t series is $(\phi_1 + 1)/2$, which exceeds ϕ_1 where $\phi_1 \in (0,1)$. Hence, where a stationary AR(1) model is the DGP, and the standard HEGY test equation is used, one may lose power in the test for the nonseasonal unit root. In practice, therefore, one may consider an additional step where there are no seasonal unit roots, i.e. a standard ADF test in a regression that includes seasonal dummies. Dickey *et al.* (1986) argue that the standard Dickey–Fuller asymptotics apply for this regression.

Two final remarks conclude this section on seasonal unit roots. The first concerns the approach in Canova and Hansen (1995) where a test procedure is developed that considers the null hypothesis of constant seasonality versus unit root seasonality. This and the HEGY method are evaluated in simulation experiments in Hylleberg (1995), where it is concluded that the two approaches seem complementary in the sense that one method is useful when the other is not, and vice versa. The second remark concerns an alternative differencing filter for seasonal series, i.e. the fractional differencing filter. Porter-Hudak (1990) and Ray (1993) consider the differencing filter $(1 - B^S)^D$ where D can take fractional values. It is found that such fractional processes can adequately describe and forecast seasonal time series.

5.2 Seasonal Cointegration

The preceding section was confined to a review of developments in univariate time series analysis. Of course, most of this analysis is usually performed prior to the construction of multivariate time series models. Once one has decided on the most appropriate differencing filter, i.e. the appropriate number of seasonal and nonseasonal unit roots in univariate time series, the next step may be to relate the differenced time series using, for example, transfer function analysis. Typically, in traditional time series analysis one constructs models for two or more $\Delta_1\Delta_4$ or Δ_4 differenced time series. However, we have seen in the previous section that probably only for a few time series are these differencing filters required, and that more often there may be only a few seasonal unit roots. Furthermore, it may be that univariate time series need the Δ_4 filter, while a linear combination of $(1 + B + B^2 + B^3)$ transformed series is already stationary. In other words, two time series each with a nonseasonal unit root and a few seasonal unit roots may have such roots in common; linear combinations of appropriately transformed time series do not possess such unit roots. This calls for an extension of the cointegration concept as defined in Engle and Granger (1987), see also Engle *et al.* (1989).

Residual based method

Engle *et al.* (1993) (EGHL) propose a test method for the presence of seasonal and nonseasonal cointegration relations. Suppose that two time series y_t and x_t have some or all unit roots at nonseasonal and/or seasonal frequencies. When there is cointegration at the zero frequency, i.e. when y_t and x_t have a common nonseasonal unit root, then the process u_t defined by

$$u_t = (1 + B + B^2 + B^3)y_t - \alpha_1(1 + B + B^2 + B^3)x_t \tag{5.16}$$

is a stationary process. Seasonal cointegration at the biannual frequency π, corresponding to unit root -1, amounts to the stationarity of the process v_t, which is defined by

$$v_t = (1 - B + B^2 - B^3)y_t - \alpha_2(1 - B + B^2 - B^3)x_t. \tag{5.17}$$

Finally, seasonal cointegration at the annual frequency $\pi/2$, corresponding to the unit roots $\pm i$, amounts to the stationarity of the process w_t, defined by

$$w_t = (1 - B^2)y_t - \alpha_3(1 - B^2)x_t - \alpha_4(1 - B^2)y_{t-1} - \alpha_5(1 - B^2)x_{t-1}. \tag{5.18}$$

In the case where all three u_t, v_t, and w_t series are stationary, a simple version of a seasonal cointegration model is

$$\Delta_4 y_t = \gamma_{11}u_{t-1} + \gamma_{21}v_{t-1} + \gamma_{31}w_{t-2} + \gamma_{41}w_{t-3} + \varepsilon_{1,t}, \tag{5.19}$$

$$\Delta_4 x_t = \gamma_{12}u_{t-1} + \gamma_{22}v_{t-1} + \gamma_{32}w_{t-2} + \gamma_{42}w_{t-3} + \varepsilon_{2,t}, \tag{5.20}$$

where γ_{1j} to γ_{4j} (for $j = 1,2$) are error correction parameters.

The test method proposed in EGHL is a two-step method, similar to Engle and Granger's approach to nonseasonal time series. The first step involves the estimation of the $\alpha_1 - \alpha_5$ parameters in (5.16), (5.17), and (5.18) via simple regressions, where such regressions may include a constant, seasonal dummies, and a trend if necessary, and a test of whether the residual processes $\hat{u}_t$, $\hat{v}_t$, and $\hat{w}_t$ are stationary. The second step is to replace the u_t, v_t, and w_t processes in (5.19) and (5.20) by their estimated counterparts, and to test the significance of the adjustment parameters. The latter step involves standard $N(0,1)$ asymptotics for the t-values for the γ_{ij} parameters ($i = 0,1,2,3,4$, $j = 1,2$), while the first step involves (extensions of the) Engle–Granger (1987) type asymptotics. For example, to test for nonseasonal cointegration, we need to check whether $\rho = 0$ in the auxiliary regression

$$(1 - B)\hat{u}_t = \rho\hat{u}_{t-1} + \sum_{i=1}^{k}\lambda_i(1 - B)\hat{u}_{t-i} + \varepsilon_t. \tag{5.21}$$

The critical values of t-test for ρ are those tabulated in Engle and Granger (1987) (see also Table 2.2 above). Similarly, to test for seasonal cointegration at frequency π, we test $\rho = 0$ in the auxiliary regression

$$(1 + B)\hat{v}_t = -\rho\hat{v}_{t-1} + \sum_{i=1}^{k}\lambda_i(1 + B)\hat{v}_{t-i} + \varepsilon_t, \tag{5.22}$$

where again we can use the Engle–Granger (1987) critical values. The test for seasonal cointegration at the frequency $\pi / 2$ is somewhat more complicated, and I refer the interested reader to EGHL for details on asymptotics and critical values.

Full system method

The presence of seasonal cointegration in a multivariate time series can also be analysed using an extension of the Johansen maximum likelihood approach (see Section 2.4). This approach is developed in Lee (1992) and Lee and Siklos (1992); see also Kunst (1993) for a recent application. This method amounts to an investigation of the ranks of certain matrices in a rewritten version of a VAR model. These matrices correspond to variables which are transformed using the various filters that are nested within the $(1 - B^4)$ filter. More precisely, consider the $m \times 1$) vector process Y_t, and assume that it can be described by the VAR(p) process

$$\mathbf{Y_t} = \Theta\mathbf{D}_t + \Phi_1\mathbf{Y}_{t-1} + \ldots + \Phi_p\mathbf{Y}_{t-p} + \mathbf{e}_t, \tag{5.23}$$

where the $\Phi_1 - \Phi_p$ are $(m \times m)$ parameter matrices, $\mathbf{D}_t$ is the (4×1) vector process $\mathbf{D}_t = (D_{1,t}, D_{2,t}, D_{3,t}, D_{4,t})$ containing the seasonal dummies, and Θ is an $(m \times 4)$ parameter matrix. Similar to the standard Johansen approach, and conditional on the assumption that $p > 4$, model (5.23) can be rewritten as

$$\begin{aligned}\Delta_4\mathbf{Y}_t = {} & \Theta\mathbf{D}_t + \Pi_1\mathbf{Y}_{1,t-1} + \Pi_2\mathbf{Y}_{2,t-1} + \Pi_3\mathbf{Y}_{3,t-2} + \Pi_4\mathbf{Y}_{3,t-1} \\ & + \Gamma_1\Delta_4\mathbf{Y}_{t-1} + \ldots + \Gamma_{p-4}\Delta_4\mathbf{Y}_{t-(p-4)} + \mathbf{e}_t,\end{aligned} \tag{5.24}$$

where

$$\begin{aligned}\mathbf{Y}_{1,t} &= (1 + B + B^2 + B^3)\mathbf{Y}_t \\ \mathbf{Y}_{2,t} &= (1 - B + B^2 - B^3)\mathbf{Y}_t \\ \mathbf{Y}_{3,t} &= (1 - B^2)\mathbf{Y}_t.\end{aligned}$$

Obviously, (5.24) is a multivariate extension of the model in (5.11). The ranks of the matrices Π_1, Π_2, Π_3, and Π_4 determine the number of cointegration relations at a certain frequency. Lee (1992) shows that seasonal cointegration at the $\pi/2$ frequency can also be checked using Π_3 only. Similar to the Johansen approach, one can now construct residual processes from regressions of $\Delta_4\mathbf{Y}_t$ and the $\mathbf{Y}_{1,t-1}$, $\mathbf{Y}_{2,t-1}$, $\mathbf{Y}_{3,t-2}$, and $\mathbf{Y}_{3,t-1}$ on lagged $\Delta_4\mathbf{Y}_t$ time series and deterministics, and construct the relevant moment matrices. Solving four eigenvalue problems results in sets of estimated eigenvalues which can be checked for their significance using the trace-test statistic (see also Section 2.4). Table 5.6 presents some simulated critical values in cases where m is at most 3. The critical values obtaining where the regression model in (5.23) does not contain seasonal dummies and the asymptotic distributions of the various test statistics are given in Lee (1992).

With the empirical time series considered in this chapter, I am not able to illustrate the practical application of the seasonal cointegration methods. Hence such an application is postponed to Chapter 9, where the seasonal cointegration model will be considered relative to an alternative model for

TABLE 5.6. Critical values for nonseasonal and nonseasonal cointegration[a]

m–r	Sample	Size	Cointegration at frequency		
			0	π	$\pi/2$
1	100	0.01	12.3	12.4	16.2
		0.05	8.6	8.6	11.9
		0.10	7.0	6.9	9.9
	200	0.01	11.9	12.0	15.7
		0.05	8.4	8.4	11.4
		0.10	6.8	6.7	9.5
2	100	0.01	24.5	24.5	30.2
		0.05	19.3	19.3	24.3
		0.10	16.9	17.0	21.6
	200	0.01	23.4	23.5	28.9
		0.05	18.5	18.8	23.5
		0.10	16.2	16.4	20.7
3	100	0.01	41.1	41.0	47.7
		0.05	34.5	34.4	40.6
		0.10	31.1	31.3	37.1
	200	0.01	39.2	39.2	46.1
		0.05	32.9	33.1	39.0
		0.10	30.0	30.0	35.7

[a] The auxiliary regression is (5.23). The m is the number of variables and r is the cointegration rank.

Source: Lee and Siklos (1992).

multivariate seasonal time series with unit roots for an additional set of empirical time series.

A final remark: in a discussion of the EGHL paper, it is mentioned in Osborn (1993) that the seasonal cointegration model for y_t as in (5.19) can be written as

$$\Delta_4 y_t = \sum_{s=1}^{4} \kappa_s (y_{t-s} - \delta_s x_{t-s}) + \varepsilon_{1,t}. \tag{5.25}$$

This expression indicates that a full cointegration model assumes long-run equilibrium relations between y_t and x_t which vary with the lag. For example, the long-run relation at $t-1$ is different from that at $t-2$.

Conclusion

In this chapter I have reviewed models and test methods that concern seasonal unit roots in univariate and multivariate time series. We found that the double-differencing filter $\Delta_1\Delta_4$, which is often considered in Box–Jenkins type procedures, assumes too many unit roots. In fact, often only a few seasonal unit roots and one nonseasonal unit root appear to be present in empirical time series.

The similarity between the seasonal unit root and seasonal cointegration models and the Box–Jenkins (and seasonal adjustment) approaches discussed in the previous two chapters is that it is assumed that nonstationary seasonal fluctuations can be removed from the data using some differencing filter. Hence in a sense it is assumed that one can identify seasonal and nonseasonal components in a single time series, and that these components can be analysed separately. In the next chapter I elaborate on the possibility that sometimes such separate components cannot be obtained.

Appendix 5.1

This appendix reproduces the proposition in Hylleberg *et al.* (1990: 221–2) which is necessary to construct the auxiliary regression (5.11).

Proposition Any (possibly infinite or rational) polynomial, $\phi(z)$, which is finite valued at the distinct, non-zero, possibly complex points, $\theta_1, \ldots, \theta_p$, can be expressed in terms of elementary polynomials and a remainder as follows:

$$\phi(z) = \sum_{k=1}^{p} \lambda_k \Delta(z) / \delta_k(z) + \Delta(z)\phi^{**}(z), \tag{5.A1}$$

where the λ_k are constants defined by

$$\lambda_k = \phi(\theta_k) / \prod_{j \neq k} \delta_j(\theta_k), \tag{5.A2}$$

$\phi^{**}(z)$ is a (possibly infinite or rational) polynomial, and

$$\delta_k(z) = 1 - (1/\theta_k)z, \tag{5.A3}$$

$$\Delta(z) = \prod_{k=1}^{p} \delta_k(z). \tag{5.A4}$$

An alternative form of (5.A1) is

$$\phi(z) = \sum_{k=1}^{p} \lambda_k \Delta(z)(1 - \delta_k(z)) / \delta_k(z) + \Delta(z)\phi^{*}(z), \tag{5.A5}$$

where $\phi^{*}(z) = \phi^{**}(z) + \Sigma\lambda_k$. From the definition of λ_k in (5.A2), it can be seen that the polynomial $\phi(z)$ will have a root at θ_k if and only if the corresponding λ_k equals zero.

Proof. See Hylleberg *et al.* (1990: 222).

6

Are Seasons, Trends, and Cycles always Independent?

When an economic time series displays seasonal variation that seems to increase with an increasing trend, we usually analyse the data after a natural logarithm transformation. An alternative strategy is to construct an explicit model for the increasing seasonal variation using, for example, seasonal dummy variables which interact with linear trends. (See Bowerman *et al.* (1990) for a comparison of the two approaches.) The use of logarithmic data seems particularly useful when an untransformed series y_t can be decomposed as $y_t = y_t^s y_t^{ns}$. A simple and stylized example is given by the equation $Y_{2,T} = 2Y_{1,T}$, where $Y_{s,T}$ $(s = 1,2)$ are the annual time series for quarters 1 and 2. Suppose $Y_{1,T}$ is an upward-trending time series; then $Y_{2,T}$ is trending as well. However, it is clear that the difference between $Y_{2,T}$ and $Y_{1,T}$, which reflects the seasonal variation, is increasing with the trend. The log transformation results in $\log Y_{2,T} = \log(2) + \log Y_{1,T}$, and hence the difference between the two annual time series in logs is constant. In current empirical time series analysis, it is standard practice to use this log transformation.

There are however economic variables for which seasonal and nonseasonal variation may not be separated just by taking logs. (See Wisniewski (1934) for an early reference to the interdependence of cyclical and seasonal variation and Hylleberg (1994) for a recent discussion.) This may be caused by, say, economic agents who can behave differently in recessions than in expansions, thereby establishing different seasonal patterns across these regimes. For example, some retail sales variables may show peaks in the clearance sales seasons, where the start and length of those seasons can depend on the current state of the economy. As another example, the growth in disposable income in several industrialized countries has created the possibility of people having more than one holiday per year, and this can have an effect on the seasonal pattern in nondurables consumption. For some economic time series, therefore, it may not be easy (or even possible) strictly to isolate seasonal and nonseasonal components since the nonseasonal and seasonal fluctuations can interact.

In this chapter some examples are given of the possible presence of an interaction between seasonal and nonseasonal components in economic variables. The investigation of this correlation in the present chapter is based on univariate time series analysis. For cross-sectional results, the reader is referred to

Miron (1994), among others. In that study it is documented that for variables such as real GNP the R^2 of the regression (3.7) is positively correlated with the variance of the estimated residuals across industrialized countries. This empirical result is taken as evidence that seasonal variation is correlated with business cycle variation across countries. In contrast, in the present chapter I focus on the correlation between seasonal and nonseasonal variation within a single time series. Notice that this assumes that one has some possibility of separating the nonseasonal from the seasonal fluctuations. In Chapters 8 and 9 I return to the empirical validity of this assumption in more detail, and question this validity.

Section 6.1 examines a simple method of checking the orthogonality of the seasonal and nonseasonal components once they are estimated using such seasonal adjustment methods as reviewed in Chapter 4. I apply the method to the US industrial production and Canadian unemployment data and find for some quarters that the estimated seasonal and nonseasonal components do not seem to be independent. This suggests that, for example, the seasonally adjusted time series can take too high or too low a value at certain seasons, which in turn may lead to biased forecasts or inappropriate policy implications. In Section 6.2 I investigate the correlation between business cycles and seasonal patterns somewhat more formally. For this purpose I draw upon some of the methods proposed in Canova and Ghysels (1994) and Ghysels (1994). Finally, in Section 6.3 I analyse the properties of various consumer confidence indices. I find evidence for pronounced seasonal fluctuations in these indices, even though the questions used to quantify consumer confidence concern annual variables. I interpret this evidence as indicating that economic agents may face difficulties in disentangling nonseasonal from seasonal fluctuations. The chapter is concluded with the suggestion that it may be useful to consider univariate and multivariate time series models that in principle can allow for the interaction between seasonal and nonseasonal fluctuations.

6.1 Correlation between Seasonal and Nonseasonal Components

In Chapter 4 I discussed the regression of the annual changes of the seasonal component $\Delta_4\hat{y}_t^s$ on the annual changes of the nonseasonal component $\Delta_4\hat{y}_t^{ns}$, which may be useful when investigating the empirical orthogonality of the two estimated components. This regression is similar to the regression proposed in Shiskin and Eisenpress (1957), which is used to check whether the time series should have been more appropriately transformed by taking natural logs. In Section 4.2 I found that there is no clear correlation between $\Delta_4\hat{y}_t^s$ and $\Delta_4\hat{y}_t^{ns}$ for three of the running sample series. Given the results summarized in the graphs in Chapter 3, where I found that a first-differenced time series may display different properties in different seasons, in the sense that growth rates in some quarters appear to be trending while they are constant in other quarters, it seems reasonable to check the correlation between $\Delta_4\hat{y}^s_t$ and $\Delta_4\hat{y}^{ns}_t$ in different seasons. This amounts to estimating the correlation between $\Delta\hat{Y}^s_{s,\mathrm{T}}$ and $\Delta\hat{Y}^{ns}_{s,\mathrm{T}}$ in season $s = 1,2,3,4$, where Δ denotes the first-order differencing filter for annual time series. (See Section 4.1 for more details on notation.) A straightforward regression of $\Delta\hat{Y}^s_{s,\mathrm{T}}$ on $\Delta\hat{Y}^{ns}_{s,\mathrm{T}}$ may not be useful in the case where the seasonal component $\hat{y}^s_t$ is approximately constant. Indeed, if $\hat{Y}^s_{s,\mathrm{T}}$ is (almost) constant, an ARMA type of model for the $\Delta\hat{Y}^s_{s,\mathrm{T}}$ series will contain a (near) unit root in the MA polynomial. It seems useful therefore to consider the following auxiliary regression:

$$\Delta\hat{Y}^s_{s,\mathrm{T}} = \mu_s + \beta_s\Delta\hat{Y}^{ns}_{s,\mathrm{T}} + \varepsilon_{s,T} + \theta_{1s}\varepsilon_{s,\mathrm{T}-1} + \ldots + \theta_{qs}\varepsilon_{s,\mathrm{T}-q}, \tag{6.1}$$

where μ_s, β_s, and θ_{1s}–θ_{qs} are parameters that vary with the season. A model like (6.1) is called a periodic regression model. In other words, for every season s ($s = 1,2,3,4$) one regresses $\hat{Y}^s_{s,\mathrm{T}} - \hat{Y}^s_{s,\mathrm{T}-1}$ on a constant and on $\hat{Y}^{ns}_{s,\mathrm{T}} - \hat{Y}^{ns}_{s,\mathrm{T}-1}$, where in every regression the error process is assumed to be an MA(q) process. Given that this auxiliary regression is considered for each season s, one may obtain different parameter estimates for the μ, β, and MA parameters. Notice that these parameters are estimated for $N - 1$ observations. To denote the seasonal variation in the parameters, I use the notation as in (6.1), where an index s for a certain parameter indicates that it can take different values in different seasons. Notice that in Chapter 4 I considered the regression as in (6.1) with $\mu_s = \mu$, $\beta_s = \beta$, and $q = 0$. The auxiliary regression in (6.1) is also considered in Franses and Ooms (1995) to analyse German and US unemployment data.

The most relevant parameter in (6.1) is β_s. The empirical significance of this parameter is an indication that the annual change in the nonseasonal component explains part of the annual change in the seasonal component. This implies for example that, when the nonseasonal component increases and β_s is positive, then the seasonal component increases too. Significant β_s parameters therefore indicate that the seasonal and nonseasonal components are correlated. When the parameters are all positively or all negatively significant in all seasons, this can be taken as evidence that the data should be transformed, for example by taking logs. When these parameters are significant in some seasons and not in others, or when the parameters are sometimes negative and sometimes positive, this is an indication that the degree of correlation between the seasonal and nonseasonal components varies throughout the year. The latter correlation structure cannot be removed simply by taking logs.

Table 6.1 reports the t-ratios of the $\hat{\beta}_s$ parameters in (6.1) for the industrial production in the USA. The first panel of the table presents the t-ratios when the data are in logs, and the second panel, the t-ratios when they are not in logs. The seasonal components for the log-transformed data are given by $\log y_t - \log\hat{y}^{ns}_t$ and for untransformed data by $y_t - \hat{y}^{ns}_t$. The final three columns in table contain the estimated t-ratios where (6.1) includes a single observation dummy variable for 1968. The inclusion of this dummy variable yields approximately

TABLE 6.1. Estimated t-ratios for $\hat{\beta}_s$ in the auxiliary regression $\Delta \hat{Y}^s_{s,T} = \mu_s + \beta_s \Delta \hat{Y}^{ns}_{s,T} + \varepsilon_{s,T} + \theta_1 \varepsilon_{s,T-1} + \ldots + \theta_q \varepsilon_{s,T-q}$: industrial production in the USA, 1960.1–1991.4

Season	q (all data)			q (without outlier[a])		
	0	1	3	0	1	3
Data in logs						
1	1.734	1.494	1.375	2.001**	1.926	1.845
2	–0.080	–0.073	0.610	–0.312	–0.312	0.071
3	–2.026**	–2.108**	–4.079**	–2.602**	–2.530**	–3.902**
4	0.361	1.082	1.084	1.050	1.038	1.125
Data not in logs						
1	0.385	0.365	–0.470	0.398	0.386	0.279
2	0.696	0.703	0.583	0.603	0.653	–0.182
3	–1.582	–1.540	0.006	–1.807	–1.772	–0.515
4	0.306	0.312	–0.451	0.241	0.308	0.093

[a] The observation for 1968 is not included in all regressions. Removal of this observation yields approximately normal estimated residuals.

** Significant at the 5 per cent level. It is assumed that the t-ratios follow asymptotic normal distributions. There may be a bias, however, since the $\hat{Y}^s_{s,T}$ and $\hat{Y}^{ns}_{s,T}$ time series are estimated time series.

normal residuals in each of the four regressions for $N - 1$ observations. Finally, the value of q in each season s is set equal to 0, 1, or 3, in order to investigate the robustness of the empirical results to variation in the dynamics of the regression model.

For log-transformed industrial production data in the first panel of Table 6.1, I find that the $\hat{\beta}_s$ appear insignificant from zero for the first, second, and fourth quarter, and for the untransformed data in the second panel I find that the $\hat{\beta}_s$ are insignificant for all seasons. This result in robust to outliers and to variation in the value of q. The only significant $\hat{\beta}_s$ values are those in the third quarter for log-transformed data, where these $\hat{\beta}_3$ values are all found to be significantly negative. The implication of this result is that when $\hat{Y}^{ns}_{s,T}$ increases in the third quarter the estimated seasonal component $\hat{Y}^s_{s,T}$ decreases. Hence in times of a positive underlying trend, which may correspond to an expansion in the US economy, the estimated seasonal component can be too small, and therefore the estimated nonseasonal component can be too large. In a sense, this implies that one can be too optimistic in the third quarter in expansion periods when one considers the seasonally adjusted US industrial production series. Moreover, when the $\hat{Y}^{ns}_{3,T}$ can be too large, the value of $\hat{Y}^{ns}_{4,T} - \hat{Y}^{ns}_{3,T}$ can be too small or can even become negative. In the latter case, one may think that a recession starts in the fourth quarter when effectively it may start in the first quarter of the next year. In Section 6.2 I return to this issue of determining business cycle turning points, and document that there are indeed many business cycle peaks in the third quarter for US industrial production.

The results in Table 6.1 suggest that it may matter whether one analyses a time series in logs or not in logs. For the US industrial production, which in fact is usually considered in logs, the seasonal and nonseasonal components do not seem to be correlated when the data are analysed in untransformed form. It may be that these results are due to the particular procedure that is used to seasonally adjust the production series. Since I have no precise information on this, I have to leave this issue for further theoretical and empirical research.

The second series analysed using the auxiliary regression in (6.1) is the unemployment series in Canada. The estimated t-ratios for β_s in (6.1) are reported in Table 6.2. For this series I observe from the second panel of the table that for untransformed data the $\hat{\beta}_s$ parameters ($s = 1,2,3,4$) are significant for all seasons and for all choices of q. For these untransformed data the $\hat{\beta}_s$ is positive in the first two quarters, while it is negative in the last two quarters. This implies that, in times of increasing unemployment (usually in recessions), the estimated seasonal component is too large in quarters 1 and 2, and hence that the seasonally adjusted data are too small, which in turn implies that one can be too optimistic in the first two quarters during recession periods. This suggests that one may have the impression that a recession is (almost) over when in fact it is not. Of course, for similar reasons one may be too pessimistic in the last two quarters in recessions. The first panel of Table 6.2 shows that the results for

TABLE 6.2. Estimated t-ratios for $\hat{\beta}_s$ in the auxiliary regression $\Delta\hat{Y}^s_{s,T} = \mu_s + \beta_s\Delta\hat{Y}^s_{s,T} + \varepsilon_{s,T} + \theta_1\varepsilon_{s,T-1} + \ldots + \theta_q\varepsilon_{s,T-q}$: unemployment in Canada, 1960.1–1987.4

Season	q (all data)			q (without outlier[a])		
	0	1	3	0	1	3
Data in logs						
1	1.310	1.321	1.169	0.012	–0.055	–0.064
2	–1.887	–1.711	–2.164**	–0.999	–0.818	–1.041
3	–1.415	–1.387	–1.499	–3.638**	–2.844**	–4.148**
4	2.461**	1.729	1.677	2.717**	1.925	2.309**
Data not in logs						
1	5.105**	5.019**	5.111**	5.652**	5.536**	5.681**
2	2.606**	2.614**	3.324**	2.883**	2.851**	2.521**
3	–7.143**	–6.915**	–7.174**	–9.537**	–9.677**	–10.979**
4	–6.936**	–9.369**	–8.334**	–6.867**	–8.944**	–7.365**

[a] The observation for 1966 is not included in all regressions. Removal of this observation yields approximately normal estimated residuals.

** Significant at the 5 per cent level. It is assumed that the t-ratios follow asymptotic normal distributions. There may be a bias, however, since the $\hat{Y}^s_{s,T}$ and $\hat{Y}^{ns}_{s,T}$ time series are estimated time series.

log-transformed time series are not as pronounced as for the untransformed time series, and that the sign of the $\hat{\beta}_s$ changes. It appears that for the log-transformed data significant t-ratios can be found in the third quarter, where these t-ratios are negative, and in the fourth quarter, where these are positive.

All in all, the empirical results reported in Tables 6.1 and 6.2 suggest that the estimated seasonal and nonseasonal components (when these are estimated using official seasonal adjustment methods) can be correlated, and that the sign of this correlation can vary with the seasons. Whether this can affect the determination of, say, business cycle turning-points will be discussed next.

6.2 Business Cycles and Changes in Seasonal Patterns

The results in the previous section indicate that the seasonal and nonseasonal components in economic time series can be correlated. Since applied analysis in macroeconomics can be concerned with the analysis of business cycle fluctuations, it is of natural interest to investigate whether this correlation can affect the determination of, say, business cycle turning points and the duration of recessions. In this section I will briefly discuss recent studies in Ghysels (1994) and Canova and Ghysels (1994) that focus on this issue, and where it is found that business cycle fluctuations do not seem independent of seasonal fluctuations. I emphasize this finding here by reconsidering the US industrial production series within the light of these two empirical studies.

Seasonality in business cycle turning points

In Ghysels (1994) it is documented that regime shifts in the US economy from a recession to an expansion and vice versa are not equally distributed throughout the year. In other words, there are some seasons in which it appears to be more likely that an economy will pull out of a recession than it is in other seasons. Furthermore, it is found that the length of a recession or expansion also depends on the season in which such a business cycle period starts. The results in Ghysels (1994) are obtained using a so-called periodic Markov switching regime model, which extends the Hamilton (1989) model with seasonally varying transition probabilities. The data considered in Ghysels (1994) are the NBER business cycle reference dates since 1857. Note that the NBER decides on turning points using a set of seasonally adjusted time series.

To emphasize the findings in Ghysels (1994), consider the industrial production data for the USA to estimate peaks and troughs in the US business cycle. Of course, this industrial production series is not the only variable that is considered by the NBER to construct the turning points. Other relevant series concern, for example, employment, monetary aggregates, and financial indices (See Watson (1994) for an analysis of some of these series.) It can however be expected that the industrial production series is somehow relatively important in determining the business cycle turning points. This seems to be reflected by the peaks and troughs that are recorded in Table 6.3. In this table I give the official NBER peaks and troughs, and those obtained when considering the first differences of the seasonally adjusted industrial production series, i.e. $\Delta_1 y_t^{ns}$. A period is considered to be a recession period when two consecutive quarters display negative growth. The quarter before a sequence of two negative growth observations is considered to be a peak, while the last observation of a recession period is considered to be a trough. Alternative methods for the determination of turning points are surveyed in Boldin (1994). The numbers in parentheses are the leads and lags relative to the NBER data. The NBER peaks and troughs can be found in Watson (1994).

The results in Table 6.3 indicate that, for the sample 1960.1–1991.4, there are eight recessions using the industrial production series, which is two more than the NBER finds. Most turning points for the industrial production series match those of the NBER, with the noticeable difference that the start of the 1974/75 recession is 1973.4 according to the NBER and 1974.3 according to the industrial production series. The key feature of the peaks obtained using the production series is that four peaks occur in the third quarter and three in the first quarter. This seems to suggest (similar to the findings in Ghysels 1994) that there is seasonal variation in business cycle peaks when using the industrial

TABLE 6.3. Peaks and troughs according to the NBER and obtained using the first differenced seasonally adjusted data for US industrial production[a]

NBER		Industrial production			
Peak	Trough	Peak		Trough	
1960.2	1961.1	1960.1	(–1)	1961.1	(0)
		1966.4		1967.2	
1969.4	1970.4	1969.3	(–1)	1970.4	(0)
1973.4	1975.1	1974.3	(+3)	1975.2	(+1)
		1979.1		1979.3	
1980.1	1980.3	1980.1	(0)	1980.3	(0)
1981.3	1982.4	1981.3	(0)	1982.4	(0)
1990.3	1991.2	1990.3	(0)	1991.1	(–1)

[a] For the US industrial production data, a period is considered to be a recession period when two consecutive quarters display negative growth. For SA data this amounts to constructing $\Delta_1 y_t^{ns} = y_t^{ns} - y_{t-1}^{ns}$, and to checking whether $\Delta_1 y_t^{ns}$ is negative. The quarter before a sequence of two such negative growth observations is considered to be a peak, while the last observation of a recession period is considered to be a trough. The numbers is parentheses are the leads and lags relative to the NBER dates. The NBER peaks and troughs can be found in Watson (1994), among others.

production series since it seems more likely that a recession starts in the second or fourth quarter than in the other quarters. Note that there does not appear such an obvious seasonal variation for the troughs in Table 6.3.

The result that a peak may be more often found in the third quarter seems to correspond to the finding in Table 6.1 that in expansions the seasonally adjusted time series may take too large a value in the third quarter, and that the $\Delta_1 y_t^{ns}$ series may then become negative in the fourth quarter. Jointly interpreting the results in Tables 6.1 and 6.3, it seems that the finding of many peaks in the third quarter may be related to the observation that the seasonal and nonseasonal components are correlated. In other words, it may be for this industrial production series that it is not appropriate to base turning points on this seasonally adjusted time series. We return to this issue in somewhat more detail in Chapter 8, where I propose an alternative dating method that may reduce the possibility of seasonally varying turning points.

Cycles and changing seasonality

Another recent study that considers the relationship between business cycle fluctuations and seasonal patterns is Canova and Ghysels (1994). In addition to formal statistical tests for predictive failure, these authors consider (versions of) the regression model

$$\phi_p(B)\Delta_1 y_t = \sum_{s=1}^{4} \delta_s D_{s,t} + \sum_{s=1}^{4} \delta^r_s D^r_{s,t} + \varepsilon_t, \tag{6.2}$$

where $D^r_{s,t}$ are seasonal dummy variables which take nonzero values in recessions only, and δ_s and δ^r_s are unknown parameters for the seasons $s = 1,2,3,4$. One may now consider testing whether the $D^r_{s,t}$ dummies have explanatory value for $\Delta_1 y_t$. However, the significance of these variables does not convey information on whether the seasonal fluctuations themselves differ across business cycle stages since, typically, $\Delta_1 y_t$ may anyhow be negative in recessions. Hence, it seems more appropriate to consider the null hypothesis that $\delta^r_s = \delta^r$ for $s = 1,2,3,4$. Under this hypothesis, the seasonal fluctuations correspond to δ_s for all the observations and $\delta_s + \delta^r_s$ in recessions, and hence the seasonal fluctuations are then similar across the cycle. Under the alternative hypothesis, $\delta^r_s \neq \delta^r$ for all r, the seasonal pattern in recessions is different from that in expansions.

Based on forecasts from (6.2) and on the statistical tests for predictive failure, Canova and Ghysels (1994) conclude not only that seasonal patterns are unstable, but also that changes in these patterns are linked with the business cycle stages. In Chapter 3 above, I found that the seasonal pattern in the US industrial production series may not be constant, and I now proceed to investigate whether these changes correspond to changes in the cycle. For this purpose, I construct four dummy variables $D^r_{s,t}$ using the official recession periods as indicated by the NBER (see Table 6.3). Some experimentation yields that an empirically adequate regression model like (6.2) contains lags of $\Delta_1 y_t$ at 1, 3, 4, and 5. The usual diagnostic measures do not indicate that this model is misspecified. Since the first six observations are not used because of taking first differences and including the fifth lag of $\Delta_1 y_t$, the NBER recession of 1960.3–1961.1 is not included in the estimation sample. The auxiliary regression model is estimated for 122 observations, which is a sample that covers five recessions. In total, these recessions concern 19 observations, of which there are five observations in the first, second, and fourth quarters, while there are only four recession observations for the third quarter. Hence the recessions seem to be equally spread over the four quarters. The F-test for the equivalence of the δ_s parameters in (6.2), i.e. $\delta_s = \delta$ for all s, obtains the insignificant value of 1.264, while the F-test for the hypothesis $\delta^r_s = \delta^r$ obtains the highly significant value of 7.626. Notice that, in order for this inference to be valid, I have to assume that the Δ_1 filter is adequate to transform the data to stationarity. Based on a standard t-test, it appears that $D^r_{1,t}$ can be deleted from the model. Imposing all apparently valid parameter restrictions, the final auxiliary regression model is

$$\begin{aligned}\Delta_1 y_t = {} & \underset{(0.002)}{0.014}D^e_t - \underset{(0.005)}{0.032}D^r_t + \underset{(0.008)}{0.034}D^r_{2,t} + \underset{(0.008)}{0.026}D^r_{3,t} + \underset{(0.069)}{0.143}\Delta_1 y_{t-1} \\ & - \underset{(0.058)}{0.197}\Delta_1 y_{t-3} + \underset{(0.060)}{0.352}\Delta_1 y_{t-4} - \underset{(0.063)}{0.335}\Delta_1 y_{t-5} + \hat{\varepsilon}_t,\end{aligned} \tag{6.3}$$

where $D^e_t = 1$ when there is an expansion (according to the NBER), and $D^e_t = 0$ elsewhere, and where D^r_t is a similar variable for recessions. The roots of the AR polynomial are –0.938, –0.076±0.815i and 0.618±0.391i. Obviously, the first root is close to –1, as can be expected given the results of the HEGY tests for seasonal unit roots in Table 5.2 above. However, a regression such as (6.2) for the $(1 - B)(1 + B)y_t$ variable gives roughly similar results as (6.3). The diagnostics for (6.3) are $F_{\text{AR},1\text{-}1} = 0.559$, $F_{\text{AR},1\text{-}4} = 0.695$, $F_{\text{ARCH},1} = 1.976$, $F_{\text{ARCH},1\text{-}4} = 1.199$, and $JB = 0.053$. Hence this model cannot be rejected by the data. From (6.3) one can calculate that the growth in industrial production in each quarter in an expansion is 1.3 per cent (since $1 / \hat{\phi}_p(1) = 0.965$). Notice that there is no difference across these growth rates, and thus there does not seem to be deterministic seasonal variation in expansion periods. However, there is stochastic seasonality because of the $\Delta_1 y_{t-4}$ and the seasonal (near) unit root. The growth rate in recessions equals –3.0 per cent in the first and fourth quarters, 0.2 per cent in the second quarter, and –0.5 per cent in the third quarter. Finally, it can be noticed that the absolute value of the parameter for D^r_t exceeds the parameter for D^e_t, which reflects the asymmetry in business cycles (see Neftçi 1984 and Hamilton 1989, among others). In sum, the results in (6.3) for US industrial production seem to confirm the findings in Canova and Ghysels (1994) that seasonal fluctuations differ across the business cycle stages.

6.3 Seasonality in Consumer Confidence

Given the empirical findings above, a natural question concerns the (economic) forces that may cause the seasonal fluctuations to change with a trend or the business cycle. In the beginning of this chapter I gave as examples retail sales, for which the seasonal fluctuations may depend on the start of the clearance sales period (which in turn may depend on the state of the economy), and the consumer spending on holidays, where in times of (expected) expansion one may consider having two holidays per year and in times of a recession at most only a single one. These two examples tentatively suggest that a possible cause for correlated seasonal and business cycle variation is that the expectations of economic agents vary with the season, and that these may be based on the current economic situation even though the expectations may concern yearly horizons. Empirical evidence on such seasonal variation in business surveys is presented in Ghysels and Nerlove (1988). Furthermore, it is worth mentioning that the US consumer confidence index is available only in seasonally adjusted form. The same applies to the data on European consumer confidence since 1989. Before that year, these data are available in unadjusted form, and I now proceed to analyse these data when observed before 1989.

Empirical evidence for some European countries

As discussed in Chapter 4, one of the main underlying assumptions of seasonal adjustment methods is that an economic time series can be viewed as the sum or a product of a seasonal and a nonseasonal component. Although the issue of seasonal adjustment is viewed mainly as a statistical problem, an important motivation for seasonal adjustment is that economic agents may prefer to make decisions on deseasonalized data. Indeed, the producer of, for example, ice cream is aware of the fact that there is a seasonal variation in the demand for this product, and therefore is likely to be interested only in the direction of the underlying trend.

It may however be interesting to study the extent to which economic agents are able to separate nonseasonal from seasonal variation. One way to investigate this is to study the results of seasonal surveys when such surveys involve questions regarding expectations of annual trends. For this purpose I analyse some time series of consumer opinion on economic and financial conditions, which are collected by several statistical agencies in European Community countries and are summarized in issues of *European Economy* (Supplement B). We shall consider these data only for the sample period 1975–88, because after that period these expectations series began to be seasonally adjusted. Since the data will not be used in subsequent chapters, they are not given in the Data Appendix but can be obtained from the author upon request.

The consumer confidence data can be divided into eight categories: (1) the general consumer confidence index; (2) the financial situation of households over the last and next 12 months; (3) the general economic situation over the last and next 12 months; (4) the price trends over the last and next 12 months; (5) the unemployment over the next 12 months only; (6) the major purchases at present and over the next 12 months; (7) the savings at present and over the next 12 months; and (8) the index of net acquisition of financial assets. In this chapter I consider only the series of expectations over the next 12 months, as well as the general confidence index.

The six expectations series of interest are weighted averages of the percentages of responses to the following questions: How do you think the financial position of your household will change over the next 12 months? How do you think the general economic situation in this country will develop over the next 12 months? What do you think about the price trends over the next 12 months? What do you think about the unemployment level in the country over the next 12 months? Do you think that there is an advantage to be gained by making major purchases (furniture, washing machines, TV sets, and so on) over the next 12 months? and Will there be savings by you or your household over the next 12 months? The general consumer confidence index is an arithmetic average of the answers to the four questions on the financial situation of households and general economic situation (past and future) together with that on the advisability of making major purchases.

The time series of the confidence indices are collected for European countries that are member of the European Community. For eight countries the observations start in May 1972: Belgium, Denmark, (West) Germany, France, Ireland, Italy, the Netherlands, and the United Kingdom. The observations for Greece, Spain, and Portugal start much later, and therefore I do not consider those series any further. In summary, I investigate seasonal patterns in seven consumer confidence indices in eight countries. It appears that for the period 1975–88 I can construct such time series which are measured three times per year.

The procedure to test for seasonality amounts to a simple regression of index z_t on a constant, two seasonal dummies, and z_{t-1}. It appears that this is a fairly adequate description of all time series considered. Since these series are based on averages of answers to questions in consumer surveys, I conjecture that it is quite unrealistic to allow for the possibility of unit roots since, for example, this would imply that an oil crisis could establish that consumers have permanent pessimistic opinions, or that, in the case of a random walk with drift, consumers may become increasingly optimistic every year. Moreover, the method of weighted-averaging the percentages of the agreement of consumers with certain answers ensures that the indices have upper and lower bounds.

The results of the F-tests for the significance of the two seasonal dummies in the regression model are displayed in Table 6.4. The results reported in this table suggest that many indices display highly significant seasonal fluctuations,

TABLE 6.4. Testing for seasonality in consumer confidence indices[a,b]

Ind.[c]	Belgium	Denmark	Germany	France	Ireland	Italy	Neth.	UK
GCC	7.463***	1.996	3.160*	0.073	0.038	2.564*	2.512*	0.310
GES	3.102*	1.468	3.864**	0.690	0.263	1.388	4.365**	2.436
FP	3.316**	0.521	2.810*	0.380	0.484	4.543**	3.493**	0.972
PT	6.176***	7.294***	1.354	2.493*	1.484	5.279***	1.490	0.403
U	8.269***	6.382***	4.437**	1.555	1.752	3.943**	4.628**	3.188*
MP[d]	4.768**	0.106	0.425	5.851***	4.519**	8.660***	5.647***	1.877
S	0.019	2.666*	0.777	12.679***	1.085	2.383	4.776**	0.177

[a] Based on 14 years of three observations (January, May, October) per year. The cells are the F-test values for the significance of the two seasonal dummy variables in $z_t = \alpha_0 + \alpha_1 D_{1,t} + \alpha_2 D_{2,t} + \beta z_{t-1} + \varepsilon_t$.

[b] From 1972 to 1983 there were only three consumer surveys per year, i.e. in January, May, and October. In 1984 and 1985 measurements took place four times a year, i.e. in January/February, March/June, July/August, and October/November. Starting from 1986, monthly data are available, which are seasonally corrected from January 1989 onwards. To obtain time series that cover several years, I decide to construct series with three data points each year, i.e. in January, May, and October. For 1984 and 1985 the two first and the last observations were used. When there were no May observations available in 1986–8, I took the observations of the months April or March. Since for some of the eight countries the surveys start only in May 1974, the analysis of seasonal patterns considers all observations previous to 1975 as starting values.

[c] The indices are General Consumer Confidence (GCC), General Economic Situation (GES), Financial Position (FP), Price Trends (PT), Unemployment (U), Major Purchases (MP), and Savings (S).

[d] The sample sizes are 27 for Belgium, 35 for Ireland, and 30 for the Netherlands.

*** Significant at the 1 per cent level.
** Significant at the 5 per cent level.
* Significant at the 10 per cent level.

even though the consumers are asked to remove seasonality since the questions involve *annual* trends. Hence it seems that consumers can have time-varying expectations. One reason for this may be that economic agents base their expectations on the current state of the economy. If this is true, one can conclude that economic agents face difficulties disentangling seasonal from nonseasonal fluctuations since they are not able to separate the current trend (which they are asked to extend) and the current seasonal component. Ghysels and Nerlove (1988) report similar findings and propose to include questions in surveys that force respondents to neglect the current economic state or season.

Conclusion

In this chapter I have documented that for some economic time series there appear to be links between the seasonal fluctuations and the trend or business cycle fluctuations. Furthermore, a brief analysis of consumer opinion surveys shows that consumers can face difficulties in separating seasonal from nonseasonal fluctuations. Of course, one may conjecture that the latter difficulties constitute precisely the main motivation for constructing seasonally adjusted time series. On the other hand, when economic agents are not able to distinguish seasonal from nonseasonal fluctuations, one may expect that this will be somehow reflected in their behaviour, and hence that economic processes can be such that these fluctuations *cannot* be separated. Ghysels (1988) considers a theoretical model that may explain such economic behaviour.

Here I take the above results as indicating that it seems useful to consider econometric time series models to describe univariate and multivariate macroeconomic phenomena, which in principle allow for possible correlation between seasonal and nonseasonal variation. An important feature of these models should then be that their nested versions can be used to describe economic time series for which there is no such correlation. In this case, tests for certain parameter restrictions may guide model selection. Given the results in Section 6.1, it also seems useful to consider time series models where the parameters can take different values in different seasons. In the sequel to this chapter, I will argue that periodic models with unit roots can yield such useful descriptions, and will therefore discuss the class of periodic autoregressive time series models. In Chapter 8, I will focus explicitly on periodic models with stochastic trends, and show that specific versions of these models may be able to describe part of the empirical regularities discussed in this chapter (as well as some of the empirical features reviewed in earlier chapters).

7
Periodic Autoregressive Time Series Models

In this chapter I consider periodic autoregressive models for univariate and multivariate time series processes. A periodic autoregression (PAR) extends a nonperiodic AR model by allowing the autoregressive parameters to vary with the seasons. In other words, the PAR model assumes that the observations in each of the seasons can be described using a different model. Such a property may be useful to describe some economic time series, since in several economic situations one may expect that economic agents behave differently in different seasons. This may lead to a belief that the memory structure (which can be reflected by autoregressive components) can vary with the season. For example, technological advances throughout the years have now made it possible to buy certain vegetables in almost all seasons, whereas these products were available in, say, only the summer season several years ago; hence one may observe an increasing trend in the consumption of these vegetables in one or more seasons and no trend in the summer season (see e.g. the graphs in Kuiper, *et al.* 1993). As another example, some recent tax measures which became effective in the first part of the year may lead to seasonal variation in inflation; as a result, economic agents may come increasingly to anticipate this inflation peak. Yet another example is that, as a result of increasing personal disposable income and increasing leisure time in developed countries, consumer spending on holidays has changed over the years. Gersovitz and MacKinnon (1978), Osborn (1988), and Todd (1990) give more concrete economic examples which concern seasonally varying utility functions and optimization schemes.

In the present chapter I do not focus in such detail on economic models, but merely study econometric properties of PAR models. In Section 7.1 I discuss notation and representation issues for periodic AR models for univariate quarterly time series. It appears that a PAR model can be represented in at least two ways. The multivariate representation appears most useful to analyse the stationarity properties of the model (see Section 7.2). To save space I consider only quarterly time series, although all material naturally extends to, say, monthly time series. Furthermore, I abstain from a detailed analysis of periodic MA models (PMA) or periodic ARMA models (PARMA), since empirical experience with PAR models indicates that low-order PAR models are often sufficiently adequate to describe periodic time series (see also McLeod

1994). In Section 7.3 I discuss model selection and estimation in PAR models. Part of this section is based on Franses and Paap (1995). It appears that standard model selection techniques can readily be applied. A detailed Monte Carlo study indicates which criteria are most useful in practice. I also discuss simple methods to select between periodic and nonperiodic AR models. In Section 7.4 I study the impact of neglecting periodicity, i.e. the effects of estimating nonperiodic models for periodic time series. Furthermore, I analyse the consequences of seasonal adjustment for parameter estimation and for a test for periodicity. Finally, in Section 7.5 I briefly discuss multivariate PAR models. Most sections contain empirical investigations using the sample series that were also analysed in the previous chapters. It should be mentioned that the explicit analysis of stochastic trends (unit roots) in PAR models is postponed until Chapter 8.

7.1 Notation and Representation

In this section I consider notation and representation issues. The notation used is a simplified form of that used in Vecchia (1985), among others, and hence in order to avoid confusion I put some effort into giving notational details. Early references to periodic AR models are Gladyshev (1961), Jones and Brelsford (1967), Pagano (1978), Troutman (1979), Cleveland and Tiao (1979), Parzen and Pagano (1979), and Andel (1983). Such models are used for forecasting in Osborn and Smith (1989), among others. The periodic models in this section are natural extensions of the nonperiodic linear models reviewed in Chapter 2. For a nonlinear periodic time series model, see Ghysels (1993), where the Markov regime switching model of Hamilton (1989) is extended to allow for periodic transition probabilities.

Univariate representation

Consider a univariate time series y_t which is observed quarterly for N years, i.e. $t = 1,2, \ldots, n$. Without loss of generality, it is assumed that $n = 4N$. A periodic autoregressive model of order p (PAR(p)) can be written as

$$y_t = \mu_s + \phi_{1s} y_{t-1} + \ldots + \phi_{ps} y_{t-p} + \varepsilon_t, \tag{7.1}$$

or

$$\phi_{p,s}(B) y_t = \mu_s + \varepsilon_t, \tag{7.2}$$

with

$$\phi_{p,s}(B) = 1 - \phi_{1s} B - \ldots - \phi_{ps} B^p, \tag{7.3}$$

where μ_s is a seasonally varying intercept term, and where $\phi_{1s} - \phi_{ps}$ are autoregressive parameters up to order p which may vary with the season s,

$s = 1,2,3,4$. The ε_t is assumed to be a standard white noise process with constant variance σ^2. This assumption may be relaxed by allowing ε_t to have seasonal variances σ_s^2. Below, I will propose a simple test statistic for such seasonal heteroskedasticity. For the moment, however, I abstain from seasonal variation in σ^2 and focus on (7.1). Since some ϕ_{is}, $i = 1,2, \ldots, p$, can take zero values, the order p in (7.1) is the maximum of all p_s, where p_s denotes the AR order per season s. Hence one may also wish to consider so-called subset periodic autoregressions.

A periodic moving average model of order q (PMA(q)) for y_t can be written as

$$y_t = \delta_s + \theta_{q,s}(B)\varepsilon_t, \tag{7.4}$$

with

$$\theta_{q,s}(B) = 1 + \theta_{1s}B + \ldots + \theta_{qs}B^q. \tag{7.5}$$

(See Cipra (1985) for a theoretical analysis of a PMA(q) model.) Naturally, one can combine (7.1) and (7.4) in a PARMA(p,q) model. (See Vecchia (1985) for an analysis of a PARMA process.) Although PMA and PARMA models may be useful to describe some time series, in this book I limit the discussion to PAR models since, first, these can be easily analysed for stochastic trends, and, second, empirical experience with economic time series shows that low-order PAR models often fit periodic time series data rather well. (See the empirical results in Section 7.2, where p is often found to be equal to 1, 2, or 3 for the 11 running example series.) Hence, here, the gain of using MA structures to deal with lengthy AR structures does not seem very large.

Notice that the μ_s in (7.1) does not necessarily reflect that the mean of y_t is also seasonally varying. This can be observed from rewriting a PAR process for y_t with constant mean μ, i.e.

$$\phi_{p,s}(B)(y_t - \mu) = \varepsilon_t \tag{7.6}$$

as (7.1). It is then clear that $\mu_s = (\phi_{p,s}(1))\mu$, and hence that μ_s is seasonally varying because the autoregressive parameters are varying. Similar results apply when a deterministic trend component $\tau_s t$ is included in (7.1), where $t = 1,2, \ldots, n$ and τ_s are seasonally varying parameters.

Obviously, the number of autoregressive parameters in a PAR(p) model for a quarterly time series is four times as much as that in a nonperiodic AR(p) model. For example, where $p = 4$ in (7.1), one needs to estimate $16 + 4 = 20$ AR parameters in the PAR model. In practice, this may become too large a number of parameters, given the possibly limited number of observations. Therefore, Gersovitz and MacKinnon (1978) propose to impose some smoothness restrictions on the autoregressive parameters. On the other hand, recent empirical results of the application of PAR models to monthly time series as in Noakes *et al.* (1985) and McLeod (1994) indicate that such restrictions may not be strictly necessary in order to obtain useful estimation results. We will also abstain from such restrictions in further analysis, although I aim to be careful in the practical application of PAR(p) models.

Finally, it should be mentioned that we can already observe from (7.1) that, for data generated from a periodic AR model, it holds that the time series cannot be decomposed into seasonal and nonseasonal components. In fact, since the autoregressive parameters show seasonal variation, and since these cannot be separated from the y_t series, it is obvious also that seasonal adjustment would not yield a nonperiodic time series. Indeed, the exposition in Chapter 4 shows that seasonal adjustment methods treat the observations in all quarters similarly. This implies that we can still fit periodic models to SA data. In Section 7.4 I will analyse this implication in more detail.

Multivariate representation

Strictly speaking, the periodic time series model in (7.1) or (7.4) is a nonstationary model since the variance and autocovariances take different values in different seasons. Consider for example the PMA(1) process

$$y_t = \varepsilon_t + \theta_{1s}\varepsilon_{t-1}, \tag{7.7}$$

for which it is easily shown that the variance and first-order autocovariance in season s, to be denoted as γ_{0s} and γ_{1s}, are given by

$$\gamma_{0s} = \sigma^2(1 + \theta_{1s}^2)$$
$$\gamma_{1s} = \sigma^2\theta_{1s},$$

and hence that the first-order autocorrelation per season s, ρ_{1s}, equals

$$\rho_{1s} = \theta_{1s} / (1 + \theta_{1s}^2). \tag{7.8}$$

Obviously, this result violates the assumption of time-invariant correlations, which is one of the assumptions for stationarity. This indicates that, also in order to facilitate the analysis of stochastic trends, a more convenient representation of a PAR(p) process is given by rewriting it in a time-invariant form. Since the PAR(p) model considers different AR(p) models for different seasons, it seems natural to rewrite it as a model for annual observations (see also Gladyshev 1961; Tiao and Grupe 1980; Osborn 1991; and Lütkepohl 1991).

In general, the PAR(p) process in (7.1) can be rewritten as an AR(P) model for the (4×1) vector process $\mathbf{Y}_T = (Y_{1,T}, Y_{2,T}, Y_{3,T}, Y_{4,T})'$, $T = 1,2, \ldots, N$, where $Y_{s,T}$ is again the observation in season s in year T, $s = 1,2,3,4$, i.e.

$$\Phi_0\mathbf{Y}_T = \mu + \Phi_1\mathbf{Y}_{T-1} + \ldots + \Phi_P\mathbf{Y}_{T-P} + \varepsilon_T, \tag{7.9}$$

or

$$\Phi(B)\mathbf{Y}_T = \mu + \varepsilon_T,$$

with

$$\Phi(B) = \Phi_0 - \Phi_1 B - \ldots - \Phi_P B^P, \tag{7.10}$$

where $\mu = (\mu_1,\mu_2,\mu_3,\mu_4)'$ and $\varepsilon_T = (\varepsilon_{1,T},\varepsilon_{2,T},\varepsilon_{3,T},\varepsilon_{4,T})'$, and $\varepsilon_{s,T}$ is the observation on the error process ε_t in season s in year T. The $\Phi_0, \Phi_1, \ldots, \Phi_P$ are (4×4) parameter matrices with elements

$$\begin{aligned}\Phi_0(i,j) &= 1 \qquad i=j\\ &= 0, \qquad j>i\\ &= -\phi_{i-j,i} \quad j<i\\ \Phi_k(i,j) &= \phi_{i+4k-j,i},\end{aligned}$$

for $i = 1,2,3,4$, $j = 1,2,3,4$, and $k = 1,2,.,P$. For the model order P in (7.9) it holds that $P = 1 + [(p-1)/4]$, where $[\cdot]$ means 'integer function of'. Hence, when p equals 1,2,3, or 4, the value of P is only 1. Note that the notation is slightly different from that in Lütkepohl (1991), where this $\mathbf{Y}_T$ vector contains $Y_{4,T}$ to $Y_{1,T}$. Furthermore, because Φ_0 is a lower triangular matrix, model (7.9) is a recursive model. This is intuitively obvious since, for example, the observation in the fourth quarter in year T always comes after that in the third quarter in the same year. As an example of (7.9), consider the PAR(2) process

$$y_t = \phi_{1s}y_{t-1} + \phi_{2s}y_{t-2} + \varepsilon_t, \tag{7.11}$$

which can be written as

$$\Phi_0\mathbf{Y}_T = \Phi_1\mathbf{Y}_{T-1} + \varepsilon_T, \tag{7.12}$$

with

$$\Phi_0 = \begin{bmatrix} 1 & 0 & 0 & 0\\ -\phi_{12} & 1 & 0 & 0\\ -\phi_{23} & -\phi_{12} & 1 & 0\\ 0 & -\phi_{24} & -\phi_{14} & 1\end{bmatrix} \text{ and } \Phi_1 = \begin{bmatrix} 0 & 0 & \phi_{21} & \phi_{11}\\ 0 & 0 & 0 & \phi_{22}\\ 0 & 0 & 0 & 0\\ 0 & 0 & 0 & 0\end{bmatrix}. \tag{7.13}$$

In order to avoid confusion with multivariate time series models, I refer to models such as (7.9) as the vector of quarters (VQ) representation. Sometimes I will refer to a VQ of order P, which then corresponds to (7.9). Notice from (7.9) and (7.13) that one can always write a nonperiodic AR model in a VQ representation. In that case, the parameters in the various rows of this representation take the same values. The example in (7.13) also indicates that a PAR(4) can still be written in VQ(1) format, since then the first row of the Φ_1 matrix does not contain zeros. Furthermore, it can be seen that such a VQ of order 1 is still useful when $p_s = s + 3$. For example, when the order of the AR polynomial in quarter 4 is 7, that in quarter 3 is 6, and so on, the Φ_1 matrix does not contain any zero valued parameters, while Φ_2 only has zeros.

There are two versions of (7.9) that will sometimes be considered below since these are useful for the analysis of unit roots and for forecasting. The first is given by simply pre-multiplying (7.9) with Φ_0^{-1}, i.e.

$$\mathbf{Y}_T = \Phi_0^{-1}\,\mu + \Phi_0^{-1}\Phi_1\mathbf{Y}_{T-1} + \ldots + \Phi_0^{-1}\Phi_P\mathbf{Y}_{T-P} + \Phi_0^{-1}\varepsilon_T. \tag{7.14}$$

Notice from (7.9) that Φ_0 is invertible. The expression in (7.14) is a VAR(P) for the $\mathbf{Y}_T$ process. When $\varepsilon_T \sim N(0,\sigma^2\mathbf{I}_4)$, it follows that

$$\Phi_0^{-1}\varepsilon_T \sim N(0, \sigma^2\Phi_0^{-1}(\Phi_0^{-1})'). \tag{7.15}$$

It is easy to see that Φ_0^{-1} for any PAR process is again a lower triangular matrix. For example, for the PAR(2) case in (7.12) it can be found that

$$\Phi_0^{-1} = \begin{bmatrix} 1 & 0 & 0 & 0 \\ \phi_{12} & 1 & 0 & 0 \\ \phi_{12}\phi_{13}+\phi_{23} & \phi_{13} & 1 & 0 \\ \phi_{12}\phi_{13}\phi_{14}+\phi_{12}\phi_{24}+\phi_{23}\phi_{14} & \phi_{13}\phi_{14}+\phi_{24} & \phi_{14} & 1 \end{bmatrix}$$

This implies that the first two columns of $\Phi_0^{-1}\Phi_1$ contain only zeros; i.e.,

$$\Phi_0^{-1}\Phi_1 = \begin{bmatrix} 0 & 0 & \phi_{21} & \phi_{11} \\ 0 & 0 & \phi_{12}\phi_{21} & \phi_{12}\phi_{11}+\phi_{22} \\ 0 & 0 & (\phi_{12}\phi_{13}+\phi_{23})\phi_{21} & \phi_{11}\phi_{12}\phi_{13}+\phi_{11}\phi_{23}+\phi_{13}\phi_{22} \\ 0 & 0 & (\phi_{12}\phi_{13}\phi_{14}+\phi_{12}\phi_{24}+\phi_{23}\phi_{14})\phi_{21} & \phi_{11}(\phi_{12}\phi_{13}\phi_{14}+\phi_{12}\phi_{24}+\phi_{23}\phi_{14}) \end{bmatrix}$$

It can be seen that for a PAR(1) model the first three columns of $\Phi_0^{-1}\Phi_1$ contain only zero valued parameters. For a PAR(3) process this holds for the first column. For the PAR(p) process with $p \geq 4$, it holds that $\Phi_0^{-1}\,\Phi_1$ does not contain columns with only zeros.

A second version of (7.9) which is sometimes useful is based on the possibility of decomposing a nonperiodic AR(p) polynomial as $(1-\alpha_1 B)(1-\alpha_2 B)\ldots(1-\alpha_p B)$. This is possible when the solutions to the characteristic equation for this AR(p) polynomial are all real-valued. Where similar results hold for a PAR(p) process, for which the characteristic equation is given in the next section, it can be useful to rewrite (7.9) as

$$\prod_{i=0}^{P} [\Xi_i(B)]\mathbf{Y}_T = \mu + \varepsilon_T, \tag{7.17}$$

where the $\Xi_i(B)$ are (4×4) matrices with elements that are polynomials in B. A simple example is again the PAR(2) process in (7.12), which may sometimes be written as

$$\Xi_1(B)\Xi_0(B)\mathbf{Y}_T = \varepsilon_T, \tag{7.18}$$

with

$$\Xi_1(B) = \begin{bmatrix} 1 & 0 & 0 & -\beta_1 B \\ -\beta_2 & 1 & 0 & 0 \\ 0 & -\beta_3 & 1 & 0 \\ 0 & 0 & -\beta_4 & 1 \end{bmatrix} \text{ and } \Xi_0(B) = \begin{bmatrix} 1 & 0 & 0 & -\alpha_1 B \\ -\alpha_2 & 0 & 0 & 0 \\ 0 & -\alpha_3 & 1 & 0 \\ 0 & 0 & -\alpha_4 & 1 \end{bmatrix}. \tag{7.19}$$

such that (7.11) can be written as

$$(1-\beta_s B)(1-\alpha_s B)y_t = \mu_s + \varepsilon_t. \tag{7.20}$$

This expression equals

$$y_t - \alpha_s y_{t-1} = \mu_s + \beta_s(y_{t-1} - \alpha_{s-1} y_{t-2}) + \varepsilon_t, \tag{7.21}$$

since the backward shift operator B also operates on α_s; i.e., $B\alpha_s = \alpha_{s-1}$ for all $s = 1,2,3,4$ and with $\alpha_0 = \alpha_4$.

The analogy of a univariate PAR process with a multivariate time series process can also be exploited to derive explicit formulae for one- and multi-step-ahead forecasting. Notice that in that case the one-step-ahead forecasts concern one-year-ahead forecast for all four $Y_{s,T}$ series. For example, when I estimate the model $\mathbf{Y}_T = \Phi_0^{-1}\Phi_1\mathbf{Y}_{T-1} + \omega_T$, where $\omega_T = \Phi_0^{-1}\varepsilon_T$, the forecast for $T+1$ is $\mathbf{Y}_{T+1} = \Phi_0^{-1}\Phi_1\mathbf{Y}_T$. Of course, if one wants to forecast one step ahead for any season other than the first quarter, one may want to use the univariate representation.

7.2 Stationarity in Periodic Autoregression

There are two approaches to investigating the stationarity properties of a PAR process, i.e. investigating whether it contains unit roots. The first is to rewrite the process (7.1) as an infinite moving average process, and to check for parameter restrictions that establish that the effects of previous shocks do not die out. For simple PMA or PAR processes, this may be a useful method. (See Cipra (1985) for an application to PMA processes.) For PAR processes of higher order, however, this may become analytically cumbersome. The second method, which recognizes the links between a PAR model and a multivariate time series model such as (7.9), is then more convenient. Bentarzi and Hallin (1994*a*) give an application of this method for PMA processes. For a PAR(p) process, an investigation of the presence of unit roots in y_t amounts to investigating the solutions to the characteristic equation of (7.9), i.e. the solution to

$$|\Phi_0 - \Phi_1 z - \ldots - \Phi_P z^P| = 0. \tag{7.22}$$

When k solutions to (7.22) are on the unit circle, the $\mathbf{Y}_T$ process, and also the y_t process, has k unit roots. Notice that the number of unit roots in y_t equals that in $\mathbf{Y}_T$, and that, for example, no additional unit roots are introduced in the multivariate representation.

This can be illustrated with several examples. First, consider the PAR(2) process in (7.11) for which the characteristic equation is

$$|\Phi_0 - \Phi_1 z| = \begin{vmatrix} 1 & 0 & -\phi_{21}z & -\phi_{11}z \\ -\phi_{12} & 1 & 0 & -\phi_{22}z \\ -\phi_{23} & -\phi_{13} & 1 & 0 \\ 0 & -\phi_{24} & -\phi_{14} & 1 \end{vmatrix} = 0 \tag{7.23}$$

which becomes

$$\begin{aligned} 1 - (&\phi_{22}\phi_{13}\phi_{14} + \phi_{22}\phi_{24} + \phi_{21}\phi_{12}\phi_{13} + \phi_{21}\phi_{23} + \phi_{11}\phi_{12}\phi_{13}\phi_{14} \\ &+ \phi_{11}\phi_{12}\phi_{24} + \phi_{11}\phi_{14}\phi_{23})z + \phi_{21}\phi_{22}\phi_{23}\phi_{24}z^2 = 0 \end{aligned} \tag{7.24}$$

Hence, when the nonlinear parameter restriction

$$\begin{aligned} &\phi_{22}\phi_{13}\phi_{14} + \phi_{22}\phi_{24} + \phi_{21}\phi_{12}\phi_{13} + \phi_{21}\phi_{23} + \phi_{11}\phi_{12}\phi_{13}\phi_{14} \\ &+ \phi_{11}\phi_{12}\phi_{24} + \phi_{11}\phi_{14}\phi_{23} - \phi_{21}\phi_{22}\phi_{23}\phi_{24} = 1 \end{aligned} \tag{7.25}$$

is imposed on the parameters of the PAR(2) model in (7.11), the PAR(2) model contains a single unit root. When for a PAR(2) process (7.22) yields two real-valued solutions, one can also analyse the characteristic equation of (7.18), which is

$$|\Xi_1(z)\Xi_0(z)| = 0. \tag{7.26}$$

Given (7.19), it is easy to see that this equation equals

$$(1 - \beta_1\beta_2\beta_3\beta_4 z)(1 - \alpha_1\alpha_2\alpha_3\alpha_4 z) = 0, \tag{7.27}$$

and hence that the PAR(2) in the format of (7.21) has one unit root when either $\beta_1\beta_2\beta_3\beta_4 = 1$ or $\alpha_1\alpha_2\alpha_3\alpha_4 = 1$, and has at most two unit roots when both products equal unity. Obviously, the maximum number of unity solutions to the characteristic equation of a PAR(p) process is equal to p.

For more examples on unit roots and also on how seasonal unit roots appear in VQ models, consider the simple PAR(1) model

$$y_t = \alpha_s y_{t-1} + \varepsilon_t. \tag{7.28}$$

Note that this notation corresponds to that in (7.21), when β_s is set equal to zero for all s, and that it does not correspond to (7.1). I choose to use (7.28) for notational convenience, and also because this notation will prove useful in subsequent chapters, where the models as written in the format of (7.21) will be considered. The PAR(1) model can be written as

$$\Phi_0 \mathbf{Y}_T = \Phi_1 \mathbf{Y}_{T-1} + \varepsilon_T, \tag{7.29}$$

with

$$\Phi_0 = \begin{bmatrix} 1 & 0 & 0 & 0 \\ -\alpha_2 & 1 & 0 & 0 \\ 0 & -\alpha_3 & 1 & 0 \\ 0 & 0 & -\alpha_4 & 1 \end{bmatrix} \text{ and } \Phi_1 = \begin{bmatrix} 0 & 0 & 0 & \alpha_1 \\ 0 & 0 & 0 & 0 \\ 0 & 0 & 0 & 0 \\ 0 & 0 & 0 & 0 \end{bmatrix}. \tag{7.30}$$

The characteristic equation is

$$|\Phi_0 - \Phi_1 z| = (1 - \alpha_1\alpha_2\alpha_3\alpha_4 z) = 0, \tag{7.31}$$

and hence the PAR(1) process has a unit root when $\alpha_1\alpha_2\alpha_3\alpha_4 = 1$. Where one or more α_s values are unequal to α, i.e. when $\alpha_s \neq \alpha$ for all s, and $\alpha_1\alpha_2\alpha_3\alpha_4 = 1$, the y_t process in (7.28) is said to be periodically integrated of order 1 (PI(1)). Periodic integration of order 2 can similarly be defined in terms of the α_s and β_s in the PAR(2) process using (7.27). A concept of periodic integration was first defined in Osborn *et al.* (1988). In the next chapter I will discuss the concept of periodic integration in more detail and give a slightly more precise definition, since it will appear to be a useful concept in practice.

Since the periodic AR(1) process nests the nonperiodic AR(1) model, i.e. since, when $\alpha_s = \alpha$ for all s, (7.28) becomes

$$y_t = \alpha y_{t-1} + \varepsilon_t, \tag{7.32}$$

it is obvious that a unit root in a PAR(1) process implies a unit root in the AR(1) process (7.32). For (7.32), the characteristic equation (7.31) is $(1 - \alpha^4 z) = 0$. Hence, when $\alpha = 1$, the $\mathbf{Y}_T$ process has a single unit root. Also, when $\alpha = -1$, the process $\mathbf{Y}_T$ has a unit root. The first case corresponds to the simple random walk process, i.e. where y_t has a nonseasonal unit root, while the second case corresponds to the process where by y_t has a seasonal unit root. In other words, both the nonseasonal and the seasonal unit root processes are nested within the PAR(1) process. This suggests a simple testing strategy, i.e. first to investigate the presence of a unit root by testing whether $\alpha_1\alpha_2\alpha_3\alpha_4 = 1$, and second to test whether $\alpha_s = 1$ or $\alpha_s = -1$ for all s. In the next chapter I will consider the issue of testing for certain unit roots in more detail.

The main implication of the above analysis of a PAR(1) process is that seasonal unit roots do not disappear when analysing y_t in the VQ representation. This is even more clear when considering the seasonally integrated process

$$y_t = y_{t-4} + \varepsilon_t, \tag{7.33}$$

for which it is easily found that it can be written as a VQ(1) model, and that its characteristic equation equals

$$|\Phi_0 - \Phi_1 z| = (1 - z)^4 = 0. \tag{7.34}$$

Obviously, for the seasonally integrated process, which has one nonseasonal unit root and three seasonal unit roots (see Chapter 5), the corresponding $\mathbf{Y}_\mathrm{T}$ process has four unit roots. Hence, the seasonal unit roots -1 and $\pm i$ appear as zero frequency unit roots when we analyse the y_t variable in $\mathbf{Y}_\mathrm{T}$ format. Notice that the seasonally integrated process assumes three more unit roots in the $\mathbf{Y}_\mathrm{T}$ process than does the periodically integrated process. There are similarities, however, between these two processes, and they will be discussed in Chapter 8. Furthermore, when there are $4 - r$ unit roots in the $\mathbf{Y}_\mathrm{T}$ process, there are r cointegrating relations between the four elements of $\mathbf{Y}_\mathrm{T}$. This will also be exploited in a test procedure for unit roots in the next chapter.

A final example in this section concerns I(2) type processes and how such processes are reflected by the number of unit roots in a VQ model for the $\mathbf{Y}_\mathrm{T}$ process. Consider first the simplest process with a nonseasonal and a seasonal unit root, i.e.

$$y_t = y_{t-2} + \varepsilon_t, \tag{7.35}$$

which can be written as a VQ(1) process with

$$\Phi_0 = \begin{bmatrix} 1 & 0 & 0 & 0 \\ 0 & 1 & 0 & 0 \\ -1 & 0 & 1 & 0 \\ 0 & -1 & 0 & 1 \end{bmatrix} \text{ and } \Phi_1 = \begin{bmatrix} 0 & 0 & 1 & 0 \\ 0 & 0 & 0 & 1 \\ 0 & 0 & 0 & 0 \\ 0 & 0 & 0 & 0 \end{bmatrix}. \tag{7.36}$$

such that

$$\Phi_0^{-1}\Phi_1 = \begin{bmatrix} 0 & 0 & 1 & 0 \\ 0 & 0 & 0 & 1 \\ 0 & 0 & 1 & 0 \\ 0 & 0 & 0 & 1 \end{bmatrix} \text{ and } \Phi_0^{-1}\Phi_1 - \mathbf{I}_4 = \begin{bmatrix} -1 & 0 & 1 & 0 \\ 0 & -1 & 0 & 1 \\ 0 & 0 & 0 & 0 \\ 0 & 0 & 0 & 0 \end{bmatrix}. \tag{7.37}$$

Clearly, the eigenvalues of $\Phi_0^{-1}\Phi_1$ are 0, 0, 1, and 1, and the rank of $\Phi_0^{-1}\Phi_1 - \mathbf{I}_4$ is 2. Hence, there are two cointegrating relations between the $Y_{s,\mathrm{T}}$ variables and

y_t has two unit roots. When however the time series is an I(2) process, i.e. when for example

$$y_t = 2y_{t-1} - y_{t-2} + \varepsilon_t \tag{7.38}$$

is a useful model to describe y_t, the various parameter matrices in the VQ(1) model are such that

$$\Phi_0^{-1}\Phi_1 = \begin{bmatrix} 0 & 0 & -1 & 0 \\ 0 & 0 & -2 & 1 \\ 0 & 0 & -3 & 0 \\ 0 & 0 & -4 & 1 \end{bmatrix} \text{ and } \Phi_0^{-1}\Phi_1 - \mathbf{I}_4 = \begin{bmatrix} -1 & 0 & -1 & 2 \\ 0 & -1 & -2 & 3 \\ 0 & 0 & -3 & 4 \\ 0 & 0 & -4 & 4 \end{bmatrix}. \tag{7.39}$$

These expressions indicate that the eigenvalues of $\Phi_0^{-1}\Phi_1$ are 0, 0, 1, and 1, similar to the process in (7.35), but that the rank of $\Phi_0^{-1}\Phi_1 - \mathbf{I}_4$ equals 3. (See Johansen (1992*b*) for details of the representation of VAR models for I(2) processes.) For practical purposes, these examples suggest that it is useful to calculate the eigenvalues of $\Phi_0^{-1}\Phi_1$ since these may convey information on the maximum number of possible unit roots in y_t.

In this and the preceding section several notation and representation issues of PAR(p) models have been discussed. Although some implications of unit roots in periodic time series models have been considered, a formal statistical analysis of such roots is postponed to the next chapter. Here I shall now proceed with the estimation of the parameters in a PAR(p) process and with model selection.

7.3 Estimation and Model Selection

Given the discussion in Chapter 2, one way to select the order p in a PAR(p) process may be to use the estimated so-called periodic autocorrelation and partial autocorrelation functions. Definitions of these functions are given in McLeod (1994) and Sakai (1982), respectively. Vecchia and Ballerini (1991) derive the asymptotic distributions of the estimated periodic autocorrelation function, and use it is a method to identify a PARMA(1,1) model for a monthly river flow series. The identification of a periodic time series model is not as straightforward as it is for nonperiodic time series models. The intuitive reason for this is that periodic models have similarities with the VARMA models discussed in Section 2.2, for which it is also found that the various autocorrelation functions may not be easy to use in practice. An alternative graphical approach is proposed in Hurd and Gerr (1991) (see also Bloomfield *et al.* 1994). A method based on rank statistics is given in Bentarzi and Hallin (1994*b*).

In this section, I consider two alternative and somewhat simpler approaches to modelling PAR(p) processes. The first is to investigate the possible usefulness of periodic time series models by checking the properties of estimated residuals from nonperiodic models. The second approach is simply to estimate a PAR(p) model, where p is selected using conventional model selection criteria, and to test whether there is indeed periodic variation in the autoregressive parameters.

There are several possibilities to test for periodicity in the estimated residuals. One is to investigate periodic correlation using some modified version of the Box–Pierce test. Another is to consider a variant of an LM test statistic. (See Ghysels and Hall (1993*a*) for some expressions of LM tests for several specific models.) The LM test proposed in Franses (1993) and used in this chapter assumes that we can fit a nonperiodic AR(k) model to the time series x_t, where $x_t = \Delta_j y_t$, which yields the residuals $\hat{v}_t$. The auxiliary regression

$$\hat{v}_t = \sum_{i=1}^{k} \eta_i x_{t-i} + \sum_{s=1}^{4} (\psi_{1s} D_{s,t} \hat{v}_{t-1} + \ldots + \psi_{ms} D_{s,t} \hat{v}_{t-m}) + u_t \tag{7.40}$$

may then be used to investigate the presence of periodicity in the estimated residuals. The F-test for $\psi_{1s} = 0, \ldots, \psi_{ms} = 0$ can be used to test the null hypothesis of no periodic autocorrelation of order m. We denote this F-test as $F_{\text{PeAR},1-m}$. Under the null hypothesis, this F-statistic asymptotically follows a standard F-distribution with $(4m, n - k - 4m)$ degrees of freedom.

A second LM type test procedure for neglected periodic variation in time series models is given by the simple auxiliary regression

$$\hat{v}_t^2 = \omega_0 + \omega_1 D_{1,t} + \omega_2 D_{2,t} + \omega_3 D_{3,t} + \lambda_t, \tag{7.41}$$

which can be used to check for seasonal heteroskedasticity. Note that, when periodic variation in the AR parameters is neglected, this variation may show up in the variance of the residual process. Under the null hypothesis of no seasonal heteroskedasticity, the F-test statistic for the significance of the ω_1, ω_2, and ω_3 parameters in (7.41), which is denoted as F_{SH}, asymptotically follows a standard $F(3,n - k)$-distribution. Of course, one may extend (7.41) to include periodic ARCH components in the case of, say, financial time series, where one may expect so-called monthly or daily effects. Given the nature of the time series considered in this book, that extension is not pursued here. The F_{SH}-test may also be useful when one wants to consider nonperiodic models with seasonal variances as in, for example, Burridge and Wallis (1990).

To illustrate the use of the two test methods for periodic patterns in the estimated residuals from nonperiodic AR models, consider the results in Table 7.1 for the 11 sample series that were also analysed in previous chapters. The choice of differencing filter for each of these series is based on the test results in Table 5.2, where the series were analysed for the presence of seasonal unit roots. Note that I do not incorporate the possible structural breaks discussed in Table 5.5 in this choice. In columns (2) and (3) of the table, the test results for nonperiodic residual autocorrelation indicate that, based on these diagnostics, the selected AR order in the nonperiodic model may seem adequate to capture the dynamics. The $F_{\text{PeAR},1-1}$ and $F_{\text{PeAR},1-2}$-test values in the next two columns indicate that there is evidence of periodic serial correlation for five of the 11 series. The F_{SH} results in column (6) suggest that the residuals from the models

TABLE 7.1. Testing for periodic patterns in the estimated residuals from nonperiodic AR(k) models for suitably transformed time series

Variable	Series[a]	k	Diagnostic test statistics[b]				
			$F_{AR,1-1}$	$F_{AR,1-4}$	$F_{PeAR,1-1}$	$F_{PeAR,1-2}$	F_{SH}
		(1)	(2)	(3)	(4)	(5)	(6)
Industrial production, USA	$\Delta_2 y_t$	6	0.353	0.932	3.077**	2.596**	1.836
Unemployment, Canada	$\Delta_2 y_t$	8	0.050	0.233	2.734**	2.631**	1.201
GNP, Germany	$\Delta_4 y_t$	1	0.818	1.909	4.525**	2.511**	3.515**
UK GDP	$\Delta_1 y_t$	4	0.814	0.393	1.856	2.266**	1.419
UK total consumption	$\Delta_4 y_t$	8	1.467	0.752	1.710	0.921	1.383
UK consumer non-durables	$\Delta_4 y_t$	6	0.077	2.305	2.474**	2.609**	2.257
UK exports	$\Delta_1 y_t$	3	0.701	0.354	1.109	1.342	0.020
UK imports	$\Delta_1 y_t$	1	0.323	1.498	1.391	1.965	0.274
UK total investment	$\Delta_1 y_t$	4	0.657	0.711	2.734**	1.599	3.812**
UK public investment	$\Delta_4 y_t$	4	0.034	0.344	1.534	0.949	1.141
UK workforce[c]	$\Delta_1 y_t$	4	0.025	2.370	1.479	1.392	1.032

[a] The transformation of the data is based on the results of testing for seasonal and nonseasonal unit roots in Table 5.2. When the Δ_1 or the Δ_2 transformation is used, the AR model includes four seasonal dummy variables. When the Δ_4 filter is used, the AR model contains only a constant.

[b] The diagnostic test statistics concern residual autocorrelation of order 1 and 1–4 ($F_{AR,1-1}$ and $F_{AR,1-4}$), periodic residual autocorrelation of order 1 and 1–2 ($F_{PeAR,1-1}$ and $F_{PeAR,1-2}$), and seasonal heteroskedasticity (F_{SH}).

[c] The model for 'Workforce' contains a dummy variable to capture a severe outlier at 1959.2.

** Significant at the 5 per cent level.

for German real GNP and UK total investment display seasonal heteroskedasticity. In sum, it seems that there are some periodic patterns in the estimated residuals.

The second simple approach to investigating periodic variation in the AR parameters is by estimating, for example, a PAR(p) model to y_t and testing whether the null hypothesis of no periodic variation can be rejected. Since I confine the analysis to PAR(p) models, an estimation method is given by considering the regression model

$$y_t = \sum_{s=1}^{4} \mu_s D_{s,t} + \sum_{s=1}^{4} \phi_{1s} D_{s,t} y_{t-1} + \ldots + \sum_{s=1}^{4} \phi_{ps} D_{s,t} y_{t-p} + \varepsilon_t. \tag{7.42}$$

Under normality of the error process ε_t and with fixed starting values, the maximum likelihood estimators of the parameters ϕ_{is}, $i = 1,2, \ldots, p$ and $s = 1,2,3,4$, are obtained from OLS estimation of (7.42). (For alternative estimation methods and asymptotic results, see Pagano (1978) and Troutman (1979). Other relevant references are Cipra (1985) for the estimation of PMA(q) models, Andel (1983) for a Bayesian analysis of PAR models, Vecchia (1985), and P. L. Anderson and Vecchia (1993) for maximum likelihood estimation of PARMA models, and Adams and Goodwin (1995) for an alternative approach.) Notice that the available sample for estimating the periodic parameters in $N = n/4$, i.e. the number of observations, can be small.

Once the parameters in a PAR(p) process have been estimated, a next step is to test for periodic variation in the autoregressive parameters. In Boswijk and Franses (1996) it is shown that the likelihood ratio test for the null hypothesis

$$H_0: \phi_{is} = \phi_i \text{ for } s = 1,2,3,4 \text{ and } i = 1,2 \ldots, p, \qquad (7.43)$$

has an asymptotic $\chi^2(3p)$ distribution, *irrespective* of whether the y_t series has nonseasonal or seasonal unit roots. Hence an F-test for this H_0, which is denoted by F_{PAR}, has the standard $F(3p, n-(4+4p))$ distribution in the case of a PAR(p) process with four seasonal intercepts. An important implication of this result is that (7.42) can be estimated for the y_t series itself; i.e., there is no need to *a priori* difference the y_t series to remove stochastic trends when one wants to test for periodicity. This suggests that, for practical purposes, it seems most convenient to start by estimating the model in (7.42) and testing the H_0 in (7.43). In a second step one may then test for unit roots. An additional advantage is that this sequence of steps allows the possibility of having a periodic differencing filter, which is useful in the case of periodic integration. Since this periodic differencing filter has to be estimated from the data, i.e. the α_s in (7.28), it is therefore not appropriate to start an analysis of periodic variation in a model for, say, $\Delta_1 y_t$ or $\Delta_4 y_t$. (See Lütkepohl (1991) for alternative test procedures for periodic variation in the AR parameters of a PAR(p) process.)

To investigate whether the asymptotic distribution of the F-test for the null hypothesis in (7.43) is useful in the case of small samples, consider the Monte Carlo simulation results displayed in Tables 7.2 and 7.3. For this Monte Carlo study I generated 120 effective observations from nonperiodic AR(1) and

TABLE 7.2. Fractiles of F-statistic for the null hypothesis $\phi_{11} = \phi_{12} = \phi_{13} = \phi_{14} = \phi$ in the PAR(1) model $y_t = \phi_{1s} y_{t-1} + u_t$ when the DGP is the nonperiodic AR(1) process $y_t = \phi y_{t-1} + \varepsilon_t$, with $\varepsilon_t \sim N(0,1)$[a]

Fractiles %	$F(3,115)$	$\phi = 0.6$	$\phi = 0.8$	$\phi = 1.0$
5	0.12	0.12	0.11	0.10
10	0.19	0.20	0.19	0.18
20	0.34	0.34	0.33	0.32
50	0.79	0.81	0.78	0.78
80	1.57	1.61	1.58	1.58
90	2.13	2.16	2.09	2.11
95	2.68	2.66	2.64	2.65
Mean	1.02	1.03	1.01	1.00
Variation	0.72	0.74	0.72	0.73

[a] Based on 5,000 Monte Carlo replications. Sample size is 120 quarterly observations.

TABLE 7.3. Fractiles of F-statistic for the null hypothesis $\phi_{11} = \phi_{12} = \phi_{13} = \phi_{14} = \phi_1$ and $\phi_{21} = \phi_{22} = \phi_{23} = \phi_{24} = \phi_2$ in a PAR(2) process $y_t = \phi_{1s}y_{t-1} + \phi_{2s}y_{t-2} + u_t$ when the DGP is the nonperiodic AR(2) process $y_t = \phi_1 y_{t-1} + \phi_2 y_{t-2} + \varepsilon_t$ with $\varepsilon_t \sim N(0,1)$[a]

Fractiles %	$F(6,110)$	Roots of the AR(2) process				
		0.6,0.8	0.6,1.0	0.8,1.0	1.0,1.0	0.6,–1.0
5	0.27	0.26	0.26	0.27	0.26	0.26
10	0.36	0.35	0.35	0.36	0.37	0.36
20	0.51	0.50	0.50	0.51	0.51	0.50
50	0.90	0.90	0.90	0.89	0.88	0.89
80	1.46	1.46	1.47	1.43	1.43	1.43
90	1.83	1.84	1.84	1.81	1.83	1.82
95	2.18	2.18	2.21	2.17	2.21	2.18
Mean	1.02	1.02	1.02	1.01	1.01	1.01
Variation	0.37	0.37	0.38	0.37	0.36	0.38

[a] Based on 5,000 Monte Carlo replications. Sample size is 120 quarterly observations.

AR(2) processes, fitted PAR(1) and PAR(2) processes, respectively, and calculated the F-test statistic for the hypothesis (7.43). Based on 5,000 replications, various fractiles are calculated and compared with the fractiles from the standard $F(3p,n^*)$ distribution, where $n^* = 120–4–4p$. A comparison of these fractiles indicates that the theoretical and empirical fractiles closely match for a sample of 120 observations. This close match also holds where the y_t series contains one or two unit roots. Hence, for further empirical analysis I will use the F_{PAR}-test for hypothesis (7.42) and the corresponding asymptotic critical values.

Order selection in periodic autoregression

The order of a periodic autoregression is not known in practice and hence it has to be estimated from the data. As for nonperiodic AR models, there are several possible criteria for model selection. One can use model selection criteria such as the Akaike information criterion (AIC),

$$\text{AIC}(p) = n\log\hat{\sigma}^2 + 8p, \tag{7.44}$$

where $\hat{\sigma}^2$ is $RSS\,/\,n$, with n is the effective sample size, or the Schwarz criterion (SC),

$$\text{SC}(p) = n\log\hat{\sigma}^2 + 4p\log n. \tag{7.45}$$

Note that in (7.44) and (7.45) account has been taken of the fact that parameters are estimated for each of the four seasons. An alternative approach is to use an F-test for the deletion of the parameters ϕ_{ls} for some value of l. The order of a PAR can be set equal to p when some or all $\phi_{ps} \neq 0$, while the $\phi_{p+1,s} = 0$ for all s. This F-test is applied to PAR models with decreasing orders, where the initial order may be set at 4 or 8. Alternatively, one may estimate PAR models of order 8, 7, etc., and test whether the residuals display periodic autocorrelation. The order p is chosen when the LM test statistic calculated from the auxiliary regression (7.40) indicates that the hypothesis of no periodic autocorrelation cannot be rejected.

To investigate the empirical performance of the four PAR order selection methods given above, a Monte Carlo study is performed using 5,000 replications. In the experiments, the nominal significance level of the tests is set at 5 per cent. I start by considering PAR processes of order 2 and estimating PAR processes of orders 4 to 1.

Table 7.4 reports the selection frequencies of each of these PAR models where the PAR(2) models have roots close to zero or close to unity. It can be seen that the SC criterion performs best where the roots are not too close to zero. With the AIC one will be inclined to opt for too high an order in about 18 per cent of the cases. The F-type tests for the significance of the $\phi_{p+1,s}$ and for the periodicity in the autocorrelation function of the residuals from a PAR show selection frequencies that are reasonably stable over the DGPs. Notice that the results where a PAR(3) is selected using the F-type tests indicates that the empirical size of the tests is close to the nominal size.

TABLE 7.4. Autoregressive order selection in periodic autoregressions with no unit roots in the vector process[a]

DGP[b]	Roots of $\Phi_0^{-1}\Phi_1$	Criterion[c]	Selected order of periodic autoregression			
			1	2	3	4
(i)	$0.06 \pm 0.11i$	SC	33.4	66.3	0.3	0.0
		AIC	2.1	79.8	12.4	5.7
		F_{PeAR}	3.9	82.1	5.2	4.6
		$F(\phi_{p+1,s}=0)$	4.2	85.7	5.4	4.7
(ii)	0.78, 0.83	SC	0.0	99.7	0.3	0.0
		AIC	0.0	82.1	12.0	5.9
		F_{PeAR}	0.0	84.9	4.7	5.4
		$F(\phi_{p+1,s}=0)$	0.0	89.4	4.8	5.8

[a] Based on 5,000 Monte Carlo replications. Sample size is 120 observations. The cells report the frequencies that a certain model is selected. The data-generating process is a periodic autoregression of order 2.

[b] The data-generating processes are all PAR(2) processes. The parameters $\phi_{11}, \phi_{21}, \phi_{12}, \ldots, \phi_{24}$ are 0.70, 0.40, 0.90, 0.30, 0.80, –0.30, 0.60, and –0.40 for (i) and 2.00, –1.10,1.80, –0.90, 1.90, –0.80, 1.89, and –0.81 for (ii).

[c] The expressions of the AIC and SC criteria are given in (7.44) and (7.45). The F_{PeAR} is the test for periodicity in the residuals, (see (7.40)). The $F(\phi_{p+1,s}=0)$ is the F-type test for the significance of the four $\phi_{p+1,s}$ parameters.

TABLE 7.5. Autoregressive order selection in periodic autoregressions with one unit root in the vector process[a]

DGP[b]	Roots of $\Phi_0^{-1}\Phi_1$	Criteria[c]	Selected order of periodic autoregression			
			1	2	3	4
(i)	1.0, 0.6	SC	0.0	99.8	0.2	0.0
		AIC	0.0	82.8	11.4	5.8
		F_{PeAR}	0.0	85.4	4.4	4.6
		$F(\phi_{p+1,s}=0)$	0.0	90.4	4.7	4.9
(ii)	1.0, 0.8	SC	0.0	99.6	0.4	0.0
		AIC	0.0	81.4	12.5	6.2
		F_{PeAR}	0.0	84.9	4.4	5.1
		$F(\phi_{p+1,s}=0)$	0.0	89.7	4.8	5.5

[a] Based on 5,000 Monte Carlo replications. Sample size is 120 observations. The cells report the frequencies that a certain model is selected. The data-generating process is a periodic autoregression of order 2.

[b] The data-generating processes are all PAR(2) processes. The parameters $\phi_{11}, \phi_{21}, \phi_{12}, \ldots, \phi_{24}$ are 1.96, –0.96, 1.93, –0.93, 1.85, –0.85, 1.79, and –1.79 for (i) and 1.70, –0.70, 1.95, –0.95, 2.20, –1.20, 2.00, and –1.00 for (ii).

[c] The expressions of the AIC and SC criteria are given in (7.44) and (7.45). The F_{PeAR} is the test for periodicity in the residuals, (see (7.40)). The $F(\phi_{p+1,s}=0)$ is the F-type test for the significance of the four $\phi_{p+1,s}$ parameters.

Similar results are reported in Table 7.5, where the DGP is again a PAR(2) process, though now with one unity solution to the characteristic equation for $\mathbf{Y}_T$. Comparing the selection frequencies with those in Table 7.4, it can be observed that the number of unit roots does not seem to affect order selection, and also that the SC and $F_{\text{PeAR},1-1}$-test perform well.

Table 7.6 displays the order selection results where a subset PAR(2) is the data-generating process, i.e. where one or more parameters ϕ_{2s} are set equal to zero. For cases (i)–(iv) in the Table 0–3 of these ϕ_{2s} are set equal to zero. The performance of the SC seems most affected. The AIC as well as the F-tests display similar results as in Tables 7.4 and 7.5.

Until now only periodic processes are generated and estimated. It is also interesting to see whether the order of an initially estimated PAR process (after which step one may consider tests for parameter restrictions) corresponds to the order of a nonperiodic seasonal time series process. Table 7.7 reports the results of selecting p in a PAR(p) processes in the case where $\Delta_4 y_t = \alpha\Delta_4 y_{t-1} + \varepsilon_t$ is the DGP for some values of α. The correct model order is 5, but, given the number of redundant parameters to be estimated when considering a PAR model, one may expect less favourable results as in Tables 7.4 and 7.5. Setting the maximum value of p at 8, one observes that when α is small, say 0.2, most criteria indicate too small a model order. When this α value increases, the performance of the order selection methods largely improves, and a performance similar to that in Tables 7.4 and 7.5 can be observed. Comparable results

TABLE 7.6. Autoregressive order selection in periodic subset autoregressions.[a]

DGP[b]	Roots of $\Phi_0^{-1}\Phi_1$	Criterion[c]	Selected order of periodic autoregression			
			1	2	3	4
(i)	0.22±0.03i	SC	3.0	96.5	0.5	0.0
		AIC	0.1	81.2	12.4	6.3
		F_{PeAR}	0.2	84.9	5.1	4.8
		$F(\phi_{p+1,s}=0)$	0.2	89.3	5.5	5.0
(ii)	0.45	SC	43.5	56.5	0.0	0.0
		AIC	4.2	78.5	11.7	5.6
		F_{PeAR}	6.9	78.3	4.5	5.2
		$F(\phi_{p+1,s}=0)$	7.4	82.5	4.7	5.4
(iii)	0.89	SC	34.5	65.4	0.1	0.0
		AIC	2.8	79.3	12.2	5.7
		F_{PeAR}	5.0	81.1	4.3	4.7
		$F(\phi_{p+1,s}=0)$	5.3	85.3	4.5	5.0
(iv)	0.70	SC	62.2	37.7	0.1	0.0
		AIC	10.7	72.8	11.0	5.4
		F_{PeAR}	15.5	70.6	4.1	4.9
		$F(\phi_{p+1,s}=0)$	16.2	74.2	4.5	5.1

[a] The data-generating process is a periodic autoregression of order 2. Based on 5,000 Monte Carlo replications. Sample size is 120 observations. The cells report the frequencies that a certain model is selected.

[b] The data-generating processes are all PAR(2) processes. The parameters $\phi_{11}, \phi_{21}, \phi_{12}, \ldots, \phi_{24}$ are 0.70, 0.55, 0.90, 0.40, 0.80, –0.45, 0.60, and –0.50 for (i); 0.70, 0.55, 0.90, 0.40, 0.80, –0.45, 0.60, and 0.00 for (ii); 0.70, 0.55, 0.90, 0.40, 0.80, 0.00, 0.60, and 0.00 for (iii); and 0.70, 0.55, 0.90, 0.00, 0.80, 0.00, 0.60, and 0.00 for (iv).

[c] See note c in Table 7.5.

emerge when the DGP is a low-order nonperiodic process with seasonal unit roots.

Table 7.8 presents the frequencies of order selection for five different seasonal unit root processes. The general impression is that in at least 75 per cent of the cases one would detect the correct model order.

Overall, I conclude that order selection does not seem to be affected by the number of unit roots, that the SC often detects the correct model order, and that the F-test for the significance of $\phi_{p+1,s}$ shows satisfactory empirical performance since its success rate appears to be reasonably stable over the various DGPs. Notice further that the model order is often correctly determined, even when the DGP is a nonperiodic process. Taking the results in Tables 7.2 and 7.3 into consideration, one can observe that fitting a PAR model does not prevent the finding of a nonperiodic AR process, if the latter is in fact the DGP. As an empirical strategy I recommend using the SC to select the model order p, provided that the F-test for $\phi_{p+1,s}=0$ does not reject the null hypothesis. This approach will also be followed in the next empirical analysis of the 11 sample series.

TABLE 7.7. Periodic autoregressive order selection when the DGP is the non-periodic ARIMA(0,1,0)$_4$×(1,0,0) process: $\Delta_4 y_t = \alpha\Delta_4 y_{t-1} + \varepsilon_t$[a]

α	Criterion[b]	Selected order of periodic autoregression							
		1	2	3	4	5	6	7	8
0.2	SC	0.0	0.0	0.0	95.7	4.2	0.1	0.0	0.0
	AIC	0.0	0.0	0.0	47.9	27.9	10.6	7.1	6.5
	F_{PeAR}	0.0	0.0	0.0	57.1	18.3	5.2	6.0	6.1
	$F(\phi_{p+1,s}=0)$	0.0	0.0	0.0	61.9	19.5	5.8	6.5	6.4
0.4	SC	0.0	0.0	0.0	64.5	35.0	0.5	0.0	0.0
	AIC	0.0	0.0	0.0	11.8	56.4	14.0	9.0	8.7
	F_{PeAR}	0.0	0.0	0.0	22.5	53.5	5.1	5.9	5.7
	$F(\phi_{p+1,s}=0)$	0.0	0.0	0.0	24.2	57.7	5.5	6.4	6.3
0.6	SC	0.0	0.0	0.1	12.0	86.9	0.8	0.1	0.0
	AIC	0.0	0.0	0.0	0.5	65.4	15.9	8.9	9.2
	F_{PeAR}	0.0	0.0	0.0	1.5	73.3	6.0	5.9	6.7
	$F(\phi_{p+1,s}=0)$	0.0	0.0	0.0	1.5	78.6	6.4	6.4	7.1
0.8	SC	0.2	0.0	0.3	0.5	98.2	0.7	0.1	0.0
	AIC	0.0	0.0	0.0	0.0	65.8	14.9	9.8	9.5
	F_{PeAR}	0.0	0.0	0.0	0.0	74.5	5.4	6.4	6.8
	$F(\phi_{p+1,s}=0)$	0.0	0.0	0.0	0.0	80.1	5.9	6.9	7.1
1.0	SC	0.1	0.0	0.0	0.0	98.8	1.0	0.1	0.0
	AIC	0.0	0.0	0.0	0.0	63.8	14.9	10.6	10.7
	F_{PeAR}	0.0	0.0	0.0	0.0	72.5	6.0	7.2	7.5
	$F(\phi_{p+1,s}=0)$	0.0	0.0	0.0	0.0	77.8	6.5	7.6	8.1

[a] Based on 5,000 Monte Carlo replications. Sample size is 120 observations. The cells report the frequencies that a certain model is selected. The correct model order is 5.
[b] See note *c* in Table 7.4.

Table 7.9 presents some diagnostic results for estimated periodic autoregressive models of order p, where the p is selected using the suggested method for the 11 sample series. It appears that the value of p can be set equal to 1 or 2 for ten variables, and that only for unemployment in Canada does one need four lags in the PAR model. The diagnostic measures for residual autocorrelation, ARCH, normality, and seasonal heteroskedasticity suggest that the PAR(p) models give approximately adequate descriptions of the data. For three variables I find evidence of seasonal heteroskedasticity. An inspection of the estimated residuals, however, indicates that this evidence is due mainly to two or three observations. Hence I will assume that the PAR model as in (7.1), without seasonal variances, is adequate. Furthermore, some models need the inclusion of one or two dummy variables to remove outlying observations. An (unreported) inspection of the models where such dummies are deleted suggests that these variables affect only the diagnostic measure for normality of the residuals. Table 7.9 also includes the results for the F_{PAR}-test statistic, and it appears that the null hypothesis of no periodicity can be rejected quite

TABLE 7.8. Periodic autoregressive order selection when the DGP is the nonperiodic process: $\lambda(B)y_t = 0.5\lambda(B)y_{t-1} + \varepsilon_t$, where $\lambda(B)$ corresponds to components of the $(1 - B^4)$ filter[a]

$\lambda(B)$	Criterion[b]	Selected order of periodic autoregression[c]							
		1	2	3	4	5	6	7	8
$(1+B)$	SC	9.6	90.3	0.1	0.0	0.0	0.0	0.0	0.0
	AIC	0.2	78.8	10.5	4.4	2.2	1.5	1.1	1.2
	F_{PeAR}	0.5	68.0	3.9	4.1	3.8	3.6	5.2	5.8
	$F(\phi_{p+1,s}=0)$	0.5	71.6	4.2	4.4	3.8	3.9	5.4	6.0
$(1-B^2)$	SC	0.0	9.9	89.7	0.3	0.0	0.0	0.0	0.0
	AIC	0.0	0.3	74.9	11.4	5.2	3.1	2.5	2.6
	F_{PeAR}	0.0	0.7	70.9	3.8	4.4	4.2	4.4	5.6
	$F(\phi_{p+1,s}=0)$	0.0	0.8	75.4	4.0	4.7	4.5	4.6	6.0
$(1+B^2)$	SC	0.0	16.5	83.1	0.4	0.0	0.0	0.0	0.0
	AIC	0.0	0.6	75.3	11.0	5.5	3.1	2.1	2.4
	F_{PeAR}	0.0	1.1	70.4	3.8	4.5	4.3	4.3	5.5
	$F(\phi_{p+1,s}=0)$	0.0	1.2	75.0	4.1	4.9	4.5	4.6	5.8
$(1-B)(1+B^2)$	SC	0.0	0.0	16.5	83.1	0.3	0.0	0.0	0.0
	AIC	0.0	0.0	0.7	69.2	13.8	7.0	4.3	4.9
	F_{PeAR}	0.0	0.0	1.8	70.3	5.1	5.7	4.8	5.3
	$F(\phi_{p+1,s}=0)$	0.0	0.0	1.9	75.5	5.7	6.1	5.1	5.7
$(1+B)(1+B^2)$	SC	0.0	0.0	34.1	65.5	0.4	0.0	0.0	0.0
	AIC	0.0	0.0	3.0	68.9	13.0	6.2	4.3	4.6
	F_{PeAR}	0.0	0.0	5.6	67.2	5.0	4.2	5.1	6.5
	$F(\phi_{p+1,s}=0)$	0.0	0.0	6.1	71.4	5.4	4.5	5.5	7.0

[a] Based on 5,000 Monte Carlo replications. Sample size is 120 observations. The cells report the frequencies that a certain model is selected.

[b] The correct order for the DGPs is 2, 3, 3, 4, 4, respectively.

[c] See note *c* in Table 7.4.

convincingly for all 11 time series. In sum, PAR(p) models seem to yield useful descriptions of the data. Notice that the order of the PAR model is much lower than the orders of the models fitted in the previous chapters and in Table 7.1. Hence, it seems that there is a trade-off between periodic parameter variation and model order when this variation is neglected. In Section 7.4 I will return to this issue in more detail.

Table 7.10 reports similar results to those of Table 7.9, although now PAR(p) models are estimated which include four deterministic trends, i.e.

$$y_t = \mu_s + \tau_s t + \phi_{1s} y_{t-1} + \ldots + \phi_{ps} y_{t-p} + \varepsilon_t, \tag{7.46}$$

where $t = 1,2, \ldots, n$, and τ_s is a seasonally varying trend parameter. This model will also be used in the next chapter when I test for the presence of unit roots. Since a PAR(p) model is fitted to the untransformed y_t variables, it may occur that the trend variables are relevant. The results in Table 7.10 indicate that the four trend variables are significant for five variables, i.e. US industrial

TABLE 7.9. Diagnostic results for periodic autoregressions of order p with constants for 11 sample series[a]

Variable	p^b	$F_{AR,1-1}$	$F_{AR,1-4}$	$F_{ARCH,1-1}$
Industrial production, USA	2	0.016	0.650	0.239
Unemployment Canada[c]	4	3.677	1.356	0.195
Real GNP, Germany	2	0.929	1.864	0.106
UK GDP[d]	2	0.004	1.393	1.346
UK total consumption[e]	1	0.521	1.693	0.002
UK consumer nondurables	1	0.072	0.436	0.305
UK exports[f]	2	1.027	0.384	5.674**
UK imports	1	1.942	1.567	4.228**
UK total investment	2	0.544	1.958	2.539
UK public investment[g]	2	1.841	0.996	0.973
UK workforce[h]	2	0.081	1.777	0.396

	$F_{ARCH,1-4}$	JB	F_{SH}	F_{PAR}
Industrial production, USA	0.945	8.149**	1.418	11.299**
Unemployment, Canada[c]	0.259	1.324	0.750	5.793**
Real GNP, Germany	1.419	3.823	2.779**	43.484**
UK GDP[d]	1.959	5.339	1.896	11.715**
UK total consumption[e]	0.949	4.606	2.379	34.314**
UK consumer nondurables	1.414	5.600	1.878	31.504**
UK exports[f]	1.590	2.740	1.271	5.629**
UK imports	1.239	0.982	0.025	4.731**
UK total investment	1.708	0.928	4.388**	7.087**
UK public investment[g]	1.354	1.236	6.085**	25.887**
UK workforce[h]	0.660	4.087	1.062	10.444**

[a] The data are all in logs, except for unemployment Canada. The model is $y_t = \mu_s + \phi_{1s}y_{t-1} + \ldots + \phi_{ps}y_{t-p} + \varepsilon_t$.

[b] The order of p is selected using the Schwarz information criterion and an F—test for the significance of the $\phi_{p+1,s}$ parameters.

[c]This model includes a dummy variable for 1982.2; when this dummy variable is deleted, the F_{PAR} statistic obtains a value of 3.102**.

[d] This model includes a dummy variable for 1971.1; when this dummy variable is deleted, the F_{PAR} statistic obtains a value of 10.375**.

[e] This model includes dummy variables for 1979.3 and 1980.2; when these dummy variables are deleted, the F_{PAR} statistic obtains a value of 30.607**.

[f] This model includes dummy variables for 1967.4 and 1968.1; when these dummy variables are deleted, the F_{PAR} statistic obtains a value of 4.425**.

[g] This model includes a dummy variable for 1963.1; when this dummy variable is deleted, the F_{PAR} statistic obtains a value of 21.028**.

[h] This model includes a dummy variable for 1959.2; when this dummy variable is deleted, the F_{PAR} statistic obtains a value of 8.679**.

** Significant at the 5 per cent level. The diagnostic tests concern first and first to fourth autocorrelations ($F_{AR,1-1}$, $F_{AR,1-4}$), first and first to fourth ARCH effects ($F_{ARCH,1-1}$, $F_{ARCH,1-4}$), normality (JB), seasonal heteroskedasticity (F_{SH}), and periodic variation in the estimated AR parameters (F_{PAR}).

TABLE 7.10. Diagnostic results for periodic autoregressions of order p, with constants and trends for 11 sample series[a]

Variable	p^b	F_T	$F_{AR,1-1}$	$F_{AR,1-4}$	$F_{ARCH,1-1}$
Industrial production USA	2	7.379**	0.005	0.611	0.411
Unemployment, Canada[c]	4	2.348	1.438	0.629	1.158
Real GNP, Germany	2	1.197	1.027	1.619	0.045
UK GDP[d]	2	6.937**	0.184	1.647	0.025
UK total consumption[e]	1	2.052	0.271	0.829	0.603
UK consumer nondurables	1	3.443**	0.207	0.442	0.041
UK exports[f]	2	2.117	0.308	0.248	6.115**
UK imports	1	5.713**	0.323	1.691	2.229
UK total investment	2	2.653	0.040	0.095	3.616
UK public investment[g]	2	9.608**	0.132	0.676	1.133
UK workforce[h]	2	1.309	0.302	2.051	0.769

Variable	$F_{ARCH,1-4}$	JB	F_{SH}	F_{PAR}
Industrial production, USA	0.332	19.169**	1.895	3.021**
Unemployment, Canada[c]	0.651	6.779**	0.869	5.627**
Real GNP, Germany	0.957	3.580	3.188**	8.366**
UK GDP[d]	1.446	1.299	0.153	5.651**
UK total consumption[e]	1.207	3.254	2.739**	0.542
UK consumer nondurables	0.543	5.494	2.244	1.849
UK exports[f]	2.780**	5.670	0.218	3.357**
UK imports	0.637	1.107	0.564	3.500**
UK total investment	2.378	1.980	6.252**	2.733**
UK public investment[g]	1.111	5.775	2.045**	3.739**
UK workforce[h]	1.482	2.499	0.782	3.474**

[a] The data are all in logs, except for unemployment, Canada. The model is $y_t = \mu_s + T_s t + \phi_{1s} y_{t-1} + \ldots + \phi_{ps} y_{t-p} + \varepsilon_t$

[b] The order of p is selected using the Schwarz information criterion and an F—test for the significance of the $\phi_{p+1,s}$ parameters.

[c] This model includes a dummy variable for 1982.2; when this dummy variable is deleted, the F_{PAR} statistic obtains a value of 5.624**.

[d] This model includes a dummy variable for 1971.1; when this dummy variable is deleted, the F_{PAR} statistic obtains a value of 5.564**.

[e] This model includes dummy variables for 1979.3 and 1980.2; when these dummy variables are deleted, the F_{PAR} statistic obtains a value of 0.222.

[f] This model includes dummy variables for 1967.4 and 1968.1; when these dummy variables are deleted, the F_{PAR} statistic obtains a value of 2.183**.

[g] This model includes a dummy variable for 1963.1; when this dummy variable is deleted, the F_{PAR} statistic obtains a value of 3.485**.

[h] This model includes a dummy variable for 1959.2; when this dummy variable is deleted, the F_{PAR} statistic obtains a value of 2.783**.

** Significant at the 5 per cent level. The diagnostic tests concern first and first to fourth autocorrelations ($F_{AR,1-1}$, $F_{AR,1-4}$), first and first to fourth ARCH effects ($F_{ARCH,1-1}$, $F_{ARCH,1-4}$), normality (JB), seasonal heteroskedasticity (F_{SH}), periodic variation in the estimated AR parameters (F_{PAR}), and the significance of the four trend variables (F_T). The simulated 5 per cent critical value of this F—test in case of a single unit root in the PAR process in 3.43. This critical value is based on a Monte Carlo experiment using 20,000 replications.

production, UK GDP, UK consumption nondurables, UK imports, and UK public investment. For nine series, the results for the F_{PAR}-test remain similar, although the inclusion of the trends for the two UK consumption series means that the null hypothesis of no periodic parameter variation cannot be rejected. This evidence is interpreted as indicating that trends, which can be either deterministic or stochastic (and this will be determined in the next chapter), can have an impact on seasonal variation. Hence it seems useful to test for unit roots in both (7.1) and (7.46). The inclusion of the trend variables does not however affect the order of the PAR model. Again, the significance of the F_{SH}-statistic is due mainly to a few observations, and therefore I do not modify (7.46) with seasonal variances.

The display of the parameter estimates in the various PAR(p) models from Tables 7.9 and/or 7.10 is postponed to Chapter 8. To have an impression though of the possible number of unit roots in the 11 time series, the nonzero eigenvalues of the matrix $\hat{\Phi}_0^{-1}\hat{\Phi}_1$ are given in Table 7.11. Notice that this matrix contains all relevant information since the value of P in the VQ representation of all 11 PAR models is equal to 1 (cf. (7.9)). The nonzero eigenvalues in the table indicate that, for ten time series, there seems to be only a single real-valued eigenvalue which is close to unity. Only the Canadian unemployment series has complex roots. Furthermore, it is clear that the inclusion of four deterministic trends in the PAR model biases the value of the almost unity eigenvalues away from the unit circle. Hence one may expect that the inclusion of trends can affect the outcomes of unit root test statistics. Given the discussion on I(2) type processes in the previous section, the main conclusion from Table 7.11 is that for most series it seems that there is only a single unit root in the time series, and hence there do not seem to be I(2) type processes among the 11 sample series.

TABLE 7.11. Nonzero eigenvalues of the estimated $\Phi_0^{-1}\Phi_1$ matrix for the periodic autoregressions for the 11 sample series[a]

Variable	Without trends	With trends
Industrial production, USA	0.938, 0.032	0.462, 0.174
Unemployment, Canada	$0.854\pm0.419i$, 0.046, –0.018	$0.296\pm0.124i$, 0.439, –0.039
Real GNP, Germany	0.965, 0.036	0.855, 0.042
UK GDP	0.982, –0.002	0.777, –0.029
UK total consumption	1.020	0.775
UK consumer nondurables	1.010	0.855
UK exports	0.980, –0.002	0.769, –0.002
UK imports	1.003	0.485
UK total investment	0.948, –0.002	1.072, 0.001
UK Public investment	1.055, 0.002	0.797, –0.005
UK workforce	0.829, 0.001	0.932, 0.002

[a] The models can contain constants and trends or only constants. The models are those given in Tables 7.9 and 7.10 and include the dummy variables to remove the outlying observations (as indicated in the relevant footnotes).

Spurious periodicity because of seasonal mean shifts

In Chapter 5, I found that the evidence in favour of seasonal unit roots may become less pronounced, or may even disappear, where one allows for seasonal mean shifts. Hence, some seemingly nonstationary seasonal variation may be due to neglected deterministic shifts. Since estimating a PAR(p) model for quarterly data involves the estimation of $4p$ or more parameters, it seems worthwhile investigating whether all parameters are indeed periodically varying or whether apparent periodicity is spurious. In fact, the occurrence of mean shifts in a nonperiodic model may (spuriously) suggest that a periodic model is useful (see Franses 1995*b*). For example, consider the simple DGP for quarterly observations

$$y_t = \sum_{s=1}^{4} \beta_s D_{s,t} + \sum_{s=1}^{4} \gamma_s D_{s,t}(I_{t \geq \lambda n}) + u_t, \tag{7.47}$$

where u_t is an error process that is not necessarily a white noise process and $I_{t \geq \lambda n}$ is an indicator function with $\lambda \in [0,1]$. The parameter λ indicates at which fraction of the sample the seasonal means of y_t shift from β_s to $\beta_s + \gamma_s$. A graph of a y_t series generated by model (7.47) would suggest that the seasonal patterns have changed. Furthermore, if one were to calculate the periodic autocorrelations along the lines suggested in, for example, Vecchia and Ballerini (1991), one would likely find that the autocorrelations of y_t vary with the seasons. The main reason for this phenomenon is that a mean shift from β_s to $\beta_s + \gamma_s$ results in an upward bias in the (seasonal) autocorrelation coefficients. (See Perron (1989) and Perron and Vogelsang (1992) for some theoretical results on the effects of mean shifts in nonperiodic models.)

To illustrate the impact of seasonal mean shifts on the possible finding of periodic variation in parameters in time series models, I conducted a limited simulation experiment. I generated observations from the DGP

$$y_t = \alpha y_{t-1} + \sum_{s=1}^{4} \beta_s D_{s,t} + \sum_{s=1}^{4} \gamma_s D_{s,t}(I_{t \geq \lambda n}) + \varepsilon_t, \tag{7.48}$$

where α is set equal to 0.8 or 0.95, and ε_t is a standard white noise process. One simple method of detecting the possible empirical usefulness of a PAR model for y_t generated from (7.48) is to test the estimated residuals from a fitted nonperiodic AR model of order k for periodic autocorrelation. An LM test for first-order periodic autocorrelation can be based on the auxiliary regression in (7.40).

Table 7.12 reports the rejection frequencies of this LM test when the sample size is either 120 or 240, the β_s values are {4,–4,4,–4}, and the γ_s values are all zero except for γ_4, which can take values such as ±8, ±6, and 0. The latter value is included to investigate the size properties of the LM test (in the case of overspecification of the AR order). In the case of seasonal mean shifts, we may expect that y_t will display some autocorrelation. Hence I estimate nonperiodic AR models of order 4 for y_t. It should be mentioned that other AR orders do not yield dramatically different conclusions. The results in Table 7.12 indicate

TABLE 7.12. Spurious periodicity because of seasonal mean shifts in non-periodic models[a,b]

α	γ_4	$n = 120$		$n = 240$	
		$\lambda = 1/4$	$\lambda = 1/2$	$\lambda = 1/4$	$\lambda = 1/2$
0.8	−8	0.258	0.662	0.978	1.000
	−6	0.059	0.168	0.424	0.824
	0	0.020	0.030	0.027	0.025
	6	0.089	0.207	0.441	0.863
	8	0.282	0.716	0.989	1.000
0.95	−8	0.094	0.376	0.862	0.999
	−6	0.051	0.079	0.294	0.683
	0	0.024	0.019	0.028	0.037
	6	0.043	0.123	0.311	0.676
	8	0.102	0.416	0.907	1.000

[a] The cells contain the rejection frequencies of the LM test for periodicity in the residuals from an AR(4) process for y_t when the DGP is $y_t = \alpha y_{t-1} + \Sigma_{s=1}^{4} \beta_s D_{st}(I_{t \geq \lambda n}) + \varepsilon_t$ with $\varepsilon_t \sim N(0,1)$, $\beta_1 = -\beta_2 = \beta_3 = -\beta_4 = 4$, and where $\gamma_1 = \gamma_2 = \gamma_3 = 0$ and $\gamma_4 = \gamma_4$. Based on 1,000 Monte Carlo replications.

[b] An expression of the auxiliary regression for the test for periodicity in the residuals in given in (7.40). Furthermore, the results for λ are similar to those for $1-\lambda$.

that the null hypothesis of no periodicity in the residuals is often rejected in cases where there is seasonal mean shift. This rejection occurs more frequently when α in (7.48) takes a lower value, when γ_4 takes larger positive or negative values, when sample size increases, and when the break occurs halfway the sample. The rows corresponding to $\gamma_4 = 0$ yield the empirical size of the LM test. Since AR models of order 4 are estimated for data generated from AR(1) models, one may expect a slight size distortion. The values in the rows for $\gamma_4 = 0$ take slightly smaller values than 0.05, and hence the empirical size of the LM test seems biased downwards in case of overspecification of the AR order. The main message from Table 7.12, however, is that one may detect periodicity in the estimated residuals when a nonperiodic AR model is fitted to a nonperiodic AR data-generating process with a single seasonal mean shift. Of course, when there is more than one such seasonal mean shift, the rejection frequency of the LM test will be even larger.

Since the finding of the seasonal unit roots $\pm i$ in the German real GNP series did not appear robust over various sample sizes (see Table 5.3), I investigated whether the finding of periodicity in Table 7.9 is robust. Again, a recursive analysis similar to that in Table 5.3 was used. The results for the F_{PAR}-test are given in Table 7.13. It appears that for most sample sizes one can reject the null hypothesis of no periodic variation in the AR parameters. However, when the sample from, say, 1974.1–1990.4 would have been considered, one would not

TABLE 7.13. Recursive testing for periodic variation in the AR parameters in a PAR(2) model with constant for German real GNP[a]

Sample	F_{PAR}	Sample	F_{PAR}
1961.1–1970.4	4.814**	1962.1–1990.4	37.516**
1961.1–1971.4	4.706**	1963.1–1990.4	34.191**
1961.1–1972.4	4.795**	1964.1–1990.4	25.048**
1961.1–1973.4	6.056**	1965.1–1990.4	21.499**
1961.1–1974.4	7.470**	1966.1–1990.4	19.723**
1961.1–1975.4	8.725**	1967.1–1990.4	19.054**
1961.1–1976.4	10.492**	1968.1–1990.4	16.632**
1961.1–1977.4	14.157**	1969.1–1990.4	12.550**
1961.1–1978.4	17.639**	1970.1–1990.4	8.632**
1961.1–1979.4	20.982**	1971.1–1990.4	4.972**
1961.1–1980.4	24.397**	1972.1–1990.4	2.909**
1961.1–1981.4	27.345**	1973.1–1990.4	2.472**
1961.1–1982.4	30.065**	1974.1–1990.4	1.956
1961.1–1983.4	33.529**	1975.1–1990.4	1.787
1961.1–1984.4	35.324**	1976.1–1990.4	1.197
1961.1–1985.4	35.493**	1977.1–1990.4	2.609**
1961.1–1986.4	36.308**	1978.1–1990.4	2.135
1961.1–1987.4	37.519**	1979.1–1990.4	2.241
1961.1–1988.4	38.750**	1980.1–1990.4	2.622**
1961.1–1989.4	40.037**	1981.1–1990.4	1.694
1961.1–1990.4	38.864**	1982.1–1990.4	1.973

[a] The cells give the value of the F_{PAR} test.

** Significant at the 5 per cent level.

have rejected the null hypothesis. On the whole, however, it seems that there is significant periodic variation in the dynamic parameters in a time series model for these GNP data.

Table 7.14 presents the F_{PAR} values where it is assumed that there is a structural break at time T_b. The regression model (7.1) is now enlarged, with four $DU_{s,t}$ variables and four $D(T_b)_{s,t}$ variables, where these are defined as in (5.13). Since the choice of the T_b in Chapter 5 sometimes had an effect on the outcomes of seasonal unit root tests, I decided to consider the same dates. For the variables not analysed in Table 5.5, I set T_b equal to 1973.4, which roughly corresponds to halfway through the sample. The results given in Table 7.14 indicate that the evidence in favour of periodic parameter variation remains significant for 10 of the 11 variables. Only for UK imports may one hypothesize a structural break around 1973.4 which reduces this evidence.

Table 7.14 also reports the F_{PAR} values in the case where PAR(p) models are estimated for the data when they are not transformed using logs. Note that for Canadian unemployment logs are now taken. The empirical results in Tables 6.1 and 6.2 above suggest that the data transformation can have an impact on

TABLE 7.14. Testing for periodic variation in the AR parameters of an estimated PAR(p) process for the 11 sample series, when there is a structural break at time T_b, and when the data are not transformed using logs[a,b]

Variable	p	Structural break		Other data transformation
		T_b	F_{PAR}	F_{PAR}
Industrial production, USA	2	1976.4	2.922**	11.031**
Unemployment Canada	4	1974.4	4.865**	7.819**
Real GNP, Germany	2	1974.4	8.014**	23.637**
UK GDP	2	1973.4	2.743**	17.328**
UK total consumption	1	1973.4	8.148**	42.236**
UK consumer nondurables	1	1973.4	4.124**	85.259**
UK exports	2	1973.4	3.347**	8.117**
UK imports	1	1973.4	2.640	6.602**
UK total investment	2	1973.4	3.246**	13.194**
UK public investment	2	1975.4	11.076**	10.996**
UK workforce	2	1973.4	3.376**	11.504**

[a] Except for Unemployment, Canada, where logs are taken. The models contain four constants. The test statistic is F_{PAR}.

[b] The models are those given in Tables 7.9 and 7.10 and include the dummy variables to remove the outlying observations (as indicated in the relevant footnotes).

** Significant at the 5 per cent level.

seasonal variation in parameters. In contrast, the F_{PAR} values in Table 7.14 clearly indicate that the apparent periodicity in the log transformed data may not be due to an inappropriate data transformation.

In this section I have discussed methods of selecting a suitable PAR model, and I have applied the most useful method (which is found from Monte Carlo evidence) to the 11 sample series. It appears that these series generally can be described by low-order periodic autoregressions, and that this finding is fairly robust to data transformations and seasonal mean shifts. In the next section I will examine the effects of neglecting periodic variation in AR parameters.

7.4 Effects of Neglecting Periodicity

Given that the periodic parameters in PAR(p) models are estimated using only N observations, and that a PAR(p) model can involve many parameters, a natural question now is what the effect would be if one neglected periodic variation in the parameters. More precisely, it is of interest to investigate the consequence of considering nonperiodic models for either y_t, $\Delta_1 y_t$, or $\Delta_4 y_t$ or for $\hat{y}_t^{ns}$ where $\hat{y}_t^{ns}$ is a somehow seasonally adjusted time series.

Theoretical results on the effects of the misspecification of periodicity are derived in Tiao and Grupe (1980) and Osborn (1991). In the first study it is shown that there always exists a time-invariant approximation to a periodic model in the sense that one can always fit an ARMA(p,q) model to data generated from a PAR process. The basis of this approximation is that, although the autocovariance function of a periodically stationary process is time-varying, it averages out to an autocovariance function, which corresponds to some stationary ARMA model. When γ_{ks} denotes the k+th order autocovariance of a stationary periodic process y_t in season s, we can define its average $\bar{\gamma}_k = \frac{1}{4} \Sigma_{s=1}^{4} \gamma_{ks}$. The time-variant approximation of y_t is an ARMA model that has $\bar{\gamma}_k$ ($k = 0,1,2,...$) as its autocovariance function. Note however that the approximation is a misspecified homogeneous model since the periodic variation in the parameter cannot be removed just by considering a nonperiodic ARMA process; see for example the results in Table 7.1, where there appears to be significant periodic correlation in the residuals from some nonperiodic AR models.

In Osborn (1991) it is shown that the orders of the time-invariant models can be higher than that of the PAR model. Hence, when neglecting periodicity, one may expect the need to include several (seasonal) lags in a nonperiodic model. Furthermore, it is shown in Osborn (1991) that misspecification may result in biased forecasts. Finally, if we neglect a periodic dynamic structure, we may expect that the estimated error process is seasonally heteroskedastic. All these theoretical results seem to be reflected by the empirical results reported in Table 7.1. Given that seasonal adjustment amounts to the application of several filters, and that some seasonal adjustment methods can be represented by linear nonperiodic SARIMA models, we may also expect that the periodic variation in the AR parameters does not disappear after seasonal adjustment, although currently available test statistics may not be able to detect such variation.

To illustrate these theoretical results, a set of Monte Carlo experiments is performed. For that purpose I consider a time series y_t which is generated by a PAR(1). In the footnote to Table 7.15 I give the six DGPs used in the simulations. DGPs (i)–(iv) are periodically stationary, while DGPs (v) and (vi) are periodically integrated. The first four DGPs are included to enable a comparison of cases with and without a unit root. All Monte Carlo simulations are based on 100 effective observations with 1,000 replications in each case. The nominal significance level is set at 5 per cent in all simulations.

When one proceeds along the standard Box–Jenkins methodology, one may use the (partial) autocorrelation functions of y_t, of $\Delta_1 y_t$ (after regression on seasonal dummies to remove possible deterministic seasonality), and of $\Delta_4 y_t$. The first three panels of Tables 7.15 and 7.16 display the estimated autocorrelations and partial autocorrelations for the six DGPs, respectively. It can be observed from Table 7.15 that the autocorrelations of y_t die out only slowly, and that its fourth autocorrelation coefficient is relatively large. For $\Delta_1 y_t$ one could be inclined to fit low-order autoregressions. The third panel of Table 7.15 indicates that the $\Delta_4 y_t$ series displays an AR(1) type of pattern, with a

TABLE 7.15. Estimated autocorrelations of y_t, $\Delta_1 y_t$, $\Delta_4 y_t$ and $\hat{y}_t^{ns}$, where $\hat{y}_t^{ns}$ is a seasonally adjusted series using the linear Census X–11 filter in Table 4.1[a,b]

Series	Lag	DGP					
		(i)	(ii)	(iii)	(iv)	(v)	(vi)
y_t	1	0.73	0.80	0.74	0.87	0.59	0.91
	2	0.65	0.64	0.60	0.75	0.33	0.86
	3	0.50	0.53	0.56	0.68	0.51	0.79
	4	0.49	0.45	0.62	0.63	0.81	0.75
	8	0.21	0.17	0.36	0.36	0.66	0.54
	12	0.06	0.03	0.19	0.18	0.53	0.38
$\Delta_1 y_t$	1	–0.31	–0.07	–0.19	–0.01	–0.11	–0.07
	2	0.12	–0.10	–0.15	–0.14	–0.42	0.06
	3	–0.22	–0.08	–0.20	–0.04	–0.13	–0.07
	4	0.19	0.00	0.37	0.05	0.59	0.03
	8	0.07	–0.02	0.20	0.01	0.46	0.02
	12	0.01	–0.03	0.08	–0.00	0.33	0.01
$\Delta_4 y_t$	1	0.61	0.65	0.65	0.70	0.67	0.72
	2	0.36	0.34	0.34	0.41	0.37	0.46
	3	0.08	0.06	0.08	0.15	0.15	0.20
	4	–0.18	–0.22	–0.12	–0.11	–0.04	–0.05
	8	–0.11	–0.11	–0.08	–0.08	–0.04	–0.03
	12	–0.07	–0.06	–0.06	–0.06	–0.03	–0.04
$\hat{y}_t^{ns}$	1	0.82	0.84	0.87	0.90	0.92	0.94
	2	0.70	0.69	0.75	0.80	0.84	0.88
	3	0.57	0.56	0.65	0.70	0.78	0.81
	4	0.45	0.43	0.57	0.61	0.72	0.74
	8	0.17	0.15	0.30	0.34	0.51	0.54
	12	0.03	0.02	0.13	0.17	0.35	0.38

[a] DGP is a PAR(1) process: $y_t = \phi_s y_{t-1} + \varepsilon_t$, $\varepsilon_t \sim N(0,1)$. The number of observations is 100.The number of replications is 1,000. The figures in the cells are average estimated autocorrelations.

[b] The data-generating processes are $y_t = \phi_s y_{t-1} + \varepsilon_t$, with ϕ_1, ϕ_2, ϕ_3 and ϕ_4 as: (i) 0.4, 1.2, 0.8, 1.6; (ii) 0.6, 0.8, 1.0, 1.2; (iii) 1.3, 0.3, 1.5, 1.3; (iv) 0.7, 0.95, 1.2, 1.0; (v) 0.5, 0.9, 1.5, $1/\phi_1\phi_2\phi_3$; and (vi) 1.1, 0.91, 1.05, $1/\phi_1\phi_2\phi_3$, such that $\phi_1\phi_2\phi_3\phi_4 = 0.614$, 0.576, 0.761, 0.798, 1, and 1, respectively. The (P)ACF of $\Delta_1 y_t$ is calculated after a regression on four seasonal dummies.

parameter possibly not too close to unity. Hence a $\Delta_1\Delta_4$ filter does not seem to be appropriate in most cases. Note that the autocorrelations in the first three panels of Table 7.15 show many similarities across the DGPs, i.e. whether the DGPs have a unit root or not.

The estimated partial autocorrelations reported in the first three panels of Table 7.16 indicate that (nonperiodic) autoregressive models seem useful to describe the time series, although the autoregressive orders vary over the

TABLE 7.16. Estimated partial autocorrelations of y_t, $\Delta_1 y_t$, $\Delta_4 y_t$ and $\hat{y}_t^{ns}$. where $\hat{y}_t^{ns}$ is a seasonally adjusted series using the linear Census X–11 filter in Table 4.1[a]

Series	Lag	DGP					
		(i)	(ii)	(iii)	(iv)	(v)	(vi)
y_t	1	0.75	0.82	0.75	0.88	0.73	0.94
	2	0.25	–0.02	0.11	–0.04	–0.00	0.14
	3	–0.10	0.04	0.16	0.15	0.51	–0.12
	4	0.17	0.03	0.35	0.00	0.60	0.10
	8	0.01	0.00	0.02	–0.01	0.09	0.04
	12	–0.01	–0.02	–0.01	–0.02	0.01	0.02
$\Delta_1 y_t$	1	–0.31	–0.07	–0.19	–0.01	–0.11	–0.07
	2	0.02	–0.11	–0.20	–0.15	–0.45	0.05
	3	–0.20	–0.09	–0.32	–0.04	–0.37	–0.05
	4	0.07	–0.03	0.25	0.02	0.42	0.01
	8	–0.05	–0.06	–0.01	–0.03	0.07	–0.01
	12	–0.07	–0.07	–0.07	–0.05	–0.03	–0.03
$\Delta_4 y_t$	1	0.62	0.67	0.65	0.71	0.69	0.73
	2	–0.04	–0.18	–0.14	–0.18	–0.17	–0.15
	3	–0.22	–0.19	–0.16	–0.17	–0.07	–0.17
	4	–0.26	–0.26	–0.16	–0.23	–0.15	–0.22
	8	–0.17	–0.16	–0.10	–0.13	–0.06	–0.12
	12	–0.13	–0.12	–0.07	–0.10	–0.03	–0.09
$\hat{y}_t^{ns}$	1	0.84	0.85	0.88	0.91	0.95	0.96
	2	0.08	–0.08	–0.03	–0.10	–0.07	–0.09
	3	–0.10	–0.04	0.01	–0.03	0.10	–0.07
	4	–0.07	–0.09	0.02	–0.09	–0.03	–0.09
	8	–0.12	–0.09	–0.06	–0.08	–0.05	–0.08
	12	–0.10	–0.08	–0.05	–0.07	–0.04	–0.08

[a] DGP is a PAR(1) process: $y_t = \phi_s y_{t-1} + \varepsilon_t$, $\varepsilon_t \sim N(0,1)$. The number of observations is 100. The number of replications is 1,000. The figures in the cells are average estimated partial autocorrelations. See note *b* in Table 7.15.

different DGPs. For example, an AR(1,2) process emerges for y_t generated by DGP (i), while an AR(1,3,4,5) results for the y_t series generated by DGP (v), where AR($p_1, p_2, \ldots$) denotes an autoregressive model with lags at $p_1, p_2, \ldots$ In Table 7.17, the orders of the tentatively identified nonperiodic models for y_t, $\Delta_1 y_t$, and $\Delta_4 y_t$ are given. In sum, reasonably low-order AR models can be identified when PAR(1) processes are the underlying DGPs.

The transformation Δ_4 can be used to remove the seasonality of a time series. Alternatively, one can transform the y_t into a seasonally adjusted series $\hat{y}_t^{ns}$ using a linearized Census X-11 filter. For quarterly data, the filter takes the form

$$\hat{y}_t^{ns} = \left(\sum_{i=1}^{-28} c_i B^i + c_0 + \sum_{i=1}^{+28} c_i B^i \right) y_t, \tag{7.49}$$

TABLE 7.17. Autoregressive (AR) models identified from the (partial) autocorrelations functions of Tables 7.15 and 7.16[a]

Series	DGP					
	(i)	(ii)	(iii)	(iv)	(v)	(vi)
y_t	1,2	1	1,4,5,	1	1,3,4,5	1
$\Delta_1 y_t$	1,3	0	1,2,3,4	0	1,2,3,4	0
$\Delta_4 y_t$	1,3,4,5	1,2,3,4,5	1,4,5	1,4,5	1	1,4,5
$\hat{y}_t^{ns}$	1	1	1	1	1	1

[a] The cells indicate the lags to be included in the corresponding AR model. For example, a cell with 1,3 implies an AR model like $y_t=\phi_1 y_{t-1}+\phi_3 y_{t-3}+\varepsilon_t$. The DGPs are given in note *b* in Table 7.15.

where the c_i are given in Table 4.1. Assuming a PAR(1) process as the DGP, the estimated autocorrelations of $\hat{y}_t^{ns}$ are displayed in the bottom part of Table 7.15. These autocorrelations die out only slowly, even when $\prod_{s=1}^{4}\phi_{1s}$ is only 0.6 as in the case of DGPs (i) and (ii). The partial autocorrelations in the fourth panel of Table 7.16 for $\hat{y}_t^{ns}$ suggest that a nonperiodic AR(1) model would be tentatively identified. Note that for the stationary DGPs (i)–(iv) the autocorrelations of the $\hat{y}_t^{ns}$ series exceed $\prod_{s=1}^{4}\phi_{1s}$, which suggests that we may more often find unit roots in $\hat{y}_t^{ns}$ series when a stationary PAR is the DGP.

This suggestion is confirmed by the parameter estimates of a nonperiodic AR(1) model based on $\hat{y}_t^{ns}$, which are displayed in the first row of Table 7.18.

TABLE 7.18. Parameter estimates and rejection frequencies of different AR(1) and PAR(1) models for the adjusted $\hat{y}_t^{ns}$ series when a PAR(1) model is the DGP[a]

	DGP					
	(i)	(ii)	(iii)	(iv)	(v)	(vi)
Parameter estimates						
Nonperiodic AR(1) models						
	0.856	0.873	0.907	0.937	0.983	0.989
Periodic AR(1) models						
$\hat{\phi}_1$	0.566	0.707	1.088	0.837	0.902	0.999
$\hat{\phi}_2$	1.013	0.843	0.593	0.941	0.982	0.979
$\hat{\phi}_3$	0.872	0.961	1.053	1.026	1.031	0.993
$\hat{\phi}_4$	1.219	1.061	1.089	0.965	1.040	0.985
$\hat{\phi}_1\hat{\phi}_2\hat{\phi}_3\hat{\phi}_4$	0.609	0.608	0.740	0.780	0.950	0.957
Rejection frequencies of F—test for periodicity in parameters						
$\hat{\phi}_t^{ns}$	99.2	62.7	99.6	29.2	41.9	0.0

[a] Based on 1,000 replications and 100 effective observations. The true values of $\phi_1\ \phi_2\ \phi_3\ \phi_4$ and the DGPs are given in note *b* of Table 7.15.

These parameter estimates must be some average of the parameters in the DGP (see Tiao and Grupe 1980). Suppose that we want to specify a periodic autoregressive process for an adjusted series $\hat{y}_t^{ns}$. The relevant parameter estimates are given in the second part of Table 7.18. At first sight, it seems that the values of the estimated parameters still differ per season. This is most clearly seen for the first three DGPs. For example, the PAR(1) parameters for DGP (i) are 0.4, 1.2, 0.8, and 1.6, while the estimated parameters for the PAR(1) model for $\hat{y}_t^{ns}$ are 0.566, 1.013, 0.872, and 1.219, respectively. Additionally, it can be seen that the individual values are closer to unity. The bottom row of Table 7.18 gives the result of tests for the periodicity of the autoregressive parameter values in a PAR(1) process for $\hat{y}_t^{ns}$. In some cases the rejection frequency is very high, while for one DGP it is zero. This indicates that we may find the $(1 - B)$ filter to be adequate in cases of PAR(1) where the $\phi_s \cong 1$. A further result that is worth noting is that the product $\prod_{s=1}^{4} \hat{\phi}_{1s}$ of the estimated parameters $\hat{\phi}_{1s}$ is almost equal to the product of the parameters of the DGPs. This implies that tests for a unit root in a periodic model may be robust to seasonal adjustment method, and that one may consider PAR models for SA series for unit root testing.

The identified autoregressive orders for nonperiodic models for the y_t, $\Delta_1 y_t$, $\Delta_4 y_t$, and $\hat{y}_t^{ns}$ are summarized in Table 7.17. In practice, one would apply diagnostic tests such as $F_{\text{AR},1\text{-}1}$ and $F_{\text{AR},1\text{-}4}$ to check the adequacy of the identified model and its parameter estimates. One can also apply the $F_{\text{PeAR},1\text{-}1}$ test for first-order periodic autocorrelation in the estimated residuals based on the auxiliary regression (7.40).

The results of applying several diagnostic tests to the residuals of the AR models from Table 7.17 are displayed in Table 7.19. The rejection frequencies for the $F_{\text{AR},1\text{-}1}$ only twice exceed a value of 25 per cent, and those for the $F_{\text{AR},1\text{-}4}$ test are all below 60 per cent. For some of the DGPs, the rejection frequencies are close to the nominal size. Especially for the y_t and $\Delta_1 y_t$ series, it will be difficult to reject a nonperiodic AR model in case a PAR(1) process is the DGP. For the $\Delta_4 y_t$ series, one may sometimes detect misspecification using the $F_{\text{AR},1\text{-}4}$-test. Notice that the $F_{\text{AR},1\text{-}4}$-test rejects no residual autocorrelation for the $\hat{y}_t^{ns}$ series in several cases. This means that one has to include $\hat{y}_{t-4}^{ns}$ in a model for $\hat{y}_t^{ns}$, which may be quite unusual for a seasonally adjusted series. On the other hand, the $F_{\text{PeAR},1\text{-}1}$-test seems quite powerful in rejecting the null hypothesis of no periodicity in the estimated residuals. Hence, underlying periodicity for the $\hat{y}_t^{ns}$ series can be detected by applying a test for periodic parameters as in Table 7.18, or for periodic error processes as in Table 7.19.

The Monte Carlo experiments for analysing the effects of misspecification can be summarized as follows. When a PAR model is the DGP, one can identify reasonably low-order nonperiodic autoregressive models for the raw series and AR(1) models for the seasonally adjusted series. Using standard diagnostic tests for detecting residual autocorrelation, one usually cannot reject the nonperiodic models. As expected, a test for periodicity in the residuals yields

TABLE 7.19. Estimated rejection frequencies of tests for first and first to fourth-order residual autocorrelation and first-order periodic autocorrelation in nonperiodic AR models for y_t, $\Delta_1 y_t$, $\Delta_4 y_t$ and $\hat{y}_t^{ns}$ [a]

Test	Series	DGP					
		(i)	(ii)	(iii)	(ix)	(v)	(vi)
$F_{AR,1-1}$	y_t	13.2	2.4	1.1	4.9	19.1	13.3
	$\Delta_1 y_t$	3.2	19.5	3.5	32.1	12.0	9.4
	$\Delta_4 y_t$	19.5	22.0	2.5	13.3	31.4	10.2
	$\hat{y}_t^{ns}$	9.5	10.8	0.7	15.8	2.0	12.0
$F_{AR,1-4}$	y_t	31.6	6.6	6.4	17.2	22.1	15.2
	$\Delta_1 y_t$	16.3	14.4	10.6	18.6	20.5	10.1
	$\Delta_4 y_t$	47.8	52.6	19.4	51.9	51.4	57.6
	$\hat{y}_t^{ns}$	27.9	35.9	9.0	36.9	12.0	48.4
$F_{PeAR,1-1}$	y_t	99.8	67.6	100.0	36.3	99.8	14.1
	$\Delta_1 y_t$	99.0	64.9	100.0	31.2	99.5	13.2
	$\Delta_4 y_t$	99.2	65.0	100.0	34.5	99.7	12.7
	$\hat{y}_t^{ns}$	99.3	55.4	99.3	25.5	99.0	12.0

[a] The cells contain the rejection frequencies at the 5 per cent level for 1,000 replications and 100 effective observations. The models for y_t and $\Delta_1 y_t$ include a constant and seasonal dummies, while those for $\Delta_4 y_t$ and $\hat{y}_t^{ns}$ contain only a constant. The DGPs are given in note *b* in Table 7.15.

the best results. Hence, a sensible strategy for NSA and SA time series seems to start with a selecting a PAR(p) process, estimating its parameters, and testing whether the autoregressive parameters are periodically varying.

Finally, in order to establish that seasonal adjustment does not remove periodicity in the AR dynamics, I estimate PAR(p) models for the officially adjusted data. For US industrial production p can be set at 2, and the F_{PAR}-test obtains a value of 2.438**. For the Canadian unemployment series one can fit a PAR(4) model, and an F_{PAR}-test value of 4.865** is obtained. For the German real GNP data I estimate a PAR(2) model, which yields an F_{PAR}-test value of 2.673**. Hence, for the officially seasonally adjusted time series, one can estimate periodic models for which the null hypothesis of no periodic variation can be rejected decisively.

7.5 Multivariate Periodic Autoregression

This final section is concerned with periodic models for multiple time series. A detailed discussion of such models is given in Lütkepohl (1991) and Ula (1993). The main focus of the section is to discuss some notation and to highlight the possible practical problems that are involved in evaluating multivariate periodic autoregressive models.

Consider two quarterly observed time series y_t and x_t, $t = 1,2, \ldots, n$, and suppose that these can be described by a periodic VAR(1) (PVAR(1)),

$$\begin{bmatrix} y_t \\ x_t \end{bmatrix} = \begin{bmatrix} \alpha_s & \beta_s \\ \gamma_s & \delta_s \end{bmatrix} \begin{bmatrix} y_{t-1} \\ x_{t-1} \end{bmatrix} + \begin{bmatrix} u_t \\ w_t \end{bmatrix} \tag{7.50}$$

where α_s, β_s, γ_s, and δ_s are seasonally varying parameters, and u_t and w_t are assumed to be independent standard white noise processes. The parameters in (7.50) can be estimated using OLS per equation, where the regressions are similar to (7.42). Obviously, the identification of the PVAR model is more complicated than that of a PAR model since it involves more than one equation. One may expect (similar to the univariate case) that estimating a PVAR model and applying the usual model selection criteria may be a satisfactory method in practice. In Chapter 9 I will follow this strategy and estimate a PVAR(2) model for four UK sample series considered in the above sections.

To investigate the stationarity properties of PVAR models, it is again most useful to stack the y_t and x_t observations into their VQ processes $\mathbf{Y}_T$ and $\mathbf{X}_T$, and to stack these into the (8×1) vector $\mathbf{Z}_T = (\mathbf{Y}_T{}', \mathbf{X}_T{}')'$. The model in (7.50) may then be written as

$$\begin{bmatrix} 1 & 0 & 0 & -\alpha_1 B & 0 & 0 & 0 & -\beta_1 B \\ -\alpha_2 & 1 & 0 & 0 & -\beta_2 & 0 & 0 & 0 \\ 0 & -\alpha_3 & 1 & 0 & 0 & -\beta_3 & 0 & 0 \\ 0 & 0 & -\alpha_4 & 1 & 0 & 0 & -\beta_4 & 0 \\ 0 & 0 & 0 & -\gamma_1 B & 1 & 0 & 0 & -\delta_1 B \\ -\gamma_2 & 0 & 0 & 0 & -\delta_2 & 1 & 0 & 0 \\ 0 & -\gamma_3 & 0 & 0 & 0 & -\delta_3 & 1 & 0 \\ 0 & 0 & -\gamma_4 & 0 & 0 & 0 & -\delta_4 & 1 \end{bmatrix} \begin{bmatrix} Y_{1,T} \\ Y_{2,T} \\ Y_{3,T} \\ Y_{4,T} \\ X_{1,T} \\ X_{2,T} \\ X_{3,T} \\ X_{4,T} \end{bmatrix} = \begin{bmatrix} u_{1,T} \\ u_{2,T} \\ u_{3,T} \\ u_{3,T} \\ w_{1,T} \\ w_{2,T} \\ w_{3,T} \\ w_{4,T} \end{bmatrix}. \tag{7.51}$$

When this model is denoted as $(\Gamma_0 - \Gamma_1 B)\mathbf{Z}_T = \mathbf{u}^*_T$, we may check the solutions to the characteristic equation $|\Gamma_0 - \Gamma_1 z| = 0$. It is however clear that this yields a very complicated expression in terms of z and z^2, which does not seem useful for the purpose of formally testing for unit roots. Naturally, when the order of the PVAR increases, the expression for the characteristic equation becomes even more complicated. This suggests that alternative methods are needed. In Chapter 9 I return to this issue in more detail.

Conclusion

In this chapter I have examined several aspects of periodic autoregressive time series models. Simulation results indicate a useful empirical order selection strategy. It seems preferable first to select this order, then to test for periodic parameter variation, and finally to test for unit roots. A discussion of the latter topic will be given in the next chapter. The simulation results further suggest, if one follows this empirical strategy, that one is unlikely to fit PAR(p) models to nonperiodic AR(k) process for some p and k. For the 11 running sample series, I find significant evidence for periodic parameter variation. In the next chapter I will also investigate how a PAR(p) model with unit roots may lead to the seemingly empirical usefulness of nonperiodic models, as noticed in Chapters 3 and 5.

8

Periodic Integration

The focus in this chapter is on unit roots in periodic autoregressions. Tests for such unit roots are proposed and evaluated using Monte Carlo simulations and several sample series. Since a periodic autoregression allows the AR parameters to take different values in different seasons, it seems natural to allow for the possibility of a periodically varying differencing filter which can be used to remove the stochastic trend. A specific periodic differencing filter corresponds to the notion of periodic integration (PI). In the empirical analysis in this chapter, I find that all 11 of the sample series appear to be somehow periodically integrated. The implication of periodic integration is that the stochastic trend and the seasonal fluctuations are not independent, in the sense that accumulations of shocks can change the seasonal pattern and that the time series cannot be decomposed in two strictly separate trend and seasonal components. The empirical analysis in Chapter 6, for example, indicated the need for models that allow for correspondence between seasonal and nonseasonal fluctuations. The conjecture to be made in this chapter is that a periodically integrated model for a univariate time series can yield a useful description of some economic time series, and that it allows for this correspondence.

In Section 8.1 I discuss several approaches to testing for unit roots in periodic autoregressive models. (Some of this material is based on parts of Boswijk and Franses (1996) and Franses (1994).) These approaches are extensively evaluated using Monte Carlo experiments, and their relative usefulness is established. Applying the tests to our 11 running sample series, we find that all series are in some sense periodically integrated. In Section 8.2, I discuss several aspects of periodic integration. Examples of these aspects are a comparison of periodic integration with seasonal integration and SARIMA models, the time-varying impact of the accumulation of shocks, the estimation of the stochastic trend, one- and multi-step-ahead forecasting, and temporal aggregation. Most aspects are illustrated using one or more of the sample series. Section 8.3 deals with the impact of neglecting periodic parameter variation when the data are generated by a periodically integrated process. As examples of such implications, one may find seasonal unit roots, and business cycle turning points based on seasonally adjusted periodic integration data can suggest that there is periodicity in the distribution of peaks and troughs.

8.1 Testing for Unit Roots in Periodic Autoregression

Consider again the periodic autoregression of order p (PAR(p)) for a quarterly observed time series y_t,

$$y_t = \mu_s + \phi_{1s}y_{t-1} + \ldots + \phi_{ps}y_{t-p} + \varepsilon_t, \tag{8.1}$$

which can be written in multivariate format as

$$\Phi_0 \mathbf{Y}_T = \mu + \Phi_1 \mathbf{Y}_{T-1} + \ldots + \Phi_P \mathbf{Y}_{T-P} + \varepsilon_T, \tag{8.2}$$

where the Φ_0, Φ_1 to Φ_P are (4×4) matrices containing the ϕ_{is} parameters. (See Section 7.1 for more details on this notation.) The number of unit roots in y_t can be checked by solving the characteristic equation

$$|\Phi_0 - \Phi_1 z - \ldots - \Phi_P z^P| = 0. \tag{8.3}$$

Given these expressions, it is clear that unit roots in y_t correspond to unit roots in the (4×1) vector process $\mathbf{Y}_T$ and to cointegration relationships between the four $Y_{s,T}$ variables. For example, a single unit root in y_t corresponds to three cointegrating relations in the $\mathbf{Y}_T$ vector process. This can be illustrated by the simple PAR(1) process

$$y_t = \alpha_s y_{t-1} + \varepsilon_t, \tag{8.4}$$

which can be written as (7.29) and (7.30), or as

$$\mathbf{Y}_T = \Phi_0^{-1} \Phi_1 \mathbf{Y}_{T-1} + \Phi_0^{-1} \varepsilon_T, \tag{8.5}$$

with

$$\Phi_0^{-1}\Phi_1 = \begin{bmatrix} 0 & 0 & 0 & \alpha_1 \\ 0 & 0 & 0 & \alpha_1\alpha_2 \\ 0 & 0 & 0 & \alpha_1\alpha_2\alpha_3 \\ 0 & 0 & 0 & \alpha_1\alpha_2\alpha_3\alpha_4 \end{bmatrix} \tag{8.6}$$

since

$$\Phi_0^{-1} = \begin{bmatrix} 1 & 0 & 0 & 0 \\ \alpha_2 & 1 & 0 & 0 \\ \alpha_2\alpha_3 & \alpha_3 & 1 & 0 \\ \alpha_2\alpha_3\alpha_4 & \alpha_3\alpha_4 & \alpha_4 & 1 \end{bmatrix}. \tag{8.7}$$

In turn, (8.5) can be written in error correction form as

$$\Delta \mathbf{Y}_T = (\Phi_0^{-1}\Phi_1 - \mathbf{I}_4)\mathbf{Y}_{T-1} + \Phi_0^{-1}\varepsilon_T, \tag{8.8}$$

where Δ is the first-differencing filter for the annual vector series and

$$\Phi_0^{-1}\Phi_1 - \mathbf{I}_4 = \begin{bmatrix} -1 & 0 & 0 & \alpha_1 \\ 0 & -1 & 0 & \alpha_1\alpha_2 \\ 0 & 0 & -1 & \alpha_1\alpha_2\alpha_3 \\ 0 & 0 & 0 & \alpha_1\alpha_2\alpha_3\alpha_4 - 1 \end{bmatrix}. \tag{8.9}$$

Obviously, when $\alpha_1\alpha_2\alpha_3\alpha_4 = 1$, and when it is assumed that the $Y_{s,T}$ are at most I(1), the rank of the matrix $\Phi_0^{1}\Phi_1 - \mathbf{I}_4$ is 3, and this implies three cointegrating relationships between the $Y_{s,T}$ series.

In general, the VQ(P) process in (8.2) can be written in error correction form as

$$\Delta Y_{\mathrm{T}} = \Phi_0^{-1}\mu + \Gamma_1 \Delta \mathbf{Y}_{\mathrm{T}-1} + \ldots + \Gamma_{P-1}\Delta \mathbf{Y}_{\mathrm{T}-(P-1)} + \Pi \mathbf{Y}_{\mathrm{T}-P} + \Phi_0^{-1}\varepsilon_{\mathrm{T}}, \tag{8.10}$$

where

$$\Gamma_i = \Phi_0^{-1}(\Phi_1 + \ldots + \Phi_i) - \mathbf{I}_4, \text{ for } i = 1,2,\ldots, P-1 \tag{8.11}$$

$$\Pi = \Phi_0^{-1}(\Phi_1 + \ldots + \Phi_P) - \mathbf{I}_4. \tag{8.12}$$

When there are r cointegration relations between the $Y_{s,\mathrm{T}}$ elements, the matrix Π has rank r, with $0 < r < 4$, and can be written as

$$\Pi = \gamma\alpha', \tag{8.13}$$

where γ and α are $(4 \times r)$ matrices. In case of three cointegration relations in (8.10), it is clear that the cointegration space can be spanned by

$$\alpha = \begin{bmatrix} -\alpha_2 & 0 & 0 \\ 1 & -\alpha_3 & 0 \\ 0 & 1 & -\alpha_4 \\ 0 & 0 & 1 \end{bmatrix}, \tag{8.14}$$

such that

$$\begin{aligned} &Y_{2,\mathrm{T}} - \alpha_2 Y_{1,\mathrm{T}} \\ &Y_{3,\mathrm{T}} - \alpha_3 Y_{2,\mathrm{T}} \\ &Y_{4,\mathrm{T}} - \alpha_4 Y_{3,\mathrm{T}} \end{aligned} \tag{8.15}$$

are the three cointegrating relations. Given these relations, it can be seen that $Y_{4,\mathrm{T}} - \alpha_4\alpha_3\alpha_2 Y_{1,\mathrm{T}}$ is stationary, and hence that $Y_{1,\mathrm{T}} - \alpha_1 Y_{4,\mathrm{T}-1}$, with $\alpha_1 = 1/\alpha_2\alpha_3\alpha_4$, is a stationary variable. This implies that, in the case of a single unit root in the $\mathbf{Y}_{\mathrm{T}}$ process, the quarterly y_t series can be transformed to a process that does not contain a stochastic trend using the differencing filter $(1 - \alpha_s B)$. The PAR (p) process in (8.1) can then be written as

$$\begin{aligned} y_t - \alpha_s y_{t-1} = \mu_s + \beta_{1s}(y_{t-1} - \alpha_{s-1} y_{t-2}) + \ldots \\ + \beta_{(p-1)s}(y_{t-(p-1)} - \alpha_{s-(p-1)} y_{t-p}) + \varepsilon_t, \end{aligned} \tag{8.16}$$

where $\alpha_{s-4k} = \alpha_s$ and $\beta_{s-4k} = \beta_s$ for $k \in \mathcal{N}$ and $s \in \{1,2,3,4\}$. We call (8.16) the periodically differenced form of (8.1).

The periodic differencing filter $(1 - \alpha_s B)$ was first defined in Osborn *et al.* (1988), who defined periodic integration of order 1 by the property that there exists a set of α_s such that $(1 - \alpha_s B)y_t$ does not contain a unit root. This definition is an example of time-varying parameter definition as introduced in Granger (1986). Notice that this definition also includes the usual filter $(1 - B)$, for which all α_s equal unity. In order to avoid confusion, and also since periodic integration is the key concept in this chapter, I define periodic integration as follows.

Definition: A quarterly time series y_t is said to be periodically integrated of order 1 (PI) when the differencing filter $(1 - \alpha_s B)$ is needed to remove the

stochastic trend from y_t, where α_s are seasonally varying parameters with the property that $\alpha_1\alpha_2\alpha_3\alpha_4 = 1$ and $\alpha_s \neq \alpha$ for all $s = 1,2,3,4$.
The definition can be extended to periodic integration of order 2 and higher; this extension is not however pursued here. (See Boswijk *et al.* (1995) for a discussion of I(2) type PAR processes.)

The definition of periodic integration indicates that PI nests the usual $(1 - B)$ filter, as well as the filter $(1 + B)$, which corresponds to the seasonal unit root at the biannual frequency. This suggests that a useful strategy in the case of three cointegrating relations in $\mathbf{Y}_T$ is first to check whether $\alpha_1\alpha_2\alpha_3\alpha_4 = 1$ and then to check whether $\alpha_s = 1$ or $\alpha_s = -1$. This strategy will be discussed next. For comparison purposes, I will also discuss a method that tests whether $(1 - B)$ or $(1 + B)$ are useful filters in PAR (p) processes right away. Finally, note that the α_s parameters in the periodic differencing filter have to be estimated from the data. Boswijk and Franses (1996) show that, under the null hypothesis that $\alpha_1\alpha_2\alpha_3\alpha_4 = 1$, these α_s parameters can be estimated superconsistently.

Testing for a single unit root

We start to consider the simple periodic first-order autoregressive model in (8.4). Define $\alpha = (\alpha_1,\alpha_2,\alpha_3,\alpha_4)$. The vector process $\mathbf{Y}_T$ is stationary if the root of the characteristic equation,

$$|\Phi_0 - \Phi_1 z| = (1 - \alpha_1\alpha_2\alpha_3\alpha_4 z) = 0, \tag{8.17}$$

is outside the unit circle, i.e. if $|g(\alpha)| < 1$, where $g(\alpha) = \alpha_1\alpha_2\alpha_3\alpha_4$. Note that the values of some α_s are allowed to exceed unity. The vector process $\mathbf{Y}_T$ is integrated if (8.17) has a unit root, so that

$$\mathbf{H}_0\colon g(\alpha) = \prod_{s=1}^{4} \alpha_s = 1 \tag{8.18}$$

holds. The goal is now to test hypothesis (8.18) against the alternative that $|g(\alpha)| < 1$, where the process y_t is said to be periodically stationary. Note that the maximum number of unit roots for the $\mathbf{Y}_T$ process in (8.17) is 1. Assuming that the errors ε_t in (8.4) are normally distributed, the ML estimators of α_s are given by the OLS estimators in the regression

$$y_t = \sum_{s=1}^{4} \alpha_s D_{s,t} y_{t-1} + \varepsilon_t. \tag{8.19}$$

Because of the orthogonality of the regressors in (8.19), we have

$$\hat{\alpha}_s = \left[\sum_{t=1}^{n} D_{s,t} y_{t-1}^2\right]^{-1} \sum_{t=1}^{n} D_{s,t} y_{t-1} y_t,$$

for $s = 1,2,3,4$. Imposing the H_0 leads to the restricted regression

$$y_t = \alpha_1 D_{1,t} y_{t-1} + \alpha_2 D_{2,t} y_{t-1} + \alpha_3 D_{3,t} y_{t-1} + (\alpha_1\alpha_2\alpha_3)^{-1} D_{4,t} y_{t-1} + \varepsilon_t, \tag{8.20}$$

which can be estimated by nonlinear least squares (NLS). A likelihood ratio test statistic may be constructed as

$$LR = n \ln(RSS_0 / RSS_1), \tag{8.21}$$

where RSS_0 and RSS_1 denote the residual sums of squares from (8.19) and (8.21), respectively. A one-sided test can be constructed as

$$LR_\tau = [\text{sign}(g(\hat{\alpha}) - 1)]LR^{1/2}, \tag{8.22}$$

where $g(\hat{\alpha})$ is evaluated under the alternative hypothesis. Alternatively, a Wald test can be based on the t-type statistic

$$\tau = (\hat{V}[g(\hat{\alpha})])^{1/2}(g(\hat{\alpha}) - 1) \tag{8.23}$$

where

$$\hat{V}[g(\hat{\alpha})] = (\partial g(\hat{\alpha}) / \partial \alpha')\, \hat{V}(\hat{\alpha})[\partial g(\hat{\alpha}) / \partial \alpha], \tag{8.24}$$

and $\hat{V}(\hat{\alpha})$ is the usual covariance matrix estimator, which is diagonal because of the orthogonality of the regressors in (8.19).

Theorem 8.1. Under the H_0 in (8.18), and as $n \Rightarrow \infty$, we have

$$N(g(\hat{\alpha}) - 1) \Rightarrow \left[\int_0^1 W(r)^2 \mathrm{d}r\right]^{-1} \int_0^1 W(r) \mathrm{d}W(r), \tag{8.25}$$

$$LR_\tau, \tau \Rightarrow \left[\int_0^1 W(r)^2 \mathrm{d}r\right]^{-1/2} \int_0^1 W(r) \mathrm{d}W(r), \tag{8.26}$$

where $W(r)$ is a standard Brownian motion process.

Proof. See Boswijk and Franses (1996).

The asymptotic distributions in Theorem 8.1 are the same as those tabulated in Fuller (1976: tables 8.5.1 and 8.5.2, respectively) for the nonperiodic AR model (see also Section 2.3 above). We see that $N(g(\hat{\alpha}) - 1)$ already has an asymptotic distribution under the null hypothesis that does not depend upon nuisance parameters; so it can be used as a test statistic alternative to τ, just like the $n(\hat{\rho} - 1)$ statistic in the nonperiodic case. Observe that $g(\hat{\alpha}) - 1$ should be scaled by N if it is to be compared with the critical values in Fuller (1976). Finally, notice that it is assumed that the ε_t process is homoskedastic.

Extensions to autoregressive models that include a constant and a trend are straightforward and similar to the standard Dickey–Fuller case, provided that one considers models such as

$$y_t = \sum_{s=1}^{4} (\mu_s D_{s,t} + \tau_s D_{s,t} t + \alpha_s D_{s,t} y_{t-1}) + \varepsilon_t, \tag{8.27}$$

where t represents a deterministic trend variable. The test statistics can be calculated along similar lines as above, and the corresponding distributions can be found in the tables in Fuller (1976) (see Boswijk and Franses (1996) for details).

To verify whether the asymptotic distributions for models as in (8.27) provide reasonable approximations, I conduct a small Monte Carlo experiment. The data-generating process is a PAR(1) process for a zero mean time series without trends like (8.4). The test statistics are the τ-test in (8.23), the LR_τ test in (8.22), and the $N(g(\hat{\alpha}) - 1)$ test in (8.25).

TABLE 8.1. The empirical size of tests for a unit root in a PAR(1)[a]

Parameters in DGP[b]			Nominal	Test statistic		
α_1	α_2	α_3	size	τ	LR_τ	$N(g(\hat{\alpha})-1)$
The regression model contains no constants and no trends						
0.5	0.9	1.5	0.05	0.06	0.05	0.02
			0.10	0.11	0.10	0.06
1.1	0.91	1.05	0.05	0.06	0.04	0.03
			0.10	0.11	0.09	0.07
The regression model contains constants and no trends						
0.5	0.9	1.5	0.05	0.09	0.04	0.02
			0.10	0.14	0.08	0.05
1.1	0.91	1.05	0.05	0.09	0.05	0.03
			0.10	0.16	0.09	0.07
The regression model contains constants and trends						
0.5	0.9	1.5	0.05	0.14	0.04	0.01
			0.10	0.19	0.07	0.03
1.1	0.91	1.05	0.05	0.17	0.04	0.01
			0.10	0.21	0.10	0.03

[a] The DGP is $y_t = \alpha_s y_{t-1} + \varepsilon_t$, under the restriction $\alpha_1\alpha_2\alpha_3\alpha_4 = 1$, where ε_t is $N(0,1)$. The effective sample size is 100, and the cells contain rejection frequencies of the null hypothesis. The number of replications is 1,000.

[b] $\alpha_4 = 1/\alpha_1\alpha_2\alpha_3$.

Table 8.1 reports on the empirical size of the tests for two generating processes with the property that $\alpha_1\alpha_2\alpha_3\alpha_4 = 1$. The maintained regression models contain trends, no trends, constants, and no constants. From the rejection frequencies, we can observe that the empirical size is usually too high for the τ test, too low for the $N(g(\hat{\alpha}) - 1)$ test, and approximately adequate for the LR_τ test.

The empirical power of the tests is reported in Table 8.2. The regression model now contains no trends and no constants. The product of the four AR(1) parameters is about 0.6 and 0.8. From this table it can be concluded that the power of the τ test is usually the highest, while that of the $N(g(\hat{\alpha}) - 1)$ test is the lowest. Note that Table 8.2 does not report size-adjusted power, and hence that the latter results are likely caused by the incorrect size of the $N(g(\hat{\alpha}) - 1)$ test. The power of the LR_τ test is quite reasonable, especially at a nominal size level of 10 per cent. In summary, I recommend the use of the LR_τ test in practice.

To generalize the likelihood ratio test for a unit root to higher-order periodic autoregressions, consider the linear regression

$$y_t = \sum_{i=1}^{p}\sum_{s=1}^{4}\phi_{is}D_{s,t}y_{t-i} + \sum_{s=1}^{4}\mu_s D_{s,t} + \sum_{s=1}^{4}\tau_s D_{s,t}t + \varepsilon_t, \tag{8.28}$$

TABLE 8.2. The empirical power of tests for a unit root in a PAR(1)[a]

Parameters in DGP				Nominal	Test statistic		
α_1	α_2	α_3	α_4	size	τ	LR_τ	$N(g(\alpha)–1)$
The product $\alpha_1\alpha_2\alpha_3\alpha_4$ is about 0.6							
0.4	1.2	0.8	1.6	0.05	93.1	86.5	78.4
				0.10	98.6	97.7	94.0
0.6	0.8	1.0	1.2	0.05	96.6	91.7	90.1
				0.10	98.4	97.9	98.3
The product $\alpha_1\alpha_2\alpha_3\alpha_4$ is about 0.8							
1.3	0.3	1.5	1.3	0.05	56.9	48.1	39.8
				0.10	74.5	70.9	63.8
0.7	0.95	1.2	1.0	0.05	49.0	38.4	29.9
				0.10	69.8	63.3	54.6

[a] The DGP is $y_t = \alpha_s y_{t-1} + \varepsilon_t$, where ε_t is $N(0,1)$ distribution. The effective sample size is 100, and the cells contain rejection frequencies of the null hypothesis. The number of replications is 1,000. The regression model contains no constants and no trends.

which can be estimated using OLS. Denote RSS_1 as the residual sum of squares of (8.28) where the $\tau_s = 0$, and RSS_2 where the four trends are included. Next, consider the nonlinear regression

$$y_t = \sum_{s=1}^{4} \alpha_s y_{t-1} + \sum_{i=1}^{p-1} \sum_{s=1}^{4} \beta_{is} D_{s,t}(y_{t-i} - \alpha_{s-j} y_{t-i-1}) + \sum_{s=1}^{4} \mu_s D_{s,t} + \sum_{s=1}^{4} \tau_s D_{s,t} t + \varepsilon_t, \quad (8.29)$$

which can be estimated using NLS under the restriction $\alpha_1\alpha_2\alpha_3\alpha_4 = 1$. Denote the residual sum of squares of (8.29) for both cases as RSS_0. The relevant LR tests for (8.29) versus (8.28) are equal to

$$LR_i = n\log(RSS_0 / RSS_i). \quad (8.30)$$

Theorem 8.2. Under the H_0 in (8.18), and as $n \Rightarrow \infty$, we have

$$LR_i \Rightarrow \left\{ \left[\int_0^1 W_i(r)^2 \mathrm{d}r \right]^{-1/2} \int_0^1 W_i(r) \mathrm{d}W(r) \right\}^2, \quad (8.31)$$

for $i = 1,2$, where $W(r)$ is a standard Brownian motion process, and

$$W_1(r) = W(r) - \int_0^1 W(t)\mathrm{d}t \quad (8.32)$$

$$W_2(r) = W_1(r) - [(r - ½)12] \int_0^1 (t - ½) W_1(t)\mathrm{d}t, \quad (8.33)$$

where the processes W_1 and W_2 may be interpreted as 'demeaned' and 'detrended' Brownian motions, respectively (see Park and Phillips 1988).

Proof. See Boswijk and Franses (1996).

The distributions (8.31) appear as special cases of the Johansen (1991) cointegration analysis. Hence for the LR_1 test, the critical values are given in table 1.1 in Osterwald-Lenum (1992) (for the case where $p - r = 1$). Since the distribution of Fuller's (1976) τ_τ test statistic has virtually no mass at the positive part of the line, critical values for the LR_2 test may be obtained by taking the square of the corresponding critical values of τ_τ. In the notes to Table 8.7 below I list the asymptotic critical values. Similar to (8.22), one may construct the one-sided LR tests $LR_{1\tau}$ and $LR_{2\tau}$. The application of these one-sided tests may be less useful in case of PAR(p) models with $p > 1$, since the α_s parameters in (8.29) are not identified under the alternative hypothesis. For illustrative purposes, I will calculate both tests in the empirical part below.

Testing for a parameter restriction

Once it has been established that the hypothesis of $\alpha_1\alpha_2\alpha_3\alpha_4 = 1$ cannot be rejected using either of the LR_i test statistics in (8.30), the next step is to investigate whether the hypotheses

$$H_0\colon \alpha_s = 1 \quad \text{for } s = 1,2,3 \tag{8.34}$$

$$H_0\colon \alpha_s = -1 \quad \text{for } s = 1,2,3, \tag{8.35}$$

are valid, which given $\alpha_1\alpha_2\alpha_3\alpha_4 = 1$, imply that either $\alpha_4 = 1$ or $\alpha_4 = -1$. The first H_0 reduces the periodic differencing filter to $(1 - B)$, while the second H_0 reduces it to the differencing filter that corresponds to a seasonal unit root -1. Where the H_0 in (8.34) cannot be rejected, it is said that the PAR(p) process contains a nonseasonal unit root. Otherwise formulated, this H_0 results in a PAR process for an I(1) time series, to be abbreviated as PARI. In Boswijk and Franses (1996) it is shown that, conditional on the restriction $\alpha_1\alpha_2\alpha_3\alpha_4 = 1$, the likelihood ratio test statistics for the hypotheses (8.34) and (8.35) have asymptotic $\chi^2(3)$ distributions under the null hypothesis.

To verify this result for small samples, consider some Monte Carlo simulations. Table 8.3 reports the empirical fractiles of the F-test for the parameter restrictions corresponding to the $(1 - B)$ and $(1 + B)$ filters in a PAR(2) process. For further use, these F-statistics are denoted $F_{(1-B)}$ and $F_{(1+B)}$. It can be observed, from a comparison of the first column of Table 8.3 with the next six columns, that the empirical and theoretical fractiles closely match. Where the hypotheses in (8.34) and (8.35) can be rejected, the PAR model is called a periodically integrated AR model of order p, (PIAR(p)).

A final step in our model selection strategy can involve a test for the significance of the β_{is} parameters in (8.29). Given that $(1 - \alpha_s B)y_t$ is a periodically stationary process, the t-tests for the significance of the β_{is} parameters asymptotically follow a standard normal distribution under the null hypothesis. However, since the α_s have to be estimated, one may expect biased distributions in small samples. Table 8.4 shows simulated fractiles of a t-test for the significance of one of the second-order terms in a PAR(2) process in the format of (8.29), while the DGP is a PIAR(1). From the fractiles in this table, it can be concluded that the left-hand side of the empirical distribution closely matches the asymptotic distribution, but that there a slight bias on the right-hand side.

TABLE 8.3. Testing for (non)seasonal unit roots in PAR(2) models[a]

Fractiles %	$F(3,111)$	Data-generating process[b]					
		Nonseasonal unit roots			Seasonal unit roots		
		(i)	(ii)	(iii)	(iv)	(v)	(vi)
5	0.12	0.13	0.11	0.11	0.10	0.12	0.11
10	0.19	0.20	0.18	0.19	0.18	0.20	0.19
20	0.34	0.34	0.33	0.33	0.32	0.32	0.33
50	0.79	0.80	0.78	0.80	0.77	0.79	0.76
80	1.57	1.58	1.55	1.54	1.54	1.55	1.52
90	2.13	2.11	2.11	2.08	2.12	2.11	2.11
95	2.69	2.62	2.64	2.67	2.69	2.69	2.65
Mean	1.02	1.02	1.00	1.00	0.99	1.00	1.00
Variation	0.72	0.71	0.70	0.70	0.72	0.70	0.72

[a] Fractiles of F-statistic for the null hypothesis that $\phi_{1s} + \phi_{2s} = 1$ or that $\phi_{1s} - \phi_{2s} = 1$, given the restriction that there is one unit root in the PAR process. Based on 5,000 Monte Carlo replications. Sample size is 120 observations.

[b] The data-generating process for (i), (ii), and (iii) is $y_t - y_{t-1} = \beta_s(y_{t-1} - y_{t-2}) + \varepsilon_t$; with β_s it is 0.96. 0.93, 0.85, 0.79 for (i), 0.70, 0.95, 1.20, 1.00 for (ii), and 1.25, 0.80, 2.00, 0.50 for (iii). The roots of $\Phi_0^{-1}\Phi_1$ are 1.0, 0.6 for (i), 1.0 0.8 for (ii), and 1.0, 1.0 for (iii). The data-generating process for (iv), (v), and (vi) is $y_t + y_{t-1} = \beta_s(y_{t-1} + y_{t-2}) + \varepsilon_t$, with β_s equal to the $-\beta_s$ above. Hence the processes (i), (ii), and (iii) correspond to (iv), (v), and (vi) with respect to the roots of $\Phi_0^{-1}\Phi_1$. A PAR(2) model is estimated, the parameters are restricted as in (7.25), and then the relevant F—test statistics are calculated.

TABLE 8.4. Simulated fractiles of t-tests for the significance of lagged periodically differenced time series in a PAR(2) process[a]

DGP[b]				U/R[b]	Fractiles (%)						
α_1	α_2	α_3	α_4		5	10	20	50	80	90	95
1.053	0.888	1.071	0.999	U	−1.65	−1.21	−0.78	0.07	0.90	1.38	1.74
				R	−1.64	−1.26	−0.79	0.05	0.91	1.33	1.70
0.957	1.022	1.031	0.991	U	−1.61	−1.20	−0.76	0.06	0.92	1.39	1.77
				R	−1.61	−1.23	−0.78	0.01	0.92	1.37	1.73
0.744	1.021	1.371	0.960	U	−1.64	−1.26	−0.78	0.06	0.89	1.34	1.71
				R	−1.63	−1.24	−0.79	0.05	0.91	1.34	1.74

[a] Based on 5,000 replications. The effective sample size is 120 observations.

[b] The DGP is $y_t = \alpha_s y_{t-1} + \varepsilon_t$ with $\alpha_4 = 1/(\alpha_1\alpha_2\alpha_3)$ and $\varepsilon_t \sim N(0,1)$. The table shows the empirical fractiles of t-tests for the significance of the β_s parameters in the model $(y_t - \alpha_s y_{t-1}) = \beta_s(y_{t-1} - \alpha_{s-1}y_{t-2}) + \varepsilon_t$ with the imposed restriction $\alpha_1\alpha_2\alpha_3\alpha_4 = 1$ (R) or, without this restriction, (U). The values of α_s are chosen such that these correspond to some values to be reported in Table 8.8.

A joint test

When we are interested only in testing whether the $(1 - B)$ filter is adequate for a PAR(p) process, we may rewrite the PAR(p) model as

$$\Delta_1 y_t = \delta_s y_{t-1} + \mu_s + \tau_s t + \sum_{j=1}^{p-1} \theta_{js} \Delta_1 y_{t-j} + \varepsilon_t. \quad (8.36)$$

The presence of a nonseasonal unit root in y_t corresponds to

$$H_0\colon \delta_s = 0 \ \text{ for } s = 1,2,3,4. \quad (8.37)$$

This null hypothesis is tested against

$$H_1\colon \delta_s \neq 0 \ \text{ for at least one } s = 1,2,3,4 \quad (8.38)$$

(see Ghysels and Hall 1993*b*). Notice that under the alternative hypothesis the y_t process can be either periodically integrated or periodically stationary. Hence, under H_1 it is unknown whether the y_t process contains a stochastic trend or not. Furthermore, notice that $\delta_1\delta_2\delta_3\delta_4 = 1$ in (8.36) implies periodic integration only when $p = 1$. The restriction for PI in (8.36) is much more complicated for other values of p. The hypothesis in (8.37) can be tested using a Wald test, which can be expressed as

$$W_i = [N/(n-k)] \sum_{s=1}^{4} [t(\delta_s)]^2, \quad (8.39)$$

where $t(\delta_s)$ represents the t-value for the test $\delta_s = 0$, k is the number of regressors in (8.36), and $i = 1,2,3$ in the case where (8.36) contains (1) no seasonal dummies and no trends, (2) only four seasonal constants, and (3) four seasonal constants and four seasonal trends. Boswijk and Franses (1996) prove that the asymptotic distribution of the W_i test in (8.39) is the sum of the square of the Dickey–Fuller distribution and a $\chi^2(3)$ distribution. Table 8.5 displays some simulated critical values of these tests. These critical values will be used in the simulation experiments in Table 8.6.

In order to test for the validity of the $(1 - B)$ filter in a PAR(p) model, another possibility is to modify the Dickey–Fuller t-test for PAR models, i.e.

$$\Delta_1 y_t = \delta y_{t-1} + \mu_s + \tau t + \sum_{j=1}^{p-1} \theta_{js} \Delta_1 y_{t-j} + \varepsilon_t, \quad (8.40)$$

TABLE 8.5. Critical values of W_i test based on 10,000 Monte Carlo replications[a]

Sample	W_1			W_2			W_3		
	20%	10%	5%	20%	10%	5%	20%	10%	5%
40	1.74	2.35	2.90	2.63	3.37	4.17	3.61	4.63	5.62
80	1.62	2.15	2.62	2.36	3.01	3.59	3.12	3.84	4.51
120	1.62	2.13	2.60	2.35	2.95	3.52	3.03	3.71	4.33
160	1.59	2.06	2.51	2.30	2.86	3.40	2.97	3.67	4.27

[a] The DGP is $y_t = y_{t-1} + \varepsilon_t$ with $\varepsilon_t \sim N(0,1)$ and the auxiliary regression is $\Delta_1 y_t = \delta_s y_{t-1} + \pi_t$, where this regression is enlarged with seasonal dummies in the W_2 case and with seasonal dummies and seasonal trends in the W_3 case. An expression for the W_i test is given in (8.39).

and to test the significance of δ using the t-test. Ghysels and Hall (1993*b*) show that, under the null hypothesis, this periodic ADF test (PADF) follows a standard Dickey–Fuller distribution. Note that the deterministic trend in the auxiliary regression (8.40) enters the equation without seasonally varying parameters.

To compare the two-step and joint test methods to investigate the adequacy of the $(1-B)$ filter, consider the simulation results reported in Table 8.6. The performance of the test statistics is investigated for six DGPs. The first panel of Table 8.6 displays the relative number of cases in which a certain decision is made where the regressions do not contain deterministic components, the next two panels concern cases where the models contain seasonal dummies and/or seasonally varying deterministic trends. Note that according to (8.40) I do not include such seasonal trends for the PADF test statistic. The lag order in the regressions is set equal to 2, as in the DGP. In the PADF test, however, the regression is enlarged with lagged $\Delta_1 y_t$ variables until the $F_{\text{PEAR},\,1-1}$-test indicates that there is no significant periodic serial correlation.

TABLE 8.6. Performance of tests for unit roots in periodic autoregressions[a]

Test method[b]	Decision	DGP[c]					
		PIAR			PARI		
		(i)	(ii)	(iii)	(iv)	(v)	(vi)
The model contains no constants and no trends							
Two step	$\Pi\alpha_s = 1$	94.74	95.18	95.18	94.14	94.62	94.68
	$(1-B)$	19.74	68.42	0.04	89.24	90.28	89.62
W_1	$(1-B)$	24.22	76.80	0.04	94.68	94.50	93.96
PADF	$(1-B)$	91.96	95.12	92.76	94.80	94.58	94.32
The model contains constants and no trends							
Two step	$\Pi\alpha_s = 1$	94.94	95.28	95.00	95.40	95.14	95.04
	$(1-B)$	44.80	83.68	0.34	90.38	90.04	90.24
W_2	$(1-B)$	55.60	91.19	0.32	94.62	94.36	94.96
PADF	$(1-B)$	93.78	96.42	95.70	95.50	96.40	95.56

[a] Based on 5,000 Monte Carlo simulations. The effective sample size is 120. The cells report the relative frequencies that a certain decision is made. All test statistics are evaluated at the 5 per cent significance level.

[b] The two-step method contains the LR test in (8.31) and the F-test for the hypothesis in (8.34); the W_i tests are given in (8.39) and the PADF test is given in (8.40).

[c] Each DGP is $y_t - \alpha_s y_{t-1} = \beta_s(y_{t-1} - \alpha_{s-1} y_{t-2}) + \varepsilon_t$, with $\alpha_1\alpha_2\alpha_3\alpha_4 = 1$ and $\varepsilon_t \sim N(0.1)$. The parameters for α_s and β_s $(s = 1,2,3,4)$ are set at (i): 1.053, 0.888, 1.071, 0.999, –0.253, –0.352, –0.081, 0.331; (ii): 0.957, 1.022, 1.032, 0.991, 0.009, –0.946, –0.398, –0.646; (iii): 0.744, 1.021, 1.371, 0.960, –0.302, –0.539, –0.118, –0.365; (iv): 1, 1, 1, 1, –0.315, –0.657, –0.211, 0.127; (v): 1, 1, 1, 1, –0.037, –0.554, –0.283, –0.663; and (vi): 1, 1, 1, 1, –0.354, –0.334, –0.027, –0.436. These parameter combinations are based on some preliminary estimates of PIAR and PARI models for three sample series.

The first three DGPs in Table 8.6 correspond to a PIAR process, while the second three correspond to a PIAR process. For the fist three DGPs the $(1-B)$ restriction is not valid. We can observe from this table that the LR_i test for the nonlinear restriction $\alpha_1\alpha_2\alpha_3\alpha_4 = 1$ performs as expected; i.e., the empirical size is about 5 per cent. The $(1-B)$ filter is selected less often by the two-step method than it is using the W_i tests. Furthermore, the PADF test always seems to opt for the $(1-B)$ filter. These results seem to hold across all DGPs. In case of the PARI process, one may wish to find the adequacy of $(1-B)$ filter. Indeed, the empirical rejection frequencies of the W_i tests and PADF correspond to the nominal size of 5 per cent. On the other hand, when we use the two-step method we may reject the $(1-B)$ hypothesis in 10 per cent of the cases with a nominal size of 5 per cent.

In sum, it seems that the empirical performance of the tests is similar when the DGP in indeed a PAR model for a $(1-B)$ transformed time series. On the other hand, when the DGP is a PI process, only the two-step method is able to detect this, while the other methods select the $(1-B)$ filter. For practical purposes, therefore, I advocate using the two-step method.

Sample series

The two-step method to investigate the presence of a unit root in PAR models proposed above is illustrated using the sample series considered in Chapter 7 for which PAR(p) models appeared to be useful. In this illustration only those series are considered for which the nonzero eigenvalues of $\hat{\Phi}_0^{-1}\hat{\Phi}_1$ indicated that we may expect at most one stochastic trend in the $\mathbf{Y}_T$ series. Hence I postpone the analysis of the Canadian unemployment data to the application of a method to test for multiple unit roots below. The orders of the PAR(p) model for each of the series can be found in Table 7.9, and the nonzero eigenvalues of $\hat{\Phi}_0^{-1}\hat{\Phi}_1$ in Table 7.11. The results for the LR_i and $LR_{i\tau}$ tests, as well as for the $F_{(1-B)}$-test, are reported in Table 8.7.

The general conclusion to be drawn from these results is that a unit root in each PAR model cannot be rejected; i.e., for all ten time series the restriction $\alpha_1\alpha_2\alpha_3\alpha_4 = 1$ cannot be rejected, while we can convincingly reject the $(1-B)$ hypothesis in (8.34) for all time series. The time series are all in logs. An investigation of the adequacy of the log transformation while testing for a unit root in PAR models may be pursued along the lines of Franses and McAleer (1995*a*). This extension is left for further research. Only for the UK import series may one maintain the hypothesis that this series is periodically trend stationary at the 10 per cent level. Generally, the results in Table 8.7 imply that the sample series can be described by periodically integrated time series models, and that the appropriate differencing filter to remove the stochastic trend is $(1-\hat{\alpha}_s B)$. Hence each series has a single stochastic trend, and this trend does not correspond to a nonseasonal unit root.

TABLE 8.7. Testing for a unit root and the $(1-B)$ restriction in PAR(p) models for ten sample series

Variable[a]	Test statistics[b]				
	With trends		Without trends		
	LR_2	$LR_{2\tau}$	LR_1	$LR_{1\tau}$	$F_{(1-B)}$
Industrial production, USA	8.992	–2.999	3.265	–1.834	17.939**
Real GNP, Germany	3.753	–1.937	4.899	–2.213	20.841**
UK GDP	3.759	–1.939	0.838	–0.915	14.809**
UK total consumption	2.877	–1.696	0.993	0.996	34.142**
UK consumer nondurables	1.582	–1.258	0.342	0.585	31.617**
UK exports	3.747	–1.936	1.153	–1.074	6.292**
UK imports	10.774*	–3.282*	0.011	0.106	4.772**
UK total investment	2.548	1.596	1.998	–1.414	10.155**
UK public investment	4.176	–2.043	0.438	0.662	20.796**
UK workforce	0.840	–0.917	0.575	–0.758	12.708**

[a] The results in this table are based on the estimated PAR models displayed and evaluated in Table 7.9. The models also contain the dummy variables for single observations indicated in the footnote to that table. The sign of $\hat{\alpha}_1\hat{\alpha}_2\hat{\alpha}_3\hat{\alpha}_4-1$ in (8.22) is based on eigenvalues calculated for the matrix $\Phi_0^{-1}\Phi_1$, which are displayed in Table 7.11.

[b] The LR_1 and LR_2 test statistics are given in the text in (8.30) and the one-sided versions of these test statistics are calculated via (8.22). The $F_{(1-B)}$-test is calculated in the case where the auxiliary regression does not contain four deterministic trends. The unrestricted estimated α_s values (cf. Table 8.8) indicate that the $(1+B)$ restriction does not seem sensible.

** Significant at the 5 per cent level.

* Significant at the 10 per cent level. The 5 and 10 per cent asymptotic critical values for the test statistics are 12.96 and 10.50 (LR_2), –3.41 and –3.12 ($LR_{2\tau}$), 9.24 and 7.52 (LR_1), and –2.86 and –2.57 ($LR_{1\tau}$) (see Osterwald-Lenum 1992, and Fuller 1976). The asymptotic distribution of the $F_{(1-B)}$-test is $F(3,n-k)$, where k is the number of regressors. Note that this latter test is evaluated under the restriction that $\alpha_1\alpha_2\alpha_3\alpha_4 = 1$.

Table 8.8 reports the parameter estimates for the various PIAR(p) models. It can be seen that the estimates for α_s under the restriction that $\alpha_1\alpha_2\alpha_3\alpha_4 = 1$ are all close to unity, although they are not equal to unity. It should be mentioned here that the time series models in this table contain at most 11 parameters for approximately 130 observations. Hence, even though the SARIMA models in Chapter 3 contain far fewer parameters, the maximum of 11 seems not unreasonably large. Furthermore, note that those SARIMA models assume five unit roots instead of the single unit root found here. Finally, when extra MA terms are added to the PAR(1) and PAR(2) models in the table, they do not appear to be significant. This suggests that there may be an empirical correspondence between PIAR and SARIMA models, and this will discussed in the next section.

Given that the estimates for α_s in Table 8.8 appear to be very close to unity, one may wonder what the impact is of setting $\alpha_s = 1$. Although a more detailed discussion of the impact of nonperiodic analysis is given in Section 8.3,

Table 8.8. Parameter estimates in PIAR(p) models for ten sample series[a]

Variable	α_1	α_2	α_3	$\alpha_4 = 1/\alpha_1\alpha_2\alpha_3$
Industrial production USA	1.004	0.981	1.047	0.969
	(0.008)	(0.008)	(0.007)	(0.008)
Real GNP, Germany	1.025	0.962	0.912	1.113
	(0.011)	(0.011)	(0.009)	(0.011)
UK GDP	1.011	0.916	1.062	1.017
	(0.011)	(0.014)	(0.014)	(0.014)
UK total consumption	1.031	0.918	1.057	1.000
	(0.009)	(0.009)	(0.009)	(0.009)
UK consumer nondurables	1.003	0.932	1.030	1.039
	(0.008)	(0.007)	(0.008)	(0.008)
UK exports	0.957	1.022	1.032	0.991
	(0.013)	(0.016)	(0.015)	(0.016)
UK imports	0.965	1.038	0.974	1.025
	(0.014)	(0.015)	(0.014)	(0.015)
UK total investment	1.053	0.888	1.071	0.999
	(0.022)	(0.019)	(0.017)	(0.020)
UK public investment	0.601	2.070	0.820	0.981
	(0.069)	(0.296)	(0.035)	(0.010)
UK workforce	0.936	1.027	1.059	0.982
	(0.012)	(0.015)	(0.012)	(0.014)

	β_1	β_2	β_3	β_4	μ_1	μ_2	μ_3	μ_4
Industrial production, USA	0.779	0.255	0.357	0.482	−0.127	0.109	−0.235	0.234
	(0.155)	(0.126)	(0.172)	(0.220)	(0.048)	(0.039)	(0.040)	(0.052)
GNP, Germany	0.338	−0.676	0.351	−0.428	0.008	0.115	0.466	−0.404
	(0.195)	(0.152)	(0.154)	(0.186)	(0.137)	(0.051)	(0.069)	(0.120)
UK GDP	0.317	−0.549	−0.411	−0.288	−0.114	0.853	−0.276	−0.348
	(0.191)	(0.133)	(0.133)	(0.178)	(0.134)	(0.123)	(0.168)	(0.149)
UK total consumption					−0.412	0.911	−0.574	0.039
					(0.099)	(0.090)	(0.100)	(0.094)
UK consumer nondurables					−0.121	0.762	−0.297	−0.362
					(0.082)	(0.077)	(0.083)	(0.082)
UK Exports	0.009	−0.649	0.398	−0.646	0.390	0.086	−0.384	−0.086
	(0.143)	(0.166)	(0.124)	(0.153)	(0.126)	(0.130)	(0.132)	(0.104)
UK imports					0.347	−0.329	0.267	−0.261
					(0.138)	(0.140)	(0.134)	(0.139)
UK total investment	−0.253	−0.352	−0.080	0.331	−0.506	0.817	−0.528	0.270
	(0.214)	(0.137)	(0.160)	(0.237)	(0.192)	(0.182)	(0.218)	(0.210)
UK public investment	0.579	−2.392	−0.058	0.054	3.383	−1.067	1.035	0.105
	(0.425)	(0.333)	(0.049)	(0.253)	(0.494)	(0.585)	(0.542)	(0.471)
UK workforce	0.698	−0.361	−0.125	0.066	0.520	−0.041	−0.626	0.220
	(0.192)	(0.152)	(0.110)	(0.152)	(0.123)	(0.144)	(0.124)	(0.131)

[a] The models are $(y_t - \alpha_s y_{t-1}) = \mu_s + \beta_s(y_{t-1} - \alpha_{s-1}y_{t-2}) + \varepsilon_t$, and the parameters are estimated under the restriction $\alpha_1\alpha_2\alpha_3\alpha_4 = 1$. The calculated standard errors are given in parentheses. Note that the t-tests for the β_s parameters asymptotically follow standard normal distributions. The α_s parameters are estimated super-consistently. This prevents a simple interpretation of the estimated standard errors. Since the μ_s are nonlinear functions of β_s and α_s, similar arguments as for α_s apply. The diagnostics for these restricted models are all similar to those reported in Table 7.9.

TABLE 8.9. Estimated nonperiodic autocorrelation function for the $(1-\alpha_s B)y_t$ and $(1-B)y_t$ series (after a regression on four seasonal dummy variables), where the α_s are given in Table 8.8

Lag	Variables					
	USA Industrial production		UK imports		UK total investment	
	$(1-\alpha_s B)y_t$	$(1-B)y_t$	$(1-\alpha_s B)y_t$	$(1-B)y_t$	$(1-\alpha_s B)y_t$	$(1-B)y_t$
1	0.444**	0.242**	–0.122	–0.200**	–0.163	–0.320**
2	0.169	0.196**	–0.156	–0.051	–0.108	0.051
3	0.073	–0.061	0.017	–0.076	0.065	–0.141
4	0.026	0.205**	0.064	0.143	0.233**	0.433**
5	–0.179**	–0.264**	–0.132	–0.191**	–0.123	–0.286**
6	–0.107	–0.032	–0.077	0.017	–0.075	0.051
7	–0.134	–0.224**	0.092	–0.008	0.043	–0.127
8	–0.209**	0.008	0.034	0.105	–0.013	0.239**
9	–0.174	–0.253**	–0.122	–0.171	–0.003	–0.185**
10	–0.010	0.044	0.073	0.131	–0.032	0.073
11	–0.057	–0.165	–0.033	–0.092	0.082	–0.081
12	–0.097	0.099	–0.103	–0.041	–0.008	0.201**
s.e.	0.089	0.089	0.086	0.086	0.086	0.086

** Significant at the 5 per cent level when normality is assumed and that s.e. = $1/n^{1/2}$.

Table 8.9 presents the estimated nonperiodic autocorrelation functions of the $(1-\hat{\alpha}_s B)y_t$ and $(1-B)y_t$ series (after correction for seasonal constants) for three of the ten sample series. Similar results can be obtained for the other seven variables.

It is clear from the estimated autocorrelations in this table that the $(1-B)y_t$ time series can display long-memory properties in the sense that the ACF does not die out quickly, while the ACF of the periodically differenced series seems to correspond with either white noise or an AR(1) type of process. Even though one would not use such ACFs to identify periodic models, the ACFs for the $(1-B)y_t$ series suggest that not all stochastic trend behaviour is removed from the data. According to the techniques discussed in Chapters 3 and 4, one would then be tempted to specify models for the $(1-B)y_t$ series that include unit roots at seasonal frequencies, while the results in Table 8.7 clearly indicate that there is only a single unit root in each time series.

These findings can also be visualized by plotting the $(1-\hat{\alpha}_s B)y_t$ and $(1-B)y_t$ time series for each of the quarters, as is done for UK nondurables consumption in Figs. 8.1 and 8.2. From these figures, where $Y_{s,\mathrm{T}}-\hat{\alpha}_s Y_{s-1,\mathrm{T}}$ and $Y_{s,\mathrm{T}}-Y_{s-1,\mathrm{T}}$ are depicted for $s=2, 3, 4$, we can see that the periodically differenced time series appear to be stationary, while the $(1-B)y_t$ time series shows trending patterns in each of these three quarters. Similar graphs can be made for the other nine series. (See also the graphs in Chapter 3 for the $(1-B)y_t$ series for industrial production in the USA and real GNP in Germany.)

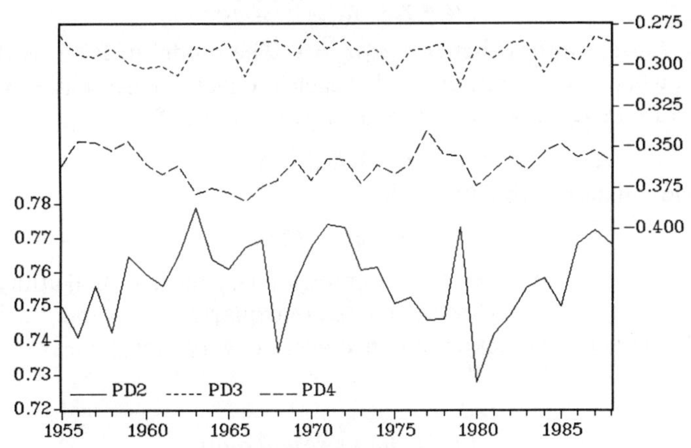

FIG. 8.1 *Periodically differenced time series for UK nondurables consumption, 1955–1988*

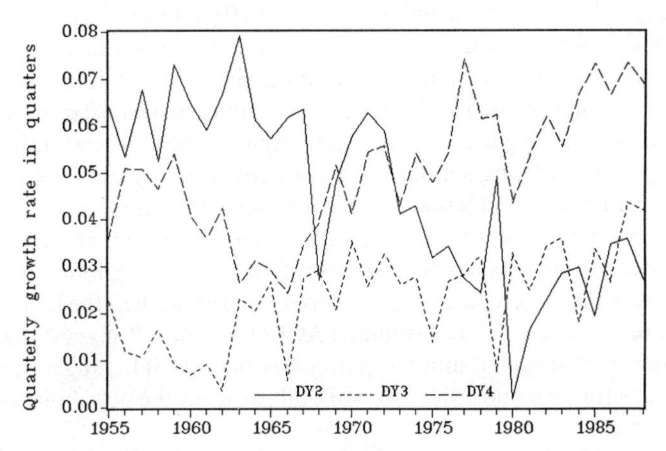

FIG. 8.2 *First -order differenced time series for UK nondurables consumption, 1955–1988*

The finding in Table 8.7 that the log of UK consumption nondurables can be described using a PIAR(1) process corresponds to the economic theoretical results in Osborn (1988). In that paper the R. E. Hall (1978) model is extended to allow for seasonality. Denote C_t as the aggregate consumption of nondurables. Assuming a periodic single-period utility function

$$U_t = C_t^{1-\gamma_s}/(1-\gamma_s), \quad 0 < \gamma_s < 1, \quad s = 1,2,3,4, \tag{8.41}$$

a periodic discount rate, and a constant real interest rate, it is derived in Osborn (1988) that $c_t = \log C_t$ can be described by

$$c_t = \mu_s + \alpha_s c_{t-1} + \varepsilon_t, \tag{8.42}$$

with

$$\alpha_s = \gamma_{s-1}/\gamma_s. \quad s = 1,2,3,4. \tag{8.43}$$

Clearly, (8.43) implies that $\alpha_1\alpha_2\alpha_3\alpha_4 = 1$. The model in R. E. Hall (1978) emerges when $\gamma_s = \gamma_{s-1}$ and $\mu_s = \mu$ for each s. Given the estimation results in Table 8.7 for (8.42), I obtain the following inequalities for the γ_s:

$$\hat{\gamma}_2 > \hat{\gamma}_3 > \hat{\gamma}_4 > \hat{\gamma}_1 \tag{8.44}$$

Since one can derive from (8.41) that

$$\partial U_t / \partial C_t = C_t^{-\gamma_s}, \tag{8.45}$$

we obtain that, for any given consumption level C_t, the marginal utility is highest in the first quarter and lowest in the second quarter. This supports the hypothesis that the utility function of consumers can vary over the seasons.

Testing for more unit roots

The above analysis is concerned with the case when one has some indication that there can be at most one unit root in the PAR(p) process. In some applications (e.g. the unemployment in Canada series) it may be useful to allow for the possible presence of more than one unit root. Since the possibility of I(2) type processes is excluded, such multiple unit roots concern the unit roots at the seasonal frequencies. Again, one may directly test for seasonal unit roots in PAR(p) processes along similar lines as (8.36) (see Ghysels *et al.* 1994*a*). A drawback of this method is that one has to assume that the $(1 - B)$ filter can remove a nonseasonal unit root. Given the results above, where I find that the $(1 - B)$ filter can be rejected against a periodic differencing filter, it is unclear what the effect is on test results for seasonal unit roots when the inappropriate $(1 - B)$ filter is assumed. The estimated ACFs in Table 8.9 suggest that (spurious) evidence of seasonal unit roots may be obtained. This suggests that it is useful to test for seasonal unit roots while allowing for the possibility of having a PIAR series.

A simple method to test for seasonal unit roots which allows for periodic and seasonal integration is given by the application of the Johansen maximum likelihood cointegration method. Consider the VQ(1) process as in (8.8), i.e.

$$\Delta \mathbf{Y}_T = \Pi \mathbf{Y}_{T-1} + \mu + \omega_T, \tag{8.46}$$

with $\omega_T = \Phi_0^{-1}\varepsilon_T$ and $\Pi = \Phi_0^{-1}\Phi_1 - \mathbf{I}_4$. The rank r of the matrix Π conveys the information on the cointegration properties of the $\mathbf{Y}_T$ vector series. The $\mathbf{Y}_T$ process is asymptotically stationary when $r = 4$. There is no cointegration between the elements of $\mathbf{Y}_T$ when $r = 0$. In the case where $0 < r < 4$, one can write $\Pi = \gamma\alpha'$, where γ and α are $(4 \times r)$ matrices, of which the matrix α contains the cointegration vectors. Johansen (1988) developed test procedures for the value of r and for linear hypotheses in terms of γ and α. For (8.46), the method amounts to the choice of the r linear combinations of elements of $\mathbf{Y}_T$ that have the largest correlation with $\Delta\mathbf{Y}_T$ (after correcting for a vector of intercepts).

The eigenvectors of the relevant canonical correlation matrix are the columns of α. The corresponding eigenvalues λ_i, where $\lambda_i \geq \lambda_{i+1}$, are used to construct statistics like trace(r) $= -N\Sigma_{i=r+1}^{4}\log(1 - \lambda_i)$ and λmax(r) $= -N\log(1 - \lambda_r)$. The trace test statistic, and the maximum eigenvalue test statistic, can be used to test for the number of cointegration vectors.

Asymptotic fractiles for these statistics are displayed in Osterwald-Lenum (1992). Preliminary Monte Carlo simulations have indicated that, for sample sizes as large as $N = 25$, these asymptotic critical values may not be appropriate. Therefore, small-sample fractiles for the statistics are calculated on the basis of 10,000 replications for a sample of 25 observations. Note that this sample size for annual data corresponds to 100 quarterly data. The relevant fractiles are displayed in Table 8.10. A comparison of these with the corresponding tables in, say, Osterwald-Lenum (1992) indicates that the critical values in small samples differ from the asymptotic ones.

To test for linear restrictions on the cointegrating vectors α, define a $(4 \times q)$-matrix $\mathbf{H}$, where $r \leq q \leq 4$, which reduces α to the parameters ϕ, or $\alpha = \mathbf{H}\phi$. For brevity, I shall denote these restrictions by their matrix $\mathbf{H}$. Assuming the validity of the restrictions $\mathbf{H}$, we can compare the corresponding eigenvalues ξ_i of

TABLE 8.10. Critical values of the Johansen cointegration tests[a]

4–r	50%	80%	90%	95%	97.5%	99%	Mean	Var.
(a) The constant term is unrestricted								
Maximal eigenvalue								
1	2.43	4.93	6.70	8.29	9.91	12.09	3.06	7.36
2	7.86	11.38	13.70	15.75	17.88	20.51	8.54	14.76
3	13.80	18.16	20.90	23.26	25.66	28.57	14.46	22.80
4	20.36	25.56	28.56	31.66	34.47	37.61	21.07	32.97
Trace								
1	2.43	4.93	6.70	8.29	9.91	12.09	3.06	7.36
2	9.78	13.99	16.56	18.90	21.26	23.70	10.45	20.64
3	21.79	27.69	31.22	34.37	37.44	40.98	22.54	42.57
4	39.32	47.10	51.59	55.92	59.60	64.33	40.09	77.08
(b) The constant term is restricted								
Maximal eigenvalue								
1	3.55	6.01	7.72	9.35	10.97	12.90	4.14	7.08
2	8.82	12.15	14.40	16.51	18.36	20.56	9.36	14.33
3	14.56	18.89	21.56	23.90	26.21	29.44	15.24	22.84
4	21.01	26.15	29.26	32.18	34.74	38.12	21.69	32.79
Trace								
1	3.55	6.01	7.72	9.35	10.97	12.90	4.14	7.08
2	11.95	16.09	18.63	20.96	22.78	25.71	12.55	20.81
3	25.01	31.01	34.44	37.85	40.56	44.60	25.74	44.16
4	43.40	51.38	55.78	59.98	63.51	67.74	44.20	78.19

[a] The sample size is 25.

the canonical correlation matrix with the λ_i via the test statistic $Q = N\Sigma_{i=1}^{r} \log[(1-\xi_i)/(1-\lambda_i)]$. Under the null hypothesis, the test statistic Q asymptotically follows a $\chi^2[r(4-q)]$ distribution. Whether this distribution is valid in small samples will be investigated below.

An application of the Johansen cointegration method to the VQ model in (8.46) gives an opportunity to gain insights into the properties of the univariate quarterly y_t series. No differencing filter is needed for y_t in the case where $r = 4$ (see also Osborn 1993). When $r = 0$, the Δ_4 filter for y_t may be appropriate. If $r = 3$, and pairs of successive $Y_{s,T}$ are cointegrated with parameters (1,–1), a transformation Δ_1 for y_t is required. This Δ_1 filter assumes the cointegration relations $(Y_{2,T} - Y_{1,T})$, $(Y_{3,T} - Y_{2,T})$, and $(Y_{4,T} - Y_{3,T})$. In terms of model (8.46), this means that the restrictions on the columns of α given by

$$\mathbf{H}_{31} = \begin{bmatrix} -1 & 0 & 0 \\ 1 & -1 & 0 \\ 0 & 1 & -1 \\ 0 & 0 & 1 \end{bmatrix} \tag{8.47}$$

cannot be rejected, and this means that y_t has a nonseasonal unit root. It is also possible to test for the presence of seasonal unit roots. When $r = 3$, one can check for the presence of root –1 by testing the restrictions

$$\mathbf{H}_{32} = \begin{bmatrix} 1 & 0 & 0 \\ 1 & 1 & 0 \\ 0 & 1 & 1 \\ 0 & 0 & 1 \end{bmatrix}. \tag{8.48}$$

If both hypotheses $\mathbf{H}_{31}$ and $\mathbf{H}_{32}$ are rejected, three general cointegration relationships between the elements of $Y_{s,T}$ have been encountered. This implies the appropriateness of the periodic differencing filter $(1 - \alpha_s B)$ with $\alpha_1\alpha_2\alpha_3\alpha_4 = 1$ and not all $\alpha_s = \alpha$.

When $r = 1$ or 2, one can proceed along similar lines to test for the presence of nonseasonal and/or specific seasonal unit roots. When $r = 2$, and one wants to test for ± 1 or $\pm i$, the **H** matrices are given by

$$\mathbf{H}_{21} = \begin{bmatrix} -1 & 0 \\ 0 & -1 \\ 1 & 0 \\ 0 & 1 \end{bmatrix} \text{ and } \mathbf{H}_{22} = \begin{bmatrix} 1 & 0 \\ 0 & 1 \\ 1 & 0 \\ 0 & 1 \end{bmatrix}. \tag{8.49}$$

When $r = 1$ and the focus is on the unit roots $(1, \pm i)$ or $(-1, \pm i)$, the matrices are

$$\mathbf{H}_{11} = \begin{bmatrix} -1 \\ 1 \\ -1 \\ 1 \end{bmatrix} \text{ and } \mathbf{H}_{12} = \begin{bmatrix} 1 \\ 1 \\ 1 \\ 1 \end{bmatrix}. \tag{8.50}$$

When r is found to be equal to 0, the $(1 - B^4)$ filter is useful.

Summarizing, an application of the Johansen cointegration method to a VAR model for the $\mathbf{Y}_T$ vector generalizes the HEGY approach since it allows for the presence of periodically varying parameters. When each of the hypotheses $\mathbf{H}$ is rejected, the time series y_t can be said to be periodically integrated. Notice that this slightly extends the definition of PI above. A possibly suitable model for such a series is a univariate periodic error correction model (PECM) where Π is replaced by $\gamma\alpha'$. The 'error' of overdifferencing is then corrected by $\alpha'\mathbf{Y}_{T-1}$, which represents linear relationships between the annual series.

Small sample evaluation of cointegration testing method

The model in (8.46) contains 16 + 14 parameters to be estimated. When some of these parameters are in fact equal to zero, as for example in (8.9), this may have an impact on the empirical performance of the cointegration method. Furthermore, nonperiodic models imply parameter restrictions on the elements of Π, and this can also effect size and power of the test strategy. Here I report on the results of a Monte Carlo study in which small periodic and nonperiodic, possibly integrated, time series are the DGPs. In the simulations only the trace test statistic will be used. This cointegration-based model selection approach will be compared with the HEGY method.

Issues of interest are whether the empirical power of the procedure is reasonably high also in cases where the model in (8.46) is overparameterized, and whether the asymptotic χ^2 distribution for the tests for restrictions is valid in samples as small as 25 annual observations. In the simulations the maintained regression model is (8.46); i.e., the model includes a constant term. The relevant critical values are obtained from Table 8.10. For comparability reasons, I consider the auxiliary regression for the HEGY method when it includes seasonal dummies and a deterministic trend. The significance level for each step of the HEGY method is set equal to 5 per cent. To gain insight into the performance of the cointegration method (which is called the VQ method), I present simulation results for the 5 and 10 per cent significance levels.

In the case where the DGP is a PIAR(1), which is a process not captured by the HEGY method, the differences between the two methods are most striking. Some simulation results relevant to this case, where only two sets of parameter values are chosen, are reported in Table 8.11. The results suggest that in about 50 per cent of the cases the VQ method selects the correct model; i.e., $r = 3$ and the restrictions $\mathbf{H}_{31}$ are rejected. One reason for this somewhat low value of the power is that the VQ model is overparameterized. This seems to be confirmed by the unreported fact that, next to the 50 per cent of the cases in which r is found to be equal to 3, generally in about 35 per cent of the cases this r is estimated to be 2. Another cause may be that the asymptotic χ^2 distribution may not apply to samples as small as 25 annual observations. In fact, at a nominal level of 5 per cent, the rejection rate of the Q test statistic is about 20 per cent. It can be expected that the results for higher-order periodic

TABLE 8.11. Monte Carlo comparison[a]

DGP	VQ[b]		HEGY[c]					
	10%	5%	1	−1	$\pm i$	–	Δ_1	Δ_4
(i) $\sigma_s = 1$	0.588	0.438	0.961	0.196	0.196	0.026	0.753	0.159
$\sigma_s = 1.25, 0.8, 0.5, 2.0$	0.595	0.441	0.960	0.225	0.215	0.023	0.727	0.179
(ii) $\sigma_s = 1$	0.619	0.488	0.948	0.859	0.272	0.020	0.117	0.247
$\sigma_s = 1.25, 0.8, 0.5, 2.0$	0.630	0.491	0.949	0.904	0.302	0.023	0.072	0.280

[a] Based on 5,000 replications of the cointegration-based VQ method and the nonperiodic HEGY method when the DGP is a periodically integrated process: $y_t = \alpha_s y_{t-1} + \varepsilon_t$, with (i) $\alpha_1 = 1.25$, $\alpha_2 = 0.8$, $\alpha_3 = 0.9$, $\alpha_4 = 1.11$, $\varepsilon_t \sim N(0, \sigma_s)$; and (ii) $\alpha_1 = 2$, $\alpha_2 = 0.5$, $\alpha_3 = 1.5$, $\alpha_4 = 0.67$, $\varepsilon_t \sim N(0, \sigma_s)$. The sample is 100 quarterly observations.

[b] The values in the cells report the number of times the correct decision is made; i.e., $r = 3$ and the Δ_1 is not appropriate.

[c] The values in the cells report the number of times the presence of the roots 1, −1, $\pm i$ cannot be rejected. The outcomes are based on the t-tests for π_1 and π_2 and the joint F-test for π_3 and π_4 in the auxiliary regression, which contains a constant, seasonal dummies and a trend (See Chapter 5). No filter (–) is chosen when all $\pi_i \neq 0$, Δ_1 when $\pi_1 = 0$ and the other π_i are not, and Δ_4 when all π_i equal zero. The test outcomes for the π_i are based on the 5 per cent significance level. The number of additional lags in the test equation is set equal to that number k, $k = 12, \ldots, 0$, for which there appeared to be no significant residual autocorrelation using the $F_{\text{AR},1\text{–}4}$ test.

autoregressive models estimated for longer time series will show an improvement of the test performance. The HEGY method indicates that in several cases one is inclined to opt for the Δ_1 filter, although often the Δ_4 filter can be found to be appropriate. Furthermore, it can be seen that a likely outcome of the HEGY method is that the root −1 or the roots $\pm i$ seem to be present. Of course, given the choice of the parameters, this may not come as a surprise. The performances of the two methods do not seem to be affected by a seasonally heteroskedastic error process.

To further investigate how the VQ method performs where the generating processes are simple periodic and nonperiodic processes implying over-parameterized VQ models, consider the results in Table 8.12. In those cases where the process y_t does not need to be differenced, i.e. the case in the upper part of the Table 8.12, the VQ method detects that the Π matrix is of full rank in a large proportion of the cases, and it sometimes performs better than the HEGY method does. As expected, the power of the method decreases when the root of the process approaches unity. For example, for an α_s parameter of 0.9 for all s, in only 12.1 per cent of the cases does the HEGY method find that the roots 1, −1, i, and $-i$ are not present jointly, and in only 14.1 per cent of the cases does the VQ approach detect that the correct $r = 4$. Again, allowing for a seasonal heteroskedastic error process does not dramatically affect the performances. Finally, when the first-order autoregressive parameter can vary with the seasons, this does not effect the outcomes to a great extent either.

TABLE 8.12. Monte Carlo evaluation of HEGY and VQ procedure[a,b]

Data generating process	HEGY	VQ	
		10%	5%
(a) $y_t = \alpha_s y_{t-1} + \varepsilon_t, \varepsilon_t \sim N(0,\sigma_s)$			
$\alpha_s = 0, \sigma_s = 1$	0.926	0.916	0.774
$\alpha_s = 0.5, \sigma_s = 1$	0.802	0.865	0.697
$\alpha_s = 0.9, \sigma_s = 1$	0.121	0.323	0.174
$\alpha = 0.5, \sigma_s = 1.1, 0.9, 1.5, 0.7$	0.789	0.875	0.699
$\alpha = 0.5, \sigma = 1.25, 0.8, 0.5, 2.0$	0.758	0.853	0.683
$\alpha_s = 0.2, 0.4, 0.6, 0.8,\ \sigma_s = 1.25, 0.8, 0.5, 2.0$	0.812	0.893	0.731
$\alpha_s = 0.6, 0.7, 0.8, 0.9,\ \sigma_s = 1.25, 0.8, 0.5, 2.0$	0.440	0.694	0.486
(b) $\Delta_1 y_t = \alpha_s \Delta_1 y_{t-1} + \varepsilon_t, \varepsilon_t \sim N(0,\sigma_s)$			
$\alpha_s = 0, \sigma_s = 1$	0.890	0.422	0.365
$\alpha_s = 0.5, \sigma_s = 1$	0.900	0.482	0.461
$\alpha_s = 0.5, \sigma_s = 1.25, 0.8, 0.5, 2.0$	0.890	0.530	0.512
$\alpha_s = 0.2, 0.4, 0.6, 0.8,\ \sigma_s = 1.25, 0.8, 0.5, 2.0$	0.889	0.460	0.427
(c) $\Delta_4 y_t = 0.5 \Delta_4 y_{t-1} + \varepsilon_t, \varepsilon_t \sim N(0,1)$	0.584	0.774	0.875

[a] The sample is 100 quarterly observations. The values in the cells report the frequencies that a method selects the correct filter.

[b] For case I, this should be no filter, or, in terms of VQ, $r = 4$. For case II this filter is Δ_1, and for case III it is Δ_4. The evaluation is based on 5,000 replications of series of length 100. All HEGY test outcomes are based on a 5 per cent significance level, while all results for a VQ model of order 1 consider 5 as well as 10 per cent significance levels. Notice that the VQ model in case III should be of order 2. Since this leads to only one non-zero parameter in Φ_2, I use the VQ(1) process.

When the Δ_1 filter for the y_t series is appropriate, the VQ method does not perform extremely well, as can be seen from part (*b*) of Table 8.12. The empirical powers in these cases are similar to those in Table 8.11. In about 45 per cent of the cases the correct filter is found. This can imply that the 20 per cent significance level for the trace test may be more useful, and that, for example, a 1 per cent level for the Q test for restrictions in the cointegration parameters may be more appropriate. Finally, when the data-generating process is a first-order autoregressive model for a series that needs a Δ_4 filter to reach stationarity, as in panel (*c*) of Table 8.12, the VQ method detects the correct filter more often than the HEGY method does. Note that the empirical success rate in this case is higher for the 5 per cent significance level. This counterintuitive result is caused by the fact that the figures in the cells correspond to one minus the size of the tests. The HEGY approach finds that in about 40 per cent of the cases the Δ_4 is not appropriate. This suggests that the size of the VQ method is not much affected by lagged $\Delta_4 y_t$ terms.

In summary, the VQ and HEGY approaches can yield similar outcomes in the cases where the data-generating processes are close to those assumed for the HEGY method. Hence, even when the multivariate time series model to which the VQ method is applied is highly overparameterized, the VQ method performs reasonably well. A suggestion for practical use of this approach is to consider also a nominal size of 20 per cent for the trace test statistic, and to test restrictions on the cointegration relations using a 1 per cent significance level. When one allows for periodically nonstationary processes, it appears that the VQ approach is a useful generalization of the HEGY method, since the latter method can only suggest the use of inappropriate filters.

Application of the VQ cointegration-based method in Franses and Romijn (1993) to several UK macroeconomic time series (including the series discussed in Table 8.7) reveals that often the rank r is estimated to be equal to 2 or 3. For example, for UK total consumption and consumption nondurables, Franses and Romijn (1993) find that $r = 2$. Given the results in Tables 8.7 and 8.8, it is clear that these findings correspond to a loss of power of the VQ method in the case of small PAR models. The main reason why the LR test statistics in (8.30) are more powerful than the Johansen tests here is that the latter tests are calculated for N observations, while the former are calculated for $n = 4N$ observations. Hence the model in (8.46) implicitly allows for seasonal heteroskedasticity, even though there may not be such seasonal variation. An application of this method to the Japanese consumption and income data analysed in Engle *et al.* (1993) is given in Franses (1994).

Canadian unemployment

To illustrate the use of the above VQ method, consider the PAR(4) for unemployment in Canada, which according to Table 7.11 may have more than one unit root. The regression model is (8.46), where a dummy variable for 1982 is included in each equation since that year contains an outlying data point (see also Table 7.9). For $N = 27$ observations, we obtain the estimates for the eigenvalues as $\hat{\lambda}_1 = 0.934$, $\hat{\lambda}_2 = 0.448$, $\hat{\lambda}_3 = 0.294$, and $\hat{\lambda}_4 = 0.149$, with the trace tests trace(3) = 4.353, trace(2) = 13.76, trace(1) = 29.804, and trace(0) = 103.234. Comparing these values with the 5 per cent critical value, there appears to be only one cointegrating relationship, while compared with the 20 per cent critical level, there are about three such relationships. (Unreported) values of the test statistics for the hypotheses $\mathbf{H}$ in (8.47)–(8.50) indicate that all of these hypotheses are rejected by the data at the 2.5 per cent level. A univariate model for Canadian unemployment may be constructed by estimating the following version of a periodic error correction model:

$$\Phi_0 \Delta \mathbf{Y}_T = (\Phi_0 \gamma)\alpha' \mathbf{Y}_{T-1} + \varepsilon_T, \tag{8.51}$$

where $\alpha' \mathbf{Y}_{T-1}$ are the estimated cointegration relations. Initially, I specify (8.51) with four such relations (although the fourth one may not reflect a stationary variable), and I delete the variables that appear not to be relevant using standard t- and F-tests. The final estimation results are

$$\Delta Y_{1,T} = \underset{(20.690)}{58.737} + \underset{(0.149)}{2.107} ECM1_{T-1} + \underset{(0.161)}{0.383} ECM2_{T-1} + \varepsilon_{1,T}$$

$$\Delta Y_{2,T} = \underset{(8.472)}{0.170} + \underset{(0.067)}{0.633} \Delta Y_{1,T} + \underset{(44.98)}{274.51} DUM82 + \varepsilon_{2,T}$$

$$\Delta Y_{3,T} = \underset{(6.765)}{5.460} + \underset{(0.102)}{1.696} \Delta Y_{2,T} - \underset{(0.092)}{0.891} \Delta Y_{1,T} + \varepsilon_{3,T}$$

$$\Delta Y_{4,T} = \underset{(16.421)}{31.517} + \underset{(0.196)}{1.257} \Delta Y_{3,T} - \underset{(0.303)}{0.692} \Delta Y_{2,T} + \underset{(0.220)}{0.504} \Delta Y_{1,T} + \underset{(62.30)}{148.43} DUM82$$
$$- \underset{(0.334)}{0.808} ECM1_{T-1} + \underset{(0.123)}{0.321} ECM2_{T-1} + \varepsilon_{4,T},$$

where the estimated standard errors are given in parentheses, and with

$$ECM1_T = -0.234\, Y_{1,T} - 0.200\, Y_{2,T} - 0.506\, Y_{3,T} + Y_{4,T}$$
$$ECM2_T = -0.770\, Y_{1,T} + Y_{2,T} + 0.590\, Y_{3,T} - 0.915\, Y_{4,T},$$

and $DUM\,82$ is a dummy variable for 1982. Given the complicated patterns in the unemployment series, this univariate PECM seems a reasonably parsimonious model. In the next chapter I will focus on multivariate PECMs, i.e. on so-called periodic cointegration models which relate two or more variables. All in all, it seems that the VQ cointegration-based method may be useful in cases where a time series displays quite complicated seasonal and trend patterns. It is my experience however that, for many regularly analysed macroeconomic time series, the two-step method based on the LR tests in (8.30) and that the F-tests for parameter variation in the differencing filter are often more useful, as can also be seen from the results in Table 8.7.

Subset periodic autoregression with a unit root

The two-step method is also useful when one wants to allow for different non-periodic filters per season. For example, when the data can be described by the process $y_t = \alpha_s y_{t-1} + \varepsilon_t$, with $\alpha_1 = \alpha_2 = 1$ and $\alpha_3 = \alpha_4 = -1$, the LR_1 test will indicate that there is one stochastic trend in the series and that the $(1 - B)$ or $(1 + B)$ filter is not appropriate. Likelihood ratio tests for specific hypotheses on the estimated α_s parameter may then reveal the variation in the sign of the parameters.

Similar arguments apply to subset PARs. If such a model fits the data for y_t, we can write it in VQ notation and construct the characteristic equation like $|\Phi_0 - \Phi_1 z| = 0$. Typically, a unit root in a subset PAR results in a quite complicated nonlinear restriction on the parameters (see Franses 1993 for two examples). However, this restriction can be evaluated using the LR tests in (8.30), once the model with the imposed nonlinear restriction is estimated using nonlinear least squares.

8.2 Periodic Integration

In the previous section it was found that for ten sample series one can fit a PIAR(1) or a PIAR(2) process; i.e., one can fit low-order PAR processes to time series that need the $(1 - \alpha_s B)$ differencing filter to remove the stochastic trend, where $\alpha_1\alpha_2\alpha_3\alpha_4 = 1$ and $\alpha_s \neq \alpha$ for $s = 1,2,3,4$. In this section I focus on some of the consequences of this finding. For illustrative purposes and for notational convenience, I will consider mainly either the PIAR(1) or PIAR(2) process for quarterly series. Needless to say, all material can be extended to higher-order PARs and monthly series.

Comparison with seasonal integration

Consider again the simple PIAR(1) process

$$y_t = \alpha_s y_{t-1} + \varepsilon_t \quad \text{with } \alpha_1\alpha_2\alpha_3\alpha_4 = 1 \tag{8.52}$$

By backward substitution, we obtain

$$y_t = y_{t-4} + \varepsilon_t + \alpha_s\varepsilon_{t-1} + \alpha_s\alpha_{s-1}\varepsilon_{t-2} + \alpha_s\alpha_{s-1}\alpha_{s-2}\varepsilon_{t-3}, \tag{8.53}$$

which shows that $\Delta_4 y_t$ is a periodic moving average process of order 3 (PMA(3)) and that $\Delta_4 y_t$ is stationary. This expression suggests that the Δ_4 filter can be used to remove the nonstationarity from y_t. However, it is clear (also from the above discussions) that the assumption of four unit roots via the Δ_4 filter amounts to overdifferencing. A simple way to see this is to write (8.53) in VQ format as $\Delta\mathbf{Y}_T = \Theta_0\varepsilon_T + \Theta_1\varepsilon_{T-1}$ with

$$\Theta_0 = \begin{bmatrix} 1 & 0 & 0 & 0 \\ \alpha_2 & 1 & 0 & 0 \\ \alpha_2\alpha_3 & \alpha_3 & 1 & 0 \\ \alpha_2\alpha_3\alpha_4 & \alpha_3\alpha_4 & \alpha_4 & 1 \end{bmatrix} \text{ and } \Theta_1 = \begin{bmatrix} 0 & \alpha_3\alpha_4\alpha_1 & \alpha_4\alpha_1 & \alpha_1 \\ 0 & 0 & \alpha_4\alpha_1\alpha_2 & \alpha_1\alpha_2 \\ 0 & 0 & 0 & \alpha_1\alpha_2\alpha_3 \\ 0 & 0 & 0 & 0 \end{bmatrix}. \tag{8.54}$$

The characteristic polynomial of the vector moving average (VMA) matrix polynomial is

$$\Theta(z) = |\Theta_0 + \Theta_1 z| = (1 - \alpha_1\alpha_2\alpha_3\alpha_4 z)^3 = (1 - z)^3 = 0. \tag{8.55}$$

This indicates that the VMA model is not invertible since it has three unit roots. Hence the Δ_4 filter for PIAR series amounts to overdifferencing. This result is even more obvious when α_s is replaced by 1.

The implication of analysing (8.52) by using a nonperiodic model will be discussed in Section 8.3. To anticipate some of the results, consider again (8.53) after taking first differences; i.e.,

$$\Delta_1\Delta_4 y_t = \varepsilon_t + (\alpha_s - 1)\varepsilon_{t-1} + \alpha_{s-1}(\alpha_s - 1)\varepsilon_{t-2} + \alpha_{s-1}\alpha_{s-2}(\alpha_s - 1)\varepsilon_{t-3} - \alpha_{s-1}\alpha_{s-2}\alpha_{s-3}\varepsilon_{t-4}. \tag{8.56}$$

When the $\alpha_s \neq 1$ but is very close to unity (as can be seen from Table 8.8), the $\alpha_s - 1$ values can become quite small, and (8.56) can be approximated by

$$\Delta_1\Delta_4 y_t = \varepsilon_t - \eta_s \varepsilon_{t-4}, \tag{8.57}$$

where η_s can take values close to unity. When (8.57) is analysed without taking account of the periodic variation in η_s, we may obtain such estimation results as reported in Chapter 3 for the Box–Jenkins SARIMA models. Notice from (8.57) that PIAR models can thus describe seasonal time series with slowly changing patterns.

Trends and constants

Similar to (8.53), one can write the periodically integrated process with four intercept terms:

$$y_t = \mu_s + \alpha_s y_{t-1} + \varepsilon_t \quad \text{with } \alpha_1\alpha_2\alpha_3\alpha_4 = 1 \tag{8.58}$$

as

$$\Delta_4 y_t = \delta_s + \varepsilon_t + \alpha_s \varepsilon_{t-1} + \alpha_s\alpha_{s-1}\varepsilon_{t-2} + \alpha_s\alpha_{s-1}\alpha_{s-2}\varepsilon_{t-3}, \tag{8.59}$$

with

$$\delta_s = \mu_s + \alpha_s \mu_{s-1} + \alpha_s\alpha_{s-1}\mu_{s-2} + \alpha_s\alpha_{s-1}\alpha_{s-2}\mu_{s-3}. \tag{8.60}$$

These expressions indicate that a PIAR(1) process can generate time series y_t that display, for example, increasing seasonal variation when $\delta_s \neq \delta$ for all $s = 1, 2, 3, 4$ because the expected value of $\Delta_4 y_t$ equals δ_s. This aspect of periodic integration can be exploited to describe seasonal time series with increasing seasonal variation without taking logs, which can be particularly useful for time series than can take negative values (like the trade balance).

Of course, one may wish to test for whether $\delta_s = \delta$ for all seasons. For this purpose it is useful to write (8.60) as

$$\begin{bmatrix} \delta_1 \\ \delta_2 \\ \delta_3 \\ \delta_4 \end{bmatrix} = \begin{bmatrix} 1 & \alpha_1\alpha_3\alpha_4 & \alpha_1\alpha_4 & \alpha_1 \\ \alpha_2 & 1 & \alpha_1\alpha_2\alpha_4 & \alpha_1\alpha_2 \\ \alpha_2\alpha_3 & \alpha_3 & 1 & \alpha_1\alpha_2\alpha_3 \\ \alpha_2\alpha_3\alpha_4 & \alpha_3\alpha_4 & \alpha_4 & 1 \end{bmatrix} \begin{bmatrix} \mu_1 \\ \mu_2 \\ \mu_3 \\ \mu_4 \end{bmatrix}, \tag{8.61}$$

which under the restriction $\alpha_1\alpha_2\alpha_3\alpha_4 = 1$ can be written as

$$\begin{bmatrix} \delta_1 \\ \delta_2 \\ \delta_3 \\ \delta_4 \end{bmatrix} = \begin{bmatrix} 1 \\ \alpha_2 \\ \alpha_2\alpha_3 \\ \alpha_2\alpha_3\alpha_4 \end{bmatrix} [1 \quad \alpha_1\alpha_3\alpha_4 \quad \alpha_1\alpha_4 \quad \alpha_1] \begin{bmatrix} \mu_1 \\ \mu_2 \\ \mu_3 \\ \mu_4 \end{bmatrix}, \tag{8.62}$$

since the rank of the matrix in (8.61) is equal to 1. Hence a test for the restriction $\delta_s = \delta$ in the PIAR(1) model, where $\alpha_1\alpha_2\alpha_3\alpha_4 = 1$ and $\alpha_s \neq 1$, can be considered only where $\delta = 0$. In fact, (8.61) shows that $\delta_2 = \alpha_2\delta_1$ and so on, and hence that δ_2 can be equal to δ_1 only when $\alpha_2 = 1$. This implies that, in a PIAR process with $\alpha_s = 1$, the δ_s in (8.60) always differ. To test for $\delta_s = 0$, it suffices to test the restriction

$$\mu_1 + \alpha_1\mu_4 + \alpha_1\alpha_4\mu_3 + \alpha_1\alpha_4\alpha_3\mu_2 = 0. \tag{8.63}$$

Under the restriction that $\alpha_1\alpha_2\alpha_3\alpha_4 = 1$, this restriction can be tested using a likelihood ratio test statistic that is asymptotically distributed, such as $\chi^2(1)$. Paap (1995) considers related tests in the case where the model in (8.58) contains deterministic trend terms. It should be mentioned that, when a PIAR(p) process with $p > 1$ is written in periodically differenced form, i.e. as a PAR $(p-1)$ process for the $(1-\alpha_s B)y_t$ series, similar results as in (8.63) can be derived. Finally, in Paap (1995) it is shown that, when (8.58) contains deterministic trend terms, the seasonal growth rates of $\Delta_4 y_t$ may be equal across the seasons.

Time-varying impact of accumulation of shocks

The fact that a periodically varying differencing filter is needed to remove the stochastic trend from a PI time series suggests that the stochastic trend and the seasonal fluctuations are somehow related. This relationship can be emphasized by calculating the impact of the accumulation of the shocks ε_t. For this purpose, I write a PIAR(2) model (which is the model found useful for seven of the 11 sample series) as

$$y_t - \alpha_s y_{t-1} = \mu_s + \beta_s(y_{t-1} - \alpha_{s-1}y_{t-2}) + \varepsilon_t, \tag{8.64}$$

with $\alpha_1\alpha_2\alpha_3\alpha_4 = 1$. In turn, this model can be written in VQ notation as

$$\Psi(B)(\Phi_0\mathbf{Y}_\mathrm{T} - \Phi_1\mathbf{Y}_{\mathrm{T}-1}) = \mu + \varepsilon_\mathrm{T}, \tag{8.65}$$

with

$$\Psi(B) = \begin{bmatrix} 1 & 0 & 0 & -\beta_1 B \\ -\beta_2 & 1 & 0 & 0 \\ 0 & -\beta_3 & 1 & 0 \\ 0 & 0 & -\beta_4 & 1 \end{bmatrix} \tag{8.66}$$

and

$$\Phi_0 = \begin{bmatrix} 1 & 0 & 0 & 0 \\ -\alpha_2 & 1 & 0 & 0 \\ 0 & -\alpha_3 & 1 & 0 \\ 0 & 0 & -\alpha_4 & 1 \end{bmatrix}, \text{ and } \Phi_1 = \begin{bmatrix} 0 & 0 & 0 & \alpha_1 \\ 0 & 0 & 0 & 0 \\ 0 & 0 & 0 & 0 \\ 0 & 0 & 0 & 0 \end{bmatrix} \tag{8.67}$$

Next, write (8.65) in a VMA representation. First, we have

$$(\Psi(B))^{-1} = (1-\beta B)^{-1}(\Omega_0 + \Omega_1 B), \tag{8.68}$$

with $\beta \equiv \beta_1\beta_2\beta_3\beta_4$,

$$\Omega_0 = \begin{bmatrix} 1 & 0 & 0 & 0 \\ \beta_2 & 1 & 0 & 0 \\ \beta_2\beta_3 & \beta_3 & 1 & 0 \\ \beta_2\beta_3\beta_4 & \beta_3\beta_4 & \beta_4 & 1 \end{bmatrix}, \text{ and } \Omega_1 = \begin{bmatrix} 0 & \beta_1\beta_3\beta_4 & \beta_1\beta_4 & \beta_1 \\ 0 & 0 & \beta_1\beta_2\beta_4 & \beta_1\beta_2 \\ 0 & 0 & 0 & \beta_1\beta_2\beta_3 \\ 0 & 0 & 0 & 0 \end{bmatrix}, \tag{8.69}$$

In case of a PIAR(1) process, $\Omega_0 = \mathbf{I}_4$, $\Omega_1 = \mathbf{0}$, $\beta = 0$, and hence $(\Psi(B))^{-1} = \mathbf{I}_4$. Using (8.68) and (8.69), the process in (8.65) can be written as

$$\Phi_0 \mathbf{Y}_T = \Phi_1 \mathbf{Y}_{T-1} + \mu^* + (1-\beta B)^{-1}(\Omega_0 \varepsilon_T + \Omega_1 \varepsilon_{T-1}), \tag{8.70}$$

where $\mu^* = (1-\beta)^{-1}(\Omega_0 + \Omega_1)\mu$. Premultiplying (8.70) with Φ_0^{-1} results in

$$\mathbf{Y}_T = \Gamma \mathbf{Y}_{T-1} + \mu^{**} + \Phi_0^{-1}(1-\beta B)^{-1}(\Omega_0 \varepsilon_T + \Omega_1 \varepsilon_{T-1}), \tag{8.71}$$

where $\Gamma = \Phi_0^{-1}\Phi_1$ and $\mu^{**} = \Phi_0^{-1}\mu^*$. Given $\alpha_1\alpha_2\alpha_3\alpha_4 = 1$, the expressions for the Γ and Φ_0^{-1} matrices are

$$\Gamma = \begin{bmatrix} 0 & 0 & 0 & \alpha_1 \\ 0 & 0 & 0 & \alpha_1\alpha_2 \\ 0 & 0 & 0 & \alpha_1\alpha_2\alpha_3 \\ 0 & 0 & 0 & 1 \end{bmatrix}, \text{ and } \Phi_0^{-1} = \begin{bmatrix} 1 & 0 & 0 & 0 \\ \alpha_2 & 1 & 0 & 0 \\ \alpha_2\alpha_3 & \alpha_3 & 1 & 0 \\ \alpha_2\alpha_3\alpha_4 & \alpha_3\alpha_4 & \alpha_4 & 1 \end{bmatrix}, \tag{8.72}$$

Note from (8.72) that Γ is an idempotent matrix; i.e., $\Gamma^m = \Gamma$ for $m = 1,2,3,\ldots$ Hence when the part $\omega_T = (1-\beta B)^{-1}(\Omega_0 \varepsilon_T + \Omega_1 \varepsilon_{T-1})$ is considered the short-run part of the model (8.71), i.e. when $\mathbf{Y}_T = \Gamma \mathbf{Y}_{T-1} + \mu^{**} + \Phi_0^{-1}\omega_T$, the long-run information is summarized in the matrix

$$\Gamma\Phi_0^{-1} = \begin{bmatrix} 1 & \alpha_1\alpha_3\alpha_4 & \alpha_1\alpha_4 & \alpha_1 \\ \alpha_2 & 1 & \alpha_1\alpha_2\alpha_4 & \alpha_1\alpha_2 \\ \alpha_2\alpha_3 & \alpha_3 & 1 & \alpha_1\alpha_2\alpha_3 \\ \alpha_2\alpha_3\alpha_4 & \alpha_3\alpha_4 & \alpha_4 & 1 \end{bmatrix}, \tag{8.73}$$

For illustrative purposes, consider the PIAR(1) case, with $\beta_s = 0$ for all s. Recursively substituting lagged $\mathbf{Y}_T$ in (8.71) yields

$$\mathbf{Y}_T = \mathbf{Y}_0^* + \Gamma\Phi_0^{-1}\mu T + \Phi_0^{-1}\varepsilon_T + \Gamma\Phi_0^{-1}\sum_{i=1}^{T-1}\varepsilon_{T-i}, \tag{8.74}$$

where $\mathbf{Y}_0^*$ is some function of the starting values $\mathbf{Y}_0$, μ, and $\Gamma\Phi_0^{-1}$.

The matrix $\Gamma\Phi_0^{-1}$ in (8.73) conveys the information on the impact of the accumulation of short-run shocks in quarter s denoted by $\Sigma_{i=1}^{T-1}\varepsilon_{s,T-i}$. Given the three cointegration relations between the elements of $\mathbf{Y}_T$, the rank of $\Gamma\Phi_0^{-1}$ is equal to 1 (cf. Engle and Granger 1987; Johansen 1991); i.e.,

$$\Gamma\Phi_0^{-1} = \begin{bmatrix} 1 \\ \alpha_2 \\ \alpha_2\alpha_3 \\ \alpha_2\alpha_3\alpha_4 \end{bmatrix} [1 \quad \alpha_1\alpha_3\alpha_4 \quad \alpha_1\alpha_4 \quad \alpha_1] \tag{8.75}$$

(see also (8.62)). Furthermore, it can be derived that $(\Gamma\Phi_0^{-1})^k = 2^k\Gamma\Phi_0^{-1}$ for $k = 1,2,3,\ldots$ Note that when all $\alpha_s = 1$ the $\Gamma\Phi_0^{-1}$ matrix contains only ones, indicating that the impact of the stochastic trend is equal for all seasons.

From the above expressions it can be observed that, for example, for a PIAR(1) process the long-run impact of the accumulation of shocks in the four seasons on the observations in the first quarter is reflected by

$$\sum_{i=1}^{T-1}\varepsilon_{1,T-i} + \alpha_1\alpha_3\alpha_4\sum_{i=1}^{T-1}\varepsilon_{2,T-i} + \alpha_1\alpha_4\sum_{i=1}^{T-1}\varepsilon_{3,T-i} + \alpha_1\sum_{i=1}^{T-1}\varepsilon_{4,T-i}. \tag{8.76}$$

Similar expressions can be derived for the other three quarters. This suggests that, given $\alpha_s \neq \alpha$ for all s, this impact varies with the seasons. This also applies to the parameters $\Gamma\Phi_0^{-1}\mu$ that correspond to the deterministic trend component. The (1×4) vector $[1, \alpha_1\alpha_3\alpha_4, \alpha_1\alpha_4, \alpha_1]$ in (8.75) conveys information on the relative importance of the accumulation of shocks in season s. If we want to test whether some trend components have the same impact, the relevant hypotheses can be formulated in terms of the α_s parameters. Under the restriction that $\alpha_1\alpha_2\alpha_3\alpha_4 = 1$, we can apply χ^2 type test statistics for hypotheses such as $\alpha_1\alpha_4 = 1$ or $\alpha_1 = 1$. (See Boswijk and Franses (1996) for formal proofs.)

Before I turn to present the estimated matrices $\Gamma\Phi_0^{-1}$ for the sample time series, I must mention that the representation as (8.74) and the form of matrix $\Gamma\Phi_0^{-1}$ can highlight the specific properties of various models for seasonal time series. For example, consider the simple nonperiodic process

$$\mathbf{y}_t = \mathbf{y}_{t-2} + \varepsilon_t, \tag{8.77}$$

which has the unit roots ± 1. It is not difficult to derive for this process that

$$\Gamma\Phi_0^{-1} = \begin{bmatrix} 1 & 0 & 1 & 0 \\ 0 & 1 & 0 & 1 \\ 1 & 0 & 1 & 0 \\ 0 & 1 & 0 & 1 \end{bmatrix}, \tag{8.78}$$

which implies that, for example, the first- and third-quarter observations $Y_{1,T}$ and $Y_{3,T}$ are driven by $\Sigma_{i=1}^{T-1}\varepsilon_{1,T-i} + \Sigma_{i=1}^{T-1}\varepsilon_{3,T-i}$. Notice that the matrix in (8.78) has rank 2, as expected from the discussion in Section 8.1. Another illuminating example is the model

$$(1 + B + B^2 + B^3)y_t = \varepsilon_t, \tag{8.79}$$

which contains the seasonal unit roots -1 and $\pm i$. For this process we obtain (after little algebra)

$$\Gamma\Phi_0^{-1} = \begin{bmatrix} 1 & 0 & 0 & -1 \\ -1 & 1 & 0 & 0 \\ 0 & -1 & 1 & 0 \\ 0 & 0 & -1 & 1 \end{bmatrix}, \tag{8.80}$$

which has rank 3 since (8.79) assumes one cointegration relation between the $Y_{s,T}$ series. Indeed, pre-multiplying $\Gamma\Phi_0^{-1}$ with the vector $[1,1,1,1]$ gives a matrix with all zeros. It is clear from (8.80) that the observations in $Y_{s,T}$ are driven by $\Sigma_{i=1}^{T-1}\varepsilon_{s,T-i} - \Sigma_{i=1}^{T-1}\varepsilon_{s-1,T-i}$ for $s = 1,2,3,4$.

In Table 8.13, I report the estimates of the matrix (8.73) for 10 of the 11 sample series. (The unemployment in Canada series is not considered because of its special structure; see (8.51).) The estimates in the various rows and columns should be interpreted as follows. When in a certain row the values of the parameters in $\Gamma\Phi_0^{-1}$ are highest, this implies that the impact of the accumulations of shocks in all seasons is highest in the season that corresponds to that row. For example, for the US industrial production series it can be observed that the

TABLE 8.13. Estimates of the impact matrix $\Gamma\Phi_0^{-1}$ for ten sample series[a]

Industrial production, USA				*Real GNP, Germany*			
1.000	1.019	0.974	1.004	1.000	1.040	1.140	1.025
0.981	1.000	0.955	0.986	0.962	1.000	1.096	0.985
1.027	1.047	1.000	1.032	0.877	0.912	1.000	0.898
0.996	1.015	0.969	1.000	0.976	1.015	1.113	1.000
UK GDP				*UK total consumption*			
1.000	1.092	1.028	1.011	1.000	1.089	1.031	1.031
0.916	1.000	0.943	0.926	0.918	1.000	0.946	0.946
0.973	1.062	1.000	0.983	0.970	1.057	1.000	1.000
0.989	1.080	1.017	1.000	0.970	1.057	1.000	1.000
UK consumer nondurables				*UK exports*			
1.000	1.073	1.042	1.003	1.000	0.978	0.948	0.957
0.932	1.000	0.971	0.934	1.022	1.000	0.969	0.978
0.960	1.030	1.000	0.962	1.055	1.032	1.000	1.009
0.997	1.070	1.039	1.000	1.045	1.023	0.991	1.000
UK imports				*UK total investment*			
1.000	0.964	0.989	0.965	1.000	1.126	1.051	1.053
1.038	1.000	1.027	0.965	0.888	1.000	0.934	0.935
1.011	0.974	1.000	0.976	0.951	1.071	1.000	1.001
1.036	1.036	1.025	1.000	0.950	1.070	0.999	1.000
UK public investment				*UK workforce*			
1.000	0.483	0.589	0.601	1.000	0.974	0.919	0.936
2.070	1.000	1.220	1.243	1.027	1.000	0.944	0.962
1.697	0.820	1.000	1.019	1.088	1.059	1.000	1.018
1.664	0.804	0.981	1.000	1.068	1.040	0.982	1.000

[a] An expression of the matrix $\Gamma\Phi_0^{-1}$ is given in (8.73). The parameter estimates for α_s are given in Table 8.8.

third-row parameters take the highest values. This means that the impact of all shocks is highest in the third quarter, and hence that the observations in this quarter are most susceptible to fluctuations in the stochastic trend. Further, when the estimated parameters in a certain column of the $\Gamma\Phi_0^{-1}$ matrix are the highest, this means that the accumulation of shocks ε_t in that particular season has the largest long-run impact. For example, for the US industrial production we can see that the second-column parameters take the largest values across the columns, and hence that the accumulation of shocks in the second quarter is a dominant component. Altogether, for this industrial production series the impact of shocks is not equal in all seasons and the third-quarter observations are most heavily affected by such shocks.

When comparing the various estimates for the long-run impact matrix in (8.73) across the ten sample series in Table 8.13, I observe some striking similarities between sets of series. For example, for GDP, total consumption, and total investment in the UK, I find that the first-row and second-column parameters take the largest values. This suggests that these series may have a common long-run correlation between seasons and trend. In Chapter 9 I will investigate whether these series have their periodic integration aspect in common using more formal tests. Furthermore, it can be observed that in four of the ten series the first-column values are largest, and that in five cases the first-row values are largest. This may reflect that the observations and shocks in the first quarter play a dominant role in driving the various macroeconomic variables.

Estimating the stochastic trend

The expressions above concern the accumulations of shocks ε_t and their long-run impact on the $Y_{s,\mathrm{T}}$ series. Of course, it may sometimes be useful to obtain an explicit expression for the stochastic trend in a PIAR process. In the next chapter we will see that this can be useful when one wants to construct multivariate models for PIAR time series. As discussed in Section 2.4, there are several methods of calculating the stochastic trend.

In the case of PIAR processes, i.e. PAR processes with a single unit root, it may be convenient to consider the Gonzalo–Granger (1995) method, which results in $\gamma_\perp'\mathbf{Y}_\mathrm{T}$ as the stochastic trend, where $\gamma_\perp$ is the (4×1) vector which is orthogonal to the adjustment parameters γ in (8.13). Consider again the PIAR(1) process $y_t = \alpha_s y_{t-1} + \varepsilon_t$ with $\alpha_1\alpha_2\alpha_3\alpha_4 = 1$, for which it holds in (8.8) that

$$\Pi = \begin{bmatrix} -1 & 0 & 0 & \alpha_1 \\ 0 & -1 & 0 & \alpha_1\alpha_2 \\ 0 & 0 & -1 & \alpha_1\alpha_2\alpha_3 \\ 0 & 0 & 0 & 0 \end{bmatrix} \qquad (8.81)$$

and that $\gamma_\perp' = (0,0,0,1)$ since for (8.81) it holds that $\gamma_\perp'\Pi = 0$. Hence, for the PIAR(1) process it appears that the $Y_{4,\mathrm{T}}$ series itself is the stochastic trend. For a PIAR(2) process, it can be shown that $\gamma_\perp' = (0,0,c_1,c_2)$, with c_1 and c_2 unknown parameters. Given that c_2 can be normalized to unity, and hence that this $\gamma_\perp'$ contains only the unknown parameter c_1 / c_2, we may apply simple techniques to estimate $\gamma_\perp'$ from an estimate of γ. The latter can simply be found using the equality $\gamma = \Pi\alpha(\alpha'\alpha)^{-1}$.

The stochastic trend from a PIAR process can be used to calculate the impulse response function (IRF), which takes into account that there a single unit root in the vector process $\mathbf{Y}_\mathrm{T}$ (see Breitung and Franses 1995). Using the empirical IRF, one may visualize that certain shocks in some seasons have a different long-run impact than similar shocks in other seasons. This can be compared to the matrix in (8.73). A less formal approach is taken in Franses (1993), where an approximation to the IRF is visualized by depicting the effect of an innovative outlier on a simulated PIAR time series, where the parameters are set equal to the estimated values from a subset PIAR process.

One-step-ahead forecasting

In Table 8.9, we saw that the estimated ACFs of $(1 - \alpha_s B)y_t$ and $(1 - B)y_t$ can be quite different, even though the estimated α_s parameters are close to unity. A natural question now is whether imposing the restriction $\alpha_s = 1$ in a PAR(p) model affects the one-step-ahead forecasts. For that purpose I will present some results from a Monte Carlo experiment and from some of the running sample series.

Table 8.14 reports the results of a Monte Carlo experiment in which the relative forecasting performance of PIAR and PARI models is compared when each of these models can be the DGP. We use the six DGPs of Table 8.6 to generate 216 observations. The first 120 observations are used to estimate a PIAR and a PARI model and the last 96 are used to evaluate forecasts. The

TABLE 8.14. Forecasting performance of PIAR and PARI models, based on 5,000 replications[a,b]

	DGP	No. of forecasts	Quarters				All
			1	2	3	4	
PIAR	(i)	96	53.62	90.76	51.80	61.80	86.96
		48	49.74	82.66	47.44	60.18	82.58
		12	54.66	45.12	65.04	47.40	73.20
	(ii)	96	64.44	50.38	46.90	40.52	51.20
		48	62.62	51.98	46.62	42.68	52.42
		12	56.76	51.52	46.90	45.72	52.74
	(iii)	96	97.80	96.64	90.40	83.24	99.90
		48	92.16	91.56	76.16	74.92	99.50
		12	69.36	70.89	53.18	57.76	94.38
PARI	(iv)	96	42.16	42.96	43.64	42.26	17.76
		48	45.54	45.20	46.84	46.36	25.30
		12	48.38	48.68	48.74	48.06	37.58
	(v)	96	40.94	42.92	40.78	43.74	17.42
		48	43.82	45.54	44.82	46.18	25.76
		12	48.02	47.86	48.78	49.30	39.22
	(vi)	96	41.56	43.32	42.00	42.18	17.22
		48	45.20	46.08	45.78	44.54	25.48
		12	48.54	48.90	48.92	48.14	37.20

[a] The cells report the relative frequencies that a PIAR model forecasts better than a PARI model, based on the mean squared prediction error of the one-step-ahead forecasts. The sample size for estimation is 120.

[b] The DGPs are given in the notes to Table 8.6. In each replication 216 observations are generated. The first 120 observations are used to estimate the models, while the last 96 observations are used for forecast evaluation. When the DGP is a PARI(2), I estimate PIAR(2) and PARI(2) models. When the DGP is a PIAR(2), I estimate PIAR(2) and PARI(k) models, where the k is found using the $F_{\mathrm{PeAR},1-1}$ test.

cells of Table 8.14 give the relative frequency that the PIAR model forecasts better than the PARI model. The first three DGPs correspond to a PIAR process. Especially for the DGPs (i) and (iii), where one of the α_s differs substantially from unity, we can observe that it does matter if one selects the wrong model. For the third DGP the PIAR model forecasts better in more than 90 per cent of the cases, while for the second DGP, where the α_s values are near unity, there is not so much difference in forecasting performance. In the second panel, I give the results in the case where a PARI process is the DGP. The outcomes for each of these DGPs is roughly the same. A PIAR model forecasts better in each quarter in about 45 per cent of the cases. However, the forecasting performance of PIAR models decreases the more forecasts one evaluates. Furthermore, note that, when one evaluates the forecasts within each season, one obtains different outcomes than when one evaluates the forecasts for all seasons jointly. In the latter case, the PIAR model may be better only in between 17 and 40 per cent of the cases. In sum, the results in Table 8.14 indicate that it seems worse to use a PARI model when a PIAR model is appropriate than it is the other way around.

To investigate whether the simulation results in Table 8.14 carry over to empirical time series, consider the forecast comparison results in Table 8.15. In this table I compare PIAR and PARI models for six of our sample series. The models are estimated until 1970.4, 1976.4, and 1982.4, and 72, 48, and 24 one-step-ahead forecasts are generated. Table 8.15 reports whether the PIAR model is better, i.e. whether the mean squared error of prediction is smaller ('–'), or whether the PARI model is better ('+'). The significance of the differences in forecasting ability is checked using a nonparametric sign-test. With a significance level of 10 per cent, we can see from the table that the differences between the two models are usually insignificant. When there is a significant difference, it is about equally likely that a PIAR or a PARI model is better. A noticeable exception is the total consumption series, for which the PIAR model is clearly better. Finally, it can be noticed that most significant differences appear in quarters 3 and 4.

Multi-step-ahead forecasting

PIAR models can also be used for multi-step-ahead forecasting. Such forecasts can be derived along similar lines as the multi-step-ahead forecasts generated from VAR models (see Lütkepohl 1991). It should be mentioned, though, that PIAR models have a feature that can appear attractive for forecasting: i.e., for the out-of-sample forecasts also, the seasonal patterns change. The periodic differencing filter $(1 - \alpha_s B)$ establishes that the differences between $\hat{Y}_{s,T+i}$ and $\hat{Y}_{s-1,T+i}$ are not constant for $s = 2,3,4$ and with increasing i. Hence the slowly changing seasonal patterns that can be observed within the sample will also

TABLE 8.15. Empirical forecasting comparison of PIAR versus PARI models for six quarterly UK macroeconomic variables[a,b]

Variable	No. of forecasts	Quarters				All
		1	2	3	4	
Total consumption	72	–	–	–	+*	+
	48	–*	–	–	–	–*
	24	–**	+	–*	–	–**
Nondurables consumption	72	–	–*	+	+**	–
	48	–	+	–	–	–
	24	+*	+	–	–**	–
Exports	72	–	+	–*	+**	+
	48	–	+	–*	+	+
	24	–	+*	–	+*	+
Imports	72	–	+	–	–	–
	48	–	–	+*	–	–
	24	–	–*	+*	–**	+*
Total investment	72	+	+*	–	–	+
	48	+	+	+	–*	+
	24	+	+	+*	–*	+
Workforce	72	+	+	–**	–	–**
	48	+	+	–**	–	+
	24	+*	+	–*	+	+

[a] The cells report whether the mean squared prediction error of one-step-ahead forecasts from the PIAR model is smaller (–) or larger (+) than that from the PARI model. Significant differences are checked using a nonparametric sign-test.

[b] The models are estimated until 1970.4, 1976.4, and 1982.4, and forecasts are generated for 1971.1–1988.4 (72), 1977.1–1988.4 (48), and 1983.1–1988.4 (24), respectively. The PIAR models are given in Table 8.8. The PARI models are of order 2, 2, 1, 1, 2, and 2, respectively.

** Significant at the 5 per cent level.
 * Significant at the 10 per cent level.

appear in the out-of-sample forecasts. Notice that this is not the case for the seasonally integrated model, i.e. when $\Delta_4 y_t = \varepsilon_t$, for which the differences between $\hat{Y}_{s,T+i}$ and $\hat{Y}_{s-1,T+i}$ remain constant. In a sense, this reflects that the seasonal integration model can describe changing seasonal patterns, but it does not extract information from the time series that can be used to generate such patterns out-of-sample.

For multi-step-ahead forecasting it may be useful to calculate the error variances of the forecasts. Here, I derive such error variances for a PIAR(2) process. To this end, it is useful to recognize that (8.71) is vector moving average process for $\mathbf{Y}_T$ (VARMA). The MA part of this model is of infinite order, as can be observed from rewriting that part as

$$\begin{aligned}
&\Phi_0^{-1}(1-\beta B)^{-1}(\Omega_0\varepsilon_T+\Omega_1\varepsilon_{T-1})\\
&\quad=(1+\beta B+\beta^2B^2+\beta^3B^3+\ldots)\Phi_0^{-1}(\Omega_0\varepsilon_T+\Omega_1\varepsilon_{T-1})\\
&\quad=\Phi_0^{-1}\Omega_0\varepsilon_T+\Phi_0^{-1}\Omega_1\varepsilon_{T-1}+\beta(\Phi_0^{-1}\Omega_0\varepsilon_{T-1}+\Phi_0^{-1}\Omega_1\varepsilon_{T-2})\\
&\qquad+\beta^2(\Phi_0^{-1}\Omega_0\varepsilon_{T-2}+\Phi_0^{-1}\Omega_1\varepsilon_{T-3})+\beta^3(\Phi_0^{-1}\Omega_0\varepsilon_{T-3}+\Phi_0^{-1}\Omega_1\varepsilon_{T-4})+\ldots\\
&\quad=\Pi_0\varepsilon_T+\Pi_1\varepsilon_{T-1}+\beta\Pi_1\varepsilon_{T-2}+\ldots+\beta^{i-1}\Pi_1\varepsilon_{T-i},
\end{aligned} \tag{8.82}$$

with

$$\begin{aligned}
\Pi_0&=\Phi_0^{-1}\Omega_0\\
\Pi_1&=\Phi_0^{-1}\Omega_1+\beta\Phi_0^{-1}\Omega_0,
\end{aligned}$$

where Π_0 is a lower triangular matrix. In the PIAR(1) case it holds that $\Pi_0 = \Phi_0^{-1}$, $\Pi_1 = 0$, and μ^{**} equals $\Phi_0^{-1}\mu$. When we combine (8.71) and (8.82), the PIAR(2) process can be written as

$$\mathbf{Y}_T=\Gamma\mathbf{Y}_{T-1}+\mu^{**}+\Pi_0\varepsilon_T+\Pi_1\varepsilon_{T-1}+\beta\Pi_1\varepsilon_{T-2}+\ldots+\beta^{i-1}\Pi_1\varepsilon_{T-i}. \tag{8.83}$$

This expression can easily be used to derive the infinite VMA representation for the $\mathbf{Y}_T$ vector process. In fact, recursively substituting lagged $\mathbf{Y}_T$ in (8.83), while taking account of the fact that $\Gamma^m = \Gamma$, yields

$$\mathbf{Y}_T=\mathbf{Y}_0^*+\Gamma\mu^{**}T+\sum_{i=0}^{T-1}\Xi_i\varepsilon_{T-i}, \tag{8.84}$$

where

$$\begin{aligned}
\mathbf{Y}_0^*&=\Gamma\mathbf{Y}_0+(\mathbf{I}_4-\Gamma)\mu^{**}\\
\Xi_0&=\Pi_0\\
\Xi_1&=(\Gamma\Pi_0+\Pi_1)=\Gamma\Xi_0+\Pi_1\\
\Xi_2&=(\Gamma\Pi_0+\Gamma\Pi_1+\beta\Pi_1)=\Gamma\Xi_1+\beta\Pi_1\\
&\ldots\\
\Xi_i&=\Gamma\Xi_{i-1}+\beta^{i-1}\Pi_1,
\end{aligned}$$

where T is the deterministic trend and $\mathbf{Y}_0$ is the starting value of the $\mathbf{Y}_T$ process. The expression in (8.84) will be used below.

Suppose that we calculate multi-step-ahead forecasts in year N; hence the starting-point is the fourth quarter in year N. For the PIAR(2) process, we have

$$\begin{aligned}
\hat{\mathbf{Y}}_{N+1}-\mathbf{Y}_{N+1}&=\Xi_0\varepsilon_{N+1}\\
\hat{\mathbf{Y}}_{N+2}-\mathbf{Y}_{N+2}&=\Xi_0\varepsilon_{N+2}+\Xi_1\varepsilon_{N+1}\\
&\ldots\\
\hat{\mathbf{Y}}_{N+h}-\mathbf{Y}_{N+h}&=\Xi_0\varepsilon_{N+h}+\Xi_1\varepsilon_{N+h-1}+\ldots+\Xi_{h-1}\varepsilon_{N+1}.
\end{aligned} \tag{8.85}$$

Given that $\varepsilon_T \sim N(0,\sigma^2\mathbf{I}_4)$, it is obvious that

$$E(\hat{\mathbf{Y}}_{N+h}-\mathbf{Y}_{N+h})=0 \quad \text{for all } h. \tag{8.86}$$

Furthermore, since ε_T is independent from ε_{T+k} for any $k \neq 0$, we easily arrive at

$$\begin{aligned}
E(\hat{\mathbf{Y}}_{N+1}-\mathbf{Y}_{N+1})^2&=\sigma^2\Xi_0\Xi_0'\\
E(\hat{\mathbf{Y}}_{N+2}-\mathbf{Y}_{N+2})^2&=\sigma^2(\Xi_0\Xi_0'+\Xi_1\Xi_1')\\
&\ldots\\
E(\hat{\mathbf{Y}}_{N+h}-\mathbf{Y}_{N+h})^2&=\sigma^2\sum_{k=0}^{h-1}\Xi_k\Xi_k'.
\end{aligned} \tag{8.87}$$

The diagonal elements of the matrices in (8.87) can now be used to calculate the forecast error variances of y_t. The assumption of normality of ε_t allows us to construct the conventional 90 or 95 per cent forecast intervals.

To highlight some specific properties of the forecast error variances from PIAR processes, consider again the PIAR(1) process, for which we obtain

$$\hat{\mathbf{Y}}_{N+1} - \mathbf{Y}_{N+1} = \Phi_0^{-1}\varepsilon_{N+1}, \tag{8.88}$$

which can be decomposed as

$$\begin{aligned}
\hat{Y}_{1,N+1} - Y_{1,N+1} &= \varepsilon_{1,N+1}\\
\hat{Y}_{2,N+1} - Y_{2,N+1} &= \alpha_2\varepsilon_{1,N+1} + \varepsilon_{2,N+1}\\
\hat{Y}_{3,N+1} - Y_{3,N+1} &= \alpha_2\alpha_3\varepsilon_{1,N+1} + \alpha_3\varepsilon_{2,N+1} + \varepsilon_{3,N+1}\\
\hat{Y}_{4,N+1} - Y_{4,N+1} &= \alpha_2\alpha_3\alpha_4\varepsilon_{1,N+1} + \alpha_3\alpha_4\varepsilon_{2,N+1} + \alpha_4\varepsilon_{3,N+1} + \varepsilon_{4,N+1}.
\end{aligned}$$

Obviously, it holds that

$$E(\hat{Y}_{s,N+1} - Y_{s,N+1}) = 0, \quad s = 1,2,3,4.$$

Furthermore, the squared prediction errors for the forecasts for observations in season s in year $N + 1$ ($SPE_{s,N+1}$) are the diagonal elements of

$$E(\hat{\mathbf{Y}}_{N+1} - \mathbf{Y}_{N+1})^2 = \sigma^2\, \Phi_0^{-1}(\Phi_0^{-1})'; \tag{8.89}$$

i.e., these are

$$\begin{aligned}
SPE_{1,N+1} &= \sigma^2\\
SPE_{2,N+1} &= (\alpha_2^2 + 1)\sigma^2\\
SPE_{3,N+1} &= (\alpha_2^2\alpha_3^2 + \alpha_3^2 + 1)\sigma^2\\
SPE_{4,N+1} &= (\alpha_2^2\alpha_3^2\alpha_4^2 + \alpha_3^2\alpha_4^2 + \alpha_4^2 + 1)\sigma^2.
\end{aligned}$$

Comparing these expressions for the SPEs with those of a nonperiodic random walk process $y_t = y_{t-1} + \varepsilon_t$, which are σ^2, $2\sigma^2$, $3\sigma^2$, and $4\sigma^2$, it is clear that a PI process allows the forecast intervals to vary with the seasons. This property reflects the seasonal heteroskedasticity in the PIAR process that is present within sample. Note that the SPE is smallest in the first quarter since all forecasts are generated from quarter 4 in year N onwards. In fact, if one generates forecasts from quarter s in year N, the SPE in quarter $s + 1$ will be smallest, $s = 1,2,3$.

It can further be derived that the PIAR(1) forecast errors for h years ahead are

$$\hat{\mathbf{Y}}_{N+h} - \mathbf{Y}_{N+h} = \Phi_0^{-1}\varepsilon_{N+h} + \Gamma\Phi_0^{-1}\sum_{j=1}^{h-1}\varepsilon_{N+j},$$

and that, for $h = 2,3,\ldots$,

$$E(\hat{\mathbf{Y}}_{N+h} - \mathbf{Y}_{N+h})^2 = \sigma^2[\Phi_0^{-1}(\Phi_0^{-1})' + (h-1)(\Gamma\Phi_0^{-1})(\Gamma\Phi_0^{-1})'], \tag{8.90}$$

and hence that the squared prediction errors $SPE_{s,N+h}$, based on quarter 4 in year N, are

$$\begin{aligned}
SPE_{1,N+h} &= \sigma^2 + (h-1)(1 + \alpha_1^2\alpha_3^2\alpha_4^2 + \alpha_1^2\alpha_4^2 + \alpha_1^2)\sigma^2\\
SPE_{2,N+h} &= (\alpha_2^2 + 1)\sigma^2 + (h-1)(\alpha_2^2 + 1 + \alpha_1^2\alpha_2^2\alpha_4^2 + \alpha_1^2\alpha_2^2)\sigma^2\\
SPE_{3,N+h} &= (\alpha_2^2\alpha_3^2 + \alpha_3^2 + 1)\sigma^2 + (h-1)(\alpha_2^2\alpha_3^2 + \alpha_3^2 + 1 + \alpha_1^2\alpha_2^2\alpha_3^2)\sigma^2\\
SPE_{4,N+h} &= (\alpha_2^2\alpha_3^2\alpha_4^2 + \alpha_3^2\alpha_4^2 + \alpha_4^2 + 1)\sigma^2 + (h-1)(\alpha_2^2\alpha_3^2\alpha_4^2 + \alpha_3^2\alpha_4^2 + \alpha_4^2 + 1)\sigma^2.
\end{aligned}$$

For PIAR processes of orders 3 and higher, one can derive the expressions for the forecast error variances along similar lines as above. A useful strategy then is to write the models in VMA format to facilitate the construction of the expressions for the error variances.

To illustrate the construction of error variances for multi-step-ahead forecasts, consider Table 8.16, which gives the estimated error variances for a PIAR(1) model for UK nondurables consumption for seven years of out-of-sample forecasts. Hence the model in Table 8.8 is re-estimated using $n - 28$ observations. It can be seen from Table 8.16 that the SPEs of the PIAR(1) process for nondurables consumption show seasonal patterns. In fact, it appears that, relative to the first and fourth quarters, the time series can be forecast more precisely in the second and third quarters.

TABLE 8.16. Forecast error variances for a PIAR(1) process for UK consumption nondurables[a]

	Quarter			
	1	2	3	4
1982	$1.000\sigma^2$	$1.869\sigma^2$	$2.983\sigma^2$	$4.222\sigma^2$
1983	$5.243\sigma^2$	$5.552\sigma^2$	$6.890\sigma^2$	$8.438\sigma^2$
1984	$9.480\sigma^2$	$9.236\sigma^2$	$10.798\sigma^2$	$12.656\sigma^2$
1985	$13.729\sigma^2$	$12.920\sigma^2$	$14.705\sigma^2$	$16.875\sigma^2$
1986	$17.972\sigma^2$	$16.604\sigma^2$	$18.614\sigma^2$	$21.093\sigma^2$
1987	$22.216\sigma^2$	$20.288\sigma^2$	$22.522\sigma^2$	$25.311\sigma^2$
1988	$26.450\sigma^2$	$23.975\sigma^2$	$26.435\sigma^2$	$29.537\sigma^2$

[a] Estimated for the sample 1955.1–1981.4; multi-step-ahead forecasts are generated for 1982.1–1988.4.

Figure 8.3 displays the multi-step-ahead forecasts yf_t for 1982.1–1988.4 generated from the PIAR(1) model. The (unreported) parameter estimates show that the parameters for this smaller sample are fairly close to those reported in Table 8.8. In addition to the yf_t series, I depict the y_t series itself and the 95 per cent forecast confidence intervals. It is clear that all 28 forecasts lie within this region, although the forecast for 1988.4 is close to the boundary.

In Table 8.17 the estimated forecast error variances are displayed for the PIAR(2) process for German real GNP. Again, I withhold the last seven years of observations for forecast evaluation. The intervals in this table show a marked seasonal pattern. It can be observed that, relative to the other quarters, the forecasts for the first and fourth quarters become increasingly less precise. This reflects the dominance of the accumulation of shocks in these quarters.

Figure 8.4 displays the 28 forecasts and forecast intervals. It is clear that the out-of-sample forecasts are well within the 95 per cent boundaries. In fact, it can be calculated that, even for a 75 per cent confidence interval, the true

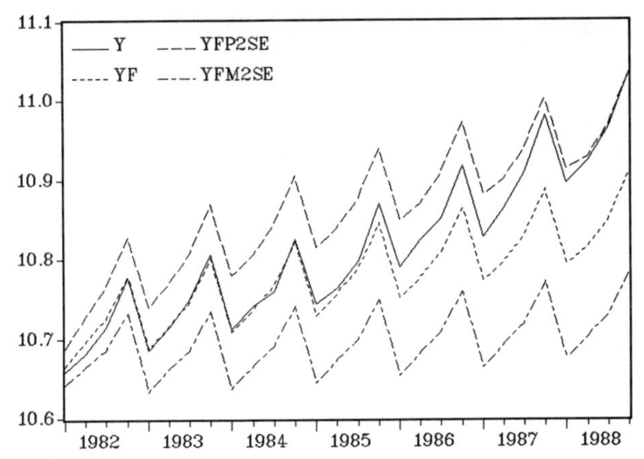

FIG. 8.3 *Out-of-sample forecasts for a PIAR (1) process: UK nondurables consumption, 1982–1989.*
The y is the time series, yf is the forecasted time series, and yf ± 2 s.e. indicates the 95 per cent forecast interval.

TABLE 8.17. Forecast error variances for a PIAR(2) process for German GNP[a]

	Quarter			
	1	2	3	4
1984	$1.000\sigma^2$	$1.084\sigma^2$	$2.546\sigma^2$	$3.649\sigma^2$
1985	$5.416\sigma^2$	$5.154\sigma^2$	$5.666\sigma^2$	$7.741\sigma^2$
1986	$9.765\sigma^2$	$9.113\sigma^2$	$8.815\sigma^2$	$11.840\sigma^2$
1987	$14.113\sigma^2$	$13.070\sigma^2$	$11.963\sigma^2$	$15.939\sigma^2$
1988	$18.461\sigma^2$	$17.027\sigma^2$	$15.112\sigma^2$	$20.038\sigma^2$
1989	$22.809\sigma^2$	$20.985\sigma^2$	$18.261\sigma^2$	$24.137\sigma^2$
1990	$27.157\sigma^2$	$24.942\sigma^2$	$21.409\sigma^2$	$28.237\sigma^2$

[a] Estimated for the sample 1960.1–1983.4; multi-step ahead forecasts are generated for 1984.1–1990.4.

observations do not exceed the boundaries. Only when the confidence interval is based on one standard error does the observation in 1987.1 not lie within this interval.

Structural mean shifts

In Chapter 7 it was argued that periodic variation in the AR parameters may be caused by neglecting shifts in mean. Furthermore, in Chapter 5 I documented for the German real GNP series that the finding of seasonal unit roots

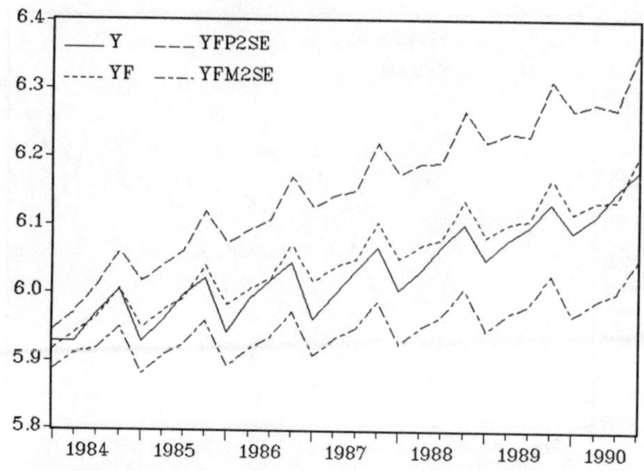

FIG. 8.4 *Out-of-sample forecasts for a PIAR(2) process: real GNP in Germany, 1984–1990.*
The y is the time series, yf is the forecasted time series, and yf ± 2 s.e. indicates the 95 per cent forecast interval.

was not robust to variation in the estimation sample. In this Chapter I find that a PIAR(2) process yields a useful description of the data. However, it may now be the case that the slowly changing seasonal patterns, which can effectively be described by a PIAR process, have been caused by deterministic shifts instead of by stochastic variation. Given that the α_s estimates are found to be very close to unity (see Table 8.8), it seems reasonable to compare the PIAR model with a model that considers the first-order-differenced time series $\Delta_1 y_t$ when it can be described using a periodic autoregression that allows for a one-time structural break at given time T_b, such as, for example,

$$(1 - B)y_t = \delta_s + \delta_s^* DU_{s,t} + \theta_s D(T_b)_{s,t} + \varepsilon_t. \tag{8.91}$$

The variables $DU_{s,t}$ and $D(T_b)_{s,t}$ are defined as in (5.13). The δ_s^* parameters have values that may be different from zero, which implies that the growth rates can change at time T_b. Franses and McAleer (1995*b*) propose several nested and nonnested procedures to select between a PIAR process and (8.91). To save space, I consider only the nested selection procedure here.

Given that the PIAR process for German GNP is of order 2, consider the nesting model

$$(1 - \alpha_s B)y_t = \beta_{1s}(1 - \alpha_{s-1}B)y_{t-1} + \kappa_s + \kappa_s^* DU_{s,t} + \theta_s^* D(T_b)_{s,t} + \varepsilon_t, \tag{8.92}$$

which is estimated subject to the restriction $\alpha_1\alpha_2\alpha_3\alpha_4 = 1$. This model may not only correspond to a convenient method of testing PIAR versus PARI with breaks; it might also be an empirically useful model. The validity of (8.92) rests in its ability to accommodate both periodic integration and a structural break. This general model, in which all variables are periodically stationary, may be estimated using nonlinear least squares, which yields a residual sum of squares RSS_G.

The PIAR model implies that there is no structural break, i.e. that the κ_s^* parameters in (8.92) are equal to zero. Since the κ_s^* are coefficients of stationary variables, the residual sum of squares of the PI model, say RSS_{PI}, can be compared with RSS_{G} from (8.91) using the standard asymptotic F-test. This test is denoted as $F_{\mathrm{PI}}(4,n-k)$, where k is the number of parameters in the general model and PIAR is the null hypothesis. The model in (8.91) implies that each of the α_s is equal to one, in which case the nonlinear restriction is satisfied automatically. Hence, comparing these models does not involve a change in the number of unit roots. When each α_s is equal to one, the κ_s and κ_s^* in (8.92) equal δ_s and δ_s^*, respectively. Denoting the residual sum of squares from estimating (8.91) by ordinary least squares as RSS_{I}, this model can be tested against (8.92) using the standard asymptotic F-test, denoted as $F_{\mathrm{I}}(3,n-k)$. The interpretation to be given to the test statistics is straightforward. A significantly large value of $F_{\mathrm{PI}}(F_{\mathrm{I}})$ leads to the rejection of the PI(I) model in favour of the general model with both periodic integration and a structural break. Insignificant test statistics in each of these three cases leads to non-rejection of the respective null hypotheses.

When T_b is not known, one should determine the value of T_b empirically. Suppose that a PI model is estimated and we want to check whether (8.91) is more appropriate. The value of T_b may be determined by applying parameter constancy tests to the PIAR model. Such a procedure necessarily involves a series of tests, which will affect the overall significance level of the tests in an unknown way. An alternative proposal is to use the likelihood ratio test statistic,

$$CHOW = n\log(RSS/(RSS_1 + RSS_2)), \tag{8.93}$$

for each subsample defined by choices for T_b, where RSS is the residual sum of squares that corresponds to the entire sample, while RSS_1 and RSS_2 correspond to the first and second subsamples. This Chow statistic is a test for coefficient stability conditional on variance equality. In the case of an unknown change point, it is possible to use the test statistic $maxCHOW$, which is the largest Chow value over all possible T_b. The distribution of $maxCHOW$ is derived in Andrews (1993) for a wide class of models, including nonlinear models.

An application of the test statistic in (8.93) to the estimated PIAR model of order 2 for German real GNP for T_b values that equal the last quarter of the years 1966–84 results in a value of 38.561 for the $maxCHOW$ statistic for $T_b =$ 1970.4. Comparing this value with the critical values in Andrews (1993) indicates that it is significant at the 5 per cent level. When model (8.92) is estimated with T_b set equal to 1970.4, the $F_{\mathrm{I}}(3,103)$-test has a value of 7.595**, while the $F_{\mathrm{PI}}(4,103)$-test obtains a value of 2.737**. This suggests that the nesting model (8.92) seems to be preferred over the two rival models. However, since the F_{PI} value of 2.737 is not significant at the 1 per cent level, there is some indication that the PIAR model seems to yield a closer description of the data than does the PARI model with breaks.

To obtain additional confidence in the usefulness of the PIAR(2) model for German real GNP, I calculate the estimates of α_s recursively, and depict them in Fig. 8.5. From this figure it is clear that $\hat{\alpha}_s$ appears fairly constant over time, and also that these estimates are different from unity.

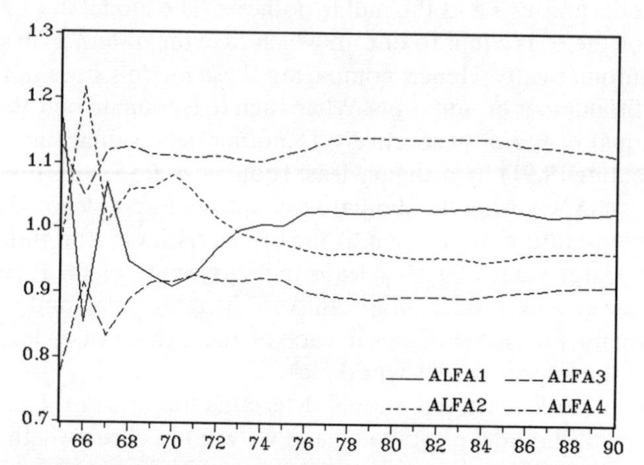

FIG. 8.5 *Forward recursive estimates of α_s in the periodic differencing filter in a PIAR(2) model for real GNP, Germany*

Temporal aggregation

The property of periodic integration can be shown to be preserved when a PIAR time series is temporally aggregated. For example, when a PIAR(1) process for y_t: $y_t = \alpha_s y_{t-1} + \varepsilon_t$ for a quarterly series is aggregated into a biannual series z_τ, where τ runs from 1 to $n/2$, i.e. $z_1 = y_1 + y_2$, $z_2 = y_3 + y_4$, . . ., it can easily be shown that this results in the PARMA(1,1) process

$$z_\tau = \eta_s z_{\tau-1} + v_\tau + \gamma_s v_{\tau-1}, \tag{8.94}$$

with

$$\eta_1 = [(\alpha_2 + 1)\alpha_1\alpha_4]/(\alpha_4 + 1)$$
$$\eta_2 = [(\alpha_4 + 1)\alpha_2\alpha_3]/(\alpha_2 + 1),$$

and hence that $\eta_1\eta_2 = 1$, provided that $\alpha_2 \neq -1$ and $\alpha_4 \neq -1$. Notice that, when $\alpha_2 = \alpha_4 = -1$, the z_τ process is white noise.

8.3 Effects of Nonperiodic Analysis

In this final section I investigate the effects of neglecting periodic parameter variation where a periodically integrated process is the DGP. Notice that I now consider the cases when nonperiodic models are fitted to periodically integrated time series. First, one may find too many seasonal unit roots, although the typical differencing filter will be equal to $(1 - B)$. Second, I study the effect of seasonal adjustment. Third, I investigate the impact of neglecting periodicity on business cycle analysis.

Seasonal and Nonseasonal Unit Roots

To illustrate the properties of nonperiodic models for PI time series, consider again the simple PIAR(1) process $y_t = \alpha_s y_{t-1} + \varepsilon_t$ with $\alpha_1\alpha_2\alpha_3\alpha_4 = 1$. Because this process is not (periodically) stationary, one cannot define its autocovariances, and hence I decide to analyse the properties of $\Delta_4 y_t$. It is easy to derive from (8.53) and (8.54) that, for example,

$$\begin{aligned}\gamma_{11} &= E(\Delta_4 y_t, \Delta_4 y_{t-1} \mid s = 1)\\ &= E(\Delta Y_{1,T}, \Delta Y_{4,T-1})\\ &= \sigma^2(\alpha_1 + \alpha_1\alpha_4^2 + \alpha_1\alpha_4^2\alpha_3^2).\end{aligned}$$

In general,

$$\gamma_{0s} = \sigma^2\{1 + \alpha_s^2[1 + \alpha_{s-1}^2(1 + \alpha_{s-2}^2)]\} \quad (8.95)$$
$$\gamma_{1s} = \sigma^2\alpha_s[1 + \alpha_{s-1}^2(1 + \alpha_{s-2}^2)] \quad (8.96)$$
$$\gamma_{2s} = \sigma^2\alpha_s\alpha_{s-1}(1 + \alpha_{s-2}^2) \quad (8.97)$$
$$\gamma_{3s} = \sigma^2\alpha_s\alpha_{s-1}\alpha_{s-2}, \quad (8.98)$$

and $\gamma_{ks} = 0$ for $k > 3$. This means that the misspecified nonperiodic model of $\Delta_4 y_t$ when y_t is a PIAR(1) is an MA(3) process. The parameters of this MA(3) model may be derived from

$$\bar{\gamma}_k = (1/4)\sum_{s=1}^{4}\gamma_{ks} \quad (8.99)$$

Notice that, when the restriction $\alpha_1\alpha_2\alpha_3\alpha_4 = 1$ does not hold, similar results as in (8.95)–(8.98) emerge for the time series $y_t - \alpha_1\alpha_2\alpha_3\alpha_4 y_{t-4}$. Hence the misspecified nonperiodic model of a PAR(1) is an ARMA(4,3) model (see also Osborn 1991).

Two simple examples may illustrate the usefulness of the expressions in (8.95)–(8.98). The first is where $\alpha_s = 1$ for all $s = 1,2,3,4$. In that case it holds that $\bar{\gamma}_0 = 4\sigma^2$, $\bar{\gamma}_1 = 3\sigma^2$, and $\bar{\gamma}_2 = 2\sigma^2$, and $\bar{\gamma}_1 = \sigma^2$ for all s. These autocovariances define the MA(3) polynomial $(1 + B + B^2 + B^3)$. Hence in this case the PAR(1) can be written as $\Delta_4 y_t = (1 + B + B^2 + B^3)\varepsilon_t$, which reduces to the simple random walk process since the $(1 + B + B^2 + B^3)$ polynomial cancels from both sides. Another example is where $\alpha_1 = \alpha_2 = -1$ and $\alpha_3 = \alpha_4 = 1$ such that all $\bar{\gamma}_k = 0$ for $k > 0$. For this PIAR model it holds that the nonperiodic model for $\Delta_4 y_t$ seems simply white noise. Hence the nonperiodic model for y_t has four unit roots, while the underlying periodic process only has one unit root.

Generally it can be shown, in the case of periodically integrated time series, that the misspecified nonperiodic model for $\Delta_4 y_t$ will be an invertible MA(q)

process. Given that many tests for unit roots are based on AR models, it is conceivable that one can fit AR models to a Δ_4-differenced PI time series (see also the simulation results in Tables 7.15, 7.16 and 7.17). In other words, when we consider nonperiodic AR models for y_t, where y_t is a periodically integrated time series, we may often find too many (seasonal) unit roots. The simulation results in Table 8.11 seem to confirm this. Furthermore, these simulation results also indicate that one often finds the nonseasonal unit root in nonperiodic models for PIAR time series.

The graphs in Fig. 8.1 and 8.2 indicate that the UK nondurables series seems most appropriately transformed using the $(1 - \alpha_s B)$ filter. In fact, the $(1 - B)$ transformed time series in Fig. 8.2 suggest that these series are not stationary. The latter nonstationarities may be picked up by one or more seasonal unit roots. Indeed, the HEGY test results in Table 5.2 point to the presence of three seasonal unit roots. Since we have found in this chapter that a PIAR(1) yields a useful description of the nondurables data, the above analysis predicts that an MA(3) for the corresponding $\Delta_4 y_t$ may seem useful. Estimation of this model results in

$$\underset{}{\Delta_4 y_t} = \underset{(0.001)}{0.025} + \hat{\varepsilon}_t + \underset{(0.062)}{0.923}\hat{\varepsilon}_{t-1} + \underset{(0.073)}{0.789}\hat{\varepsilon}_{t-2} + \underset{(0.053)}{0.712}\hat{\varepsilon}_{t-3}, \qquad (8.100)$$

which passes the usual diagnostic checks except for normality. It should be mentioned that the inclusion of two dummy variables to remove outliers results in about the same estimation outcomes as in (8.100). The $F_{\text{PeAR},1-1}$-test is significant at the 10 per cent level. The roots of the MA polynomial are the reciprocals of -0.913 and $-0.005 \pm 0.883i$, which are obviously close to -1 and $\pm i$, respectively. Hence the MA polynomial in (8.100) is close to $(1 + B + B^2 + B^3)$, which suggests that the Δ_4 filter may yield an overdifferenced process. Note, however, that the $(1 - B)$ filter is not appropriate either. This again emphasizes that the concept of periodic integration seems a concept in between nonseasonal and seasonal integration.

To compare the performance of the PIAR(1) model and the nonperiodic MA model in (8.100), a forecasting experiment for various horizons is performed. For this purpose I estimate the PIAR(1) model in Table 8.8 and (8.100) for the samples 1955.1–1970.4, 1955.1–1976.4, and 1955.1–1982.4, and I generate one-step-ahead forecasts for the remaining periods. For each of these periods I calculate the mean squared prediction error. The results are displayed in Table 8.18. From this table it is clear that the PIAR model outperforms the nonperiodic model and that most of the forecasting gain is obtained in the second quarter. This corresponds to the finding that the α_2 parameter in that season is less close to unity than the other α_s parameters.

Nonseasonal unit roots

There are several possible implications of neglecting periodicity when testing for nonseasonal unit roots using, for example, the ADF test. The first is already

TABLE 8.18. Comparison of one-step-ahead forecasts from a PIAR(1) model for y_t and a nonperiodic MA(3) model for $\Delta_4 y_t$, where y_t is UK nondurables consumption[a]

Estimation sample	Forecast period	Quarter				All
		1	2	3	4	
1955.2–1970.4	1971.1–1988.4	–	–**	–**	+**	–
1955.2–1976.4	1977.1–1988.4	–**	–*	–	–	–**
1955.2–1982.4	1983.1–1988.4	–	–*	–	–*	–*

[a] A '+' ('–') indicates that the mean squared prediction error of the PIAR model is larger (smaller) than that of the nonperiodic MA model. Statistical significance relates to a non-parametric sign-test applied to the squared forecast errors.
** Significant at the 5 per cent level.
* Significant at the 10 per cent level.

mentioned above, i.e. that one may often find the adequacy of the $(1 - B)$ filter in the case of periodic integration. Hence, if there is a single unit root in the $\mathbf{Y}_T$ process and the differencing filter is $(1 - \alpha_s B)$ with $\alpha_s \neq \alpha$ for all s, an ADF test applied to a nonperiodic AR model for y_t is likely to indicate the presence of a nonseasonal unit root. The empirical validity of this conjecture can be observed from the HEGY test results in Table 5.2, where it is found that all 11 sample series have a nonseasonal unit root, while the results in Section 8.1 indicate that all series are more appropriately modelled as some PIAR process.

In the case of a periodically stationary process, one may expect to find more nonseasonal unit roots when an ADF test is considered for a nonperiodic model. This can simply be concluded from the fact that the ACF of a nonperiodic model for a PAR process can be derived from (8.99). For example, in the case of a PAR(1) process with parameters α_s, one may expect that the first-order nonperiodic autocorrelation of y_t is about an average of these α_s, i.e. that $\rho_1 \cong \frac{1}{4}\Sigma_{s=1}^{4}\alpha_s$. For several choices of α_s, it is obvious that this ρ_1 is much closer to unity than $\alpha_1\alpha_2\alpha_3\alpha_4$ is. Consider for example the ACF and PACF in Tables 7.15 and 7.16, where these are calculated for several PAR(1) time series, and where the estimated ρ_1 is closer to unity than the product of the α_s. All this suggests that in a sense the ADF test loses power. In practice, therefore, it seems worthwhile first to check for periodic variation in the AR parameters, and if there is such variation to use one of the *LR* tests given in Section 8.2.

Although seasonal adjustment does not affect the intrinsic periodicity in a time series, as can be observed from the fact that one can still estimate PAR models for the official SA series in Section 7.4, one may expect that the data-filtering method will cause the α_s parameters in the $(1 - \alpha_s B)$ filter in SA data to get closer to unity (see also Table 7.18). Hence the application of seasonal adjustment to a PIAR series is likely to lead to the finding of a nonseasonal unit root. To show that seasonal adjustment does not affect the unit root properties of the time series to the same extent, consider the empirical performance

TABLE 8.19. Estimated rejection frequencies of the LR_1 test for a unit root in a PAR(1) model for y_t and for $\hat{y}_t^{ns}$ [a]

Variable	Nominal size(%)	DGP[b]					
		(i)	(ii)	(iii)	(iv)	(v)	(vi)
y_t	5	86.5	91.7	48.1	38.4	4.7	4.3
	10	97.7	97.9	70.9	63.3	9.5	9.3
$\hat{y}_t^{ns}$	5	86.0	84.6	56.2	30.6	4.8	3.3
	10	97.1	96.6	78.4	53.6	10.0	7.9

[a] The latter series is the seasonally adjusted series obtained using the linear approximation to the Census X-11 filter in Table 4.1. The cells contain rejection frequencies for 1,000 replications and sample size 100.

[b] The DGPs (i)–(vi) are given below Table 7.15. The first four DGPs are periodically stationary processes, and the last two are periodically integrated. Hence the first four columns refer to the empirical power, while the last two refer to the empirical size.

of the LR_1 test in Table 8.19, when it is applied to the NSA and SA time series, when the data are generated from stationary and nonstationary PAR processes. These results indicate that the linear Census X-11 filter does not seem to affect the empirical size and power properties of the LR_1 test. This suggests that it seems appropriate in practice to use this test even for SA time series.

To illustrate the practical usefulness of the two-step procedure, which includes testing for a unit root and then for certain parameter restrictions, consider the PAR(2) models for the official SA time series for US industrial production and German real GNP. For the first series we obtain an LR_1 test value of 2.852 and an $LR_{1\tau}$ value of –1.689. Hence there appears to be a unit root in the (4×1) vector process $\mathbf{Y}_T^{ns}$. Conditional on this unit root, I obtain a value of 0.028 for the $F_{(1-B)}$ test statistic. Hence, while the NSA time series is periodically integrated, the SA time series can be described using a PARI(2) process. To confirm the periodicity in the latter time series, I find that the F_{PAR} test obtains the 5 per cent significant value of 3.443.

For the SA German real GNP series one can also fit a PAR(2) process. The LR_1 test in this case obtains a value of 0.031, and an $LR_{1\tau}$ test, the value of –0.176. In contrast to the US industrial production series, the $F_{(1-B)}$ test statistic obtains the highly significant value of 4.557. Hence for this SA time series there is still significant evidence that the time series can be described using a PIAR(2) process. Notice that such a result also emerges from some of the simulation experiments reported in Table 7.18.

Business Cycles and Seasonal Cycles

Business cycle analysis is typically performed using nonperiodic models. It seems interesting therefore to study the empirical findings for certain business cycle patterns within the light of the possibility that several economic time series are periodically integrated. First, I analyse the auxiliary models that are used in Beaulieu *et al.* (1992) and Miron (1994) (see also (3.7); i.e.,

$$\Delta_1 y_t = \delta_1 D_{1,t} + \delta_2 D_{2,t} + \delta_3 D_{3,t} + \delta_4 D_{4,t} + u_t. \tag{8.101}$$

In these studies, the R^2 of this auxiliary regression is taken as a measure of the 'seasonal cycle', and the variance of the estimated residuals $\hat{u}_t$ is taken as as measure for the 'business cycle'. When this regression (8.101) is fitted to several industrial production and other macroeconomic series across several countries, Beaulieu *et al.* (1992) find that the R^2 and the residual variance are positively correlated. This positive correlation is then interpreted as meaning that countries or industries with large business cycles also have large seasonal cycles.

The previous analysis of several empirical macroeconomic time series has indicated that the model in (8.101) usually does not yield an adequate data description. This is because the first-order differencing filter is regularly rejected against the periodic differencing filter $(1 - \alpha_s B)$. The impact of using (8.101) instead of a more appropriate PIAR model can simply be established. For that purpose it is useful to define a measure of variation in the values of α_s, to be denoted as $v = \text{variance}(\alpha_s)$. We can say that when v is large there is much seasonal variation in the data, and also that the link between the stochastic trend and this seasonal variation is strong. The strength of this link can be interpreted in the sense that the elements of the $\Gamma\Phi_0^{-1}$ matrix take values that are far from unity. In terms of the regression (8.101), we may expect that, when the $(1 - B)$ filter is imposed on a PIAR process with a large v value, the R^2 of the regression will be large. Moreover, given that the $(1 - B)$ filter becomes increasingly inappropriate when v increases, we may also expect that the variance of $\hat{u}_t$ increases with v

To investigate this conjecture, I perform a limited simulation experiment and report the results in Table 8.20. We generate data from a PIAR(1) process where the parameters are 1.05, 1/1.05, $1+\beta$, and $1/(1 + \beta)$ for the α_s, $s = 1,2,3,4$, respectively. When the value of β increases, the v increases. The regression (8.101) is performed when β ranges from 0.10 to 0.19. In Table 8.20 the R^2 and the variance of $\hat{u}_t$ from this regression are reported. It is clear from the numbers in this table that one can generate a positive relation between the R^2 and the variance of $\hat{u}_t$. This suggests that PIAR processes can also generate patterns using misspecified regression models such as (8.101).

Determining business cycle peaks and troughs

A further natural question for business cycle analysis is whether seasonally adjusting a PIAR time series can have any impact on determining business cycle peaks and troughs. For the seasonally adjusted US industrial production series in Table 6.3, it is found that there are often business cycle peaks in the first and

TABLE 8.20. Correlation between 'seasonal cycle' and business cycle'[a]

β	'Seasonal cycle': R^2 of regression	'Business cycle': variance of $\hat{u}_t$
0.10	0.535	1.224
0.11	0.575	1.254
0.12	0.616	1.287
0.13	0.658	1.321
0.14	0.701	1.356
0.15	0.745	1.393
0.16	0.789	1.432
0.17	0.833	1.472
0.18	0.877	1.513
0.19	0.922	1.555

[a] The DGP is $y_t = \alpha_s y_{t-1} + \varepsilon_t$, with $\varepsilon_t \sim N(0,1)$, $\alpha_1 = 1.05$, $\alpha_2 = 1/\alpha_1$, $\alpha_3 = 1+\beta$, $\alpha_4 = 1/\alpha_3$, and the auxiliary regression model is $\Delta_1 y_t = \delta_1 D_{1,t} + \delta_2 D_{2,t} + \delta_3 D_{3,t} + \delta_4 D_{4,t} + u_t$. Based on one simulation experiment of 120 observations.

third quarters. In this chapter I establish that the unadjusted production series can be described by a PIAR(2) process. It may now be interesting to see whether such seasonal variation in business cycle peaks can be caused by PIAR processes when these are analysed in SA form. A limited Monte Carlo experiment may shed some light on this issue.

The official Census X–11 seasonal adjustment procedure is a nonlinear procedure, since it not only filters the NSA series using sets of moving average filters, but also removes outlying observations in each round of applying the moving average filters. It is thus difficult to consider in a small simulation experiment the application of exactly the same adjustment filter as is used to construct the adjusted industrial production series considered in Chapter 4. Hence I rely on the linear approximation of this filter as it is given in Table 4.1. Theoretically, the application of a linear filter to a time series generated from a PIAR process does not remove periodicity. This is because this filter treats the observations in all seasons in a similar fashion. However, the application of a linear filter smooths the original time series, and it may therefore be expected that, for example, the α_s values in the differencing filter $(1 - \alpha_s B)$ are smoothed towards unity (see Table 7.18). It is not expected that the stochastic trend is affected by the moving average filter. In sum, a seasonally adjusted PIAR time series will still contain a stochastic trend, but in practice one may find less evidence of periodic parameter variation.

Given that the quarterly NSA industrial production index can be described using a PIAR(2), with parameter estimates given in Table 8.8, I generate data from a PIAR(2) process where the parameters take values as in Table 8.8 and where the ε_t process is drawn from a normal distribution with mean zero and standard deviation 0.017. Although the experiment can be built up along several lines, the following strategy is adopted. First, 2,400 drawings from a $N(0,(0.017)^2)$ distribution are generated. With the starting values x_1 and $x_2 = 0$, 2,398 observations from a PIAR(2) process are generated. Only the last 2,057 observations are used for further analysis and the first 341 observations are thus considered to be starting values. An application of the seasonal adjustment filter in Table 4.1 to the raw x_t series yields 2,001 seasonally adjusted series $\hat{x}_t^{ns}$. First-differencing these $\hat{x}_t^{ns}$ data gives 2,000 observations on $\Delta_1 \hat{x}_t^{ns}$, and these are used to determine peaks and troughs.

The results of this limited Monte Carlo experiment are reported in Table 8.21, where the frequencies of peaks and troughs occurring in certain seasons are reported. For this experiment I obtain 196 turning points. Of course, one may generate larger samples of $\hat{x}_t^{ns}$ or repeat the exercise several times. It can however be expected that the conclusions to be drawn from Table 8.21 are not likely to change dramatically. The theoretical frequency of a peak or a trough in each of the seasons is 49 since there are 196 turning points. The results in Table 8.21 show that the probability of a peak is highest in the first and third quarters, and that the probability of a trough is highest in the first quarter.

It is clear that these simulation results closely correspond to the empirical outcomes for SA and NSA industrial production data. As shown above, the NSA time series can be described using a PIAR(2) process. The estimation results for that process predict that peaks and troughs are more likely to occur in certain seasons. Even after seasonally adjusting such a process, one still finds seasonal variation in the turning points. One prediction from the simulation experiment is that a peak is likely to occur in the first and third quarters. This

TABLE 8.21. Dating turning points in a Census X–11 seasonally adjusted time series[a]

	Quarter			
	1	2	3	4
Peak	58	52	58	28
	(49)	(49)	(49)	(49)
Trough	69	47	46	34
	(49)	(49)	(49)	(49)

[a] The data are generated by a periodically integrated process. The cells report the frequency that a certain quarter is assigned to be a peak or a trough. The data-generating process is a PIAR(2) process with parameters as in the estimated PIAR(2) process for US industrial production. The data are seasonally adjusted using the linearized version of Census X-11, as given in Table 4.1. The numbers in parentheses are the theoretical values. The sample length is 2,000 observations in only one simulation experiment.

clearly corresponds to the findings in Table 6.3, where four peaks occur in quarter 3, while three occur in quarter 1.

The natural question now is whether there is an alternative method for dating turning points in the case where a PIAR process is adequate, i.e. when seasons and stochastic trend are interdependent. As is illustrated above, a PIAR process assumes that the stochastic trend can be removed using the periodic differencing filter $(1 - \alpha_s B)$, where the α_s parameters are restricted via $\alpha_1\alpha_2\alpha_3\alpha_4 = 1$. The time series $(1 - \hat{\alpha}_s B)y_t$ obviously displays deterministic seasonal variation because of the μ_s, and it is therefore important to remove this variation prior to business cycle analysis. A regression of $(1 - \hat{\alpha}_s B)y_t$ on four seasonal dummies results in a mean zero time series, which should be corrected for the overall growth of the time series y_t. Since a periodic model assumes different models for each of the seasons, it seems appropriate to take the mean of $\Delta_4 y_t$ as a measure for the overall growth. This $\Delta_4 y_t$ process is a stationary process. Although this mean is not constant over the seasons (see (8.2.8)), its average seems a reasonable measure for the growth in each of the quarters. We denote this corrected differenced process as $\Delta_p^y y_t$.

An application of this simple dating method to the unadjusted industrial production index, where a recession is defined as two consecutive quarters with negative values for $\Delta_p^y y_t$, yields the peaks and troughs as they are reported in the last two columns in Table 8.22. The main differences between the turning points from the SA data and those from the PIAR-based method concern the peaks for the 1969–70 and 1974–5 recessions. With the SA data one would assign 1969.3 and 1974.3 as peaks, while the PIAR method yields 1969.4 and 1974.2, which are exactly equal or one-quarter closer to the NBER peaks in 1969.4 and 1973.4, respectively. Hence, by using the PIAR method, not only does one obtain peaks that are close to the NBER dates, but also, two peaks in

TABLE 8.22. Peaks and troughs according to the NBER, and obtained using the first differenced SA data and a PIAR(2) model for the NSA data: US industrial production, 1960.1–1991.4

NBER		SA[a]		PIAR[a]	
Peak	Trough	Peak	Trough	Peak	Trough
1960.2	1961.1	1960.1 (–1)	1961.1 (0)	1960.1 (–1)	1961.1 (0)
		1966.4	1967.2	1966.4	1967.2
1969.4	1970.4	1969.3 (–1)	1970.4 (0)	1969.4 (0)	1970.4 (0)
1973.4	1975.1	1974.3 (+3)	1975.2 (+1)	1974.2 (+2)	1975.2 (+1)
		1979.1	1979.3		
1980.1	1980.3	1980.1 (0)	1980.3 (0)	1980.1 (0)	1980.3 (0)
1981.3	1982.4	1981.3 (0)	1982.4 (0)	1981.3 (0)	1982.4 (0)
1990.3	1991.2	1990.3 (0)	1991.1 (–1)	1990.3 (0)	1991.1 (–1)

[a] The numbers in parentheses are the leads and lags relative to the NBER dates.

quarter 3 now become assigned to a different quarter. Furthermore, the PIAR method does not yield a (spurious) peak in 1979.1 as does the SA method. It seems that the seasonal variation in peaks disappears when one uses the PIAR method, and hence it seems worthwhile to consider PIAR models even though this may result in estimating several parameters.

A further advantage of this PIAR-based method is that it does not require any forecasts for the raw series that are needed to construct the actual seasonally adjusted time series when one relies on (variants of) the Census X–11 method. On the other hand, a possible drawback of using $\Delta_p^y y_t$ may be that one has to estimate a time series model that is nonlinear in the parameters and can involve many parameters being estimated. This may be viewed as a serious drawback for monthly time series. However, (unreported) experience with several US monthly time series leads me to conclude that, for monthly time series also, one can fit simple PIAR processes and derive simple measures to determine peaks and troughs. Finally, given the interdependence of trend and seasons in the PIAR model, it seems useful to re-estimate the model once additional observations become available. Significant changes in the determination of peaks and troughs may then be interpreted in terms of parameter stability or the overall empirical adequacy of the PIAR model.

To summarize this section, I can conclude that nonperiodic models for time series that are periodically integrated may lead to several results. One may find too many seasonal unit roots, the power of the ADF test may decrease, and there may be seasonal variation in business cycle turning points. Notice that the latter finding does not imply that there is not such variation. In fact, when a PIAR process is empirically adequate, one has obtained evidence of such seasonal variation in peaks and troughs. Finally, Franses and Ooms (1995) show that the observed regularity in Section 6.3, i.e. correlated seasonal and nonseasonal components in SA data, can be generated by PIAR processes. The main conclusion, however, is that the reasonably simple PIAR model may yield a useful data description.

Conclusion

In this chapter I considered testing for unit roots in PAR models. On the basis of our proposed model selection strategy, it was found that our sample series all appear to be periodically integrated, i.e. that there is only a single stochastic trend driving these series and that seasonal fluctuations change only slowly over time. Periodic integration implies that the accumulation of shocks is interdependent with the seasonal pattern, and hence that one cannot meaningfully separate seasonal from nonseasonal components. Using Monte Carlo simulations, I have observed that PI time series can generate the empirical nonperiodic models found useful in earlier chapters. It should be noted that PIAR processes, together with the results in Chapter 7, can explain the outcomes obtained when neglecting periodicity. Moreover, PIAR models can improve forecasts and may yield more appropriate business cycle turning points. In sum, PIAR models are not only found to be more useful empirically, but they can also generate the various documented empirical regularities in seasonal time series.

This chapter focused on univariate time series models. Although economic theories may predict univariate PIAR processes (see Osborn 1988), univariate time series analysis usually serves as a first step towards multivariate modelling. In the next chapter I will consider such multivariate periodic models with unit roots.

9

Periodic Cointegration

This chapter deals with empirical periodic models for multivariate time series with one or more stochastic trends. In the previous two chapters we have seen that for univariate variables periodic time series models are most easily analysed in the vector of quarters (VQ) form. This amounts to investigating the properties of a quarterly observed time series y_t via an analysis of the properties of the (4×1) vector process $\mathbf{Y}_T$, which contains the observations in each of the quarters. As noted in Section 7.5, the analysis of the stationarity property of a periodic VAR (PVAR) model can be quite involved, even for the simple PVAR(1) model (see equation (7.51)). In this chapter I propose two reasonably simple alternative approaches to analysing a multivariate time series using a periodic model.

The outline of this chapter is as follows. In Section 9.1 I consider some notation and representation issues of multivariate periodic models. It is shown that several empirically useful models impose restrictions on the rank of the matrix containing the long-run relationships. Since it does not seem tractable in practice to formally investigate this rank with methods such as those given in Chapter 8, I continue in Section 9.2 with the analysis of common aspects in periodic time series. This approach starts with an analysis of univariate time series. In the case of periodically integrated time series, a useful common aspect can be called 'common periodic integration', indicating that for a set of variables it holds that they have similar relationships between the trend and the seasonal fluctuations. This section establishes that the UK GDP, total consumption, total investment, and exports series are common periodically integrated; i.e., a linear combination of the four variables is stationary and nonperiodic. This leads to the construction of a multivariate error correction model for these variables. To assess the stationarity of the PVAR model for the four variables, I propose to use a sequential method. In a first step we extract the stochastic trend from each of the series, and in a second step we test for cointegration among the four stochastic trends. The empirical usefulness of this strategy is evaluated using some Monte Carlo experiments. Franses and Paap (1995) document additional results. These experiments also suggest a useful method of extracting trends from a univariate periodically integrated time series.

In Section 9.3 I consider an alternative approach to the construction of a multivariate model for periodic time series. Part of the material is drawn from

Boswijk and Franses (1995). This approach starts with an analysis of a periodic cointegration model (PCM), i.e. an error correction model where the cointegrating parameters and the adjustment parameters are allowed to vary with the seasons. The model then involves so-called periodic cointegration relationships. We consider estimation methods and illustrate their practical usefulness by an analysis of the consumption of nondurables and disposable income in Sweden. The source of the data is given in the Data Appendix. We find that a reasonably simple PCM describes the data, and that significant error correction occurs only in the second and fourth quarters. When the same data are considered using seasonal cointegration techniques, which were discussed in Chapter 5, we do not find any long-run relations at the seasonal frequencies. Furthermore, when we consider the seasonally adjusted time series we find only weak evidence for zero frequency cointegration.

Section 9.4 establishes, via some simulation experiments, that business cycle turning points based on seasonally adjusted time series generated from a PCM can display significant seasonal variation in peaks and troughs.

9.1 Representation of Periodic VAR Models

Consider two quarterly observed time series, y_t and x_t, $t = 1,2,\ldots, n$, which can be represented by the VQ processes $\mathbf{Y}_{\mathrm{T}}$ and $\mathbf{X}_{\mathrm{T}}$, which contain the observations y_t and x_t when stacked as $(Y_{1,\mathrm{T}}, Y_{2,\mathrm{T}}, Y_{3,\mathrm{T}}, Y_{4,\mathrm{T}})'$ and $(X_{1,\mathrm{T}}, X_{2,\mathrm{T}}, X_{3,\mathrm{T}}, X_{4,\mathrm{T}})'$. One possible representation of a periodic VAR model for the bivariate time series $(y_t, x_t)'$ amounts to a periodic extension of the usual VAR process (see Section 2.2); i.e.,

$$\begin{bmatrix} y_t \\ x_t \end{bmatrix} = \begin{bmatrix} \phi_{1s,1} & \phi_{1s,2} \\ \phi_{1s,3} & \phi_{1s,4} \end{bmatrix} \begin{bmatrix} y_{t-1} \\ x_{t-1} \end{bmatrix} + \ldots + \begin{bmatrix} \phi_{ps,1} & \phi_{ps,2} \\ \phi_{ps,3} & \phi_{ps,4} \end{bmatrix} \begin{bmatrix} y_{t-p} \\ x_{t-p} \end{bmatrix} + \begin{bmatrix} u_t \\ w_t \end{bmatrix}, \tag{9.1}$$

where u_t and w_t are white noise error processes and $\phi_{is,j}$ are parameters that correspond to lag i, season s, and matrix element j. Kleibergen and Franses (1995) propose an alternative representation of a periodic VAR, which in some respects is somewhat more convenient. In this section I will continue to use (9.1) (see Lütkepohl 1991). The number of AR parameters in the PVAR(p) process in (9.1) is $16p$. In general, a PVAR(p) process for m variables concerns $4m^2p$ parameters. For example, a PVAR process of order 4 for four variables involves 256 parameters. Obviously, for practical purposes this number of parameters is usually too many.

To assess whether the bivariate process $z_t = (y_t, x_t)'$ has one or more stochastic trends, it is again useful to write (9.1) in the VQ format

$$\Lambda_0 \mathbf{Z}_{\mathrm{T}} = \Lambda_1 \mathbf{Z}_{\mathrm{T}-1} + \ldots + \Lambda_{\mathrm{P}} \mathbf{Z}_{\mathrm{T-P}} + \mathbf{v}_{\mathrm{T}}, \tag{9.2}$$

where $\Lambda_0, \Lambda_1, \ldots, \Lambda_{\mathrm{P}}$ are (8×8) matrices, $\mathbf{Z}_{\mathrm{T}}$ is the (8×1) vector process

$$\mathbf{Z}_{\mathrm{T}} = (Y_{1,\mathrm{T}}, Y_{2,\mathrm{T}}, Y_{3,\mathrm{T}}, Y_{4,\mathrm{T}}, X_{1,\mathrm{T}}, X_{2,\mathrm{T}}, X_{3,\mathrm{T}}, X_{4,\mathrm{T}})', \tag{9.3}$$

and $\mathbf{v}_T$ is an (8×1) vector process containing the stacked u_t and w_t error processes. As for the univariate case, it holds that $P = 1 + [(p-1)/4]$. For practical purposes this means that a PVAR(4) can be written as a VQ process of order 1. Notice that a PVAR(4) for two variables involves 64 parameters, which corresponds to the 64 parameters in the matrix $\Lambda_0^{-1}\Lambda_1$ in the rewritten version of (9.2), i.e.

$$\mathbf{Z}_T = \Lambda_0^{-1}\Lambda_1\mathbf{Z}_{T-1} + \Lambda_0^{-1}\mathbf{v}_T. \tag{9.4}$$

The presence of unit roots in the $\mathbf{Z}_T$ process can be investigated by solving the characteristic equation $|\Lambda_0 - \Lambda_1 z| = 0$. This is similar to investigating the rank of the matrix Π in the error correction form of (9.4), i.e.

$$\Delta\mathbf{Z}_T = \Pi\mathbf{Z}_{T-1} + \Lambda_0^{-1}\mathbf{v}_T. \tag{9.5}$$

See also Section 7.2 for univariate periodic time series.

The general model in (9.5) nests several bivariate models which can be considered for practical purposes. For notational convenience, I discuss only the models for the y_t time series variable. These models can all be viewed as imposing certain restrictions on the cointegrating vectors which are contained in the Π matrix in (9.5). When the rank of Π is zero, there are no such cointegration relations between the elements of $\mathbf{Z}_T$, and hence one may consider the model

$$\phi_s(B)\Delta_4 y_t = \gamma_s(B)\Delta_4 x_t + \varepsilon_t, \tag{9.6}$$

where $\phi_s(B)$ and $\gamma_s(B)$ are polynomials in B with periodic parameters, ε_t denotes a white noise process (possibly with seasonally varying variance), and the Δ_4 filter for quarterly series corresponds to the Δ filter for the annual time series $\mathbf{Z}_T$. Notice that model (9.6) assumes that there are eight stochastic trends in the bivariate system. On the other end, the rank of Π is 8, and one may specify a model as (9.6) in levels.

Typically, it can be expected for macroeconomic time series that the rank of Π is somewhere in between 0 and 8. To give some examples of possibly useful models, consider the following cases. When the rank of Π is equal to 7 and the cointegration relations are $Y_{s,T} - Y_{s-1,T}$ for $s = 2,3,4$ and $Y_{s,T} - X_{s,T}$ for $s = 1,2,3,4$, which in turn imply the stationary relations $X_{s,T} - X_{s-1,T}$ for $s = 2,3,4$, one may consider the error correction model for y_t as

$$\phi_s(B)\Delta_1 y_t = \gamma_s(B)\Delta_1 x_t + \psi_s(y_{t-1} - x_{t-1}) + \varepsilon_t, \tag{9.7}$$

where $(y_{t-1} - x_{t-1})$ is an error correction variable. A slightly modified version of (9.7) in case of univariate periodically integrated time series is

$$\phi_s(B)(y_t - \alpha_s y_{t-1}) = \gamma_s(B)(x_t - \phi_s x_{t-1}) + \psi_s(y_{t-1} - \kappa_s x_{t-1}) + \varepsilon_t, \tag{9.8}$$

which is estimated under the restrictions $\alpha_1\alpha_2\alpha_3\alpha_4 = 1$ and $\phi_1\phi_2\phi_3\phi_4 = 1$. Note that the seven cointegration relations between the elements of $\mathbf{Z}_T$ imply that only one of the κ_s values is unrestricted. Furthermore, when the α_s and ϕ_s are all different for all s, $\kappa_s = \kappa$ for all $s = 1,2,3,4$ only when $\alpha_s = \phi_s$ for all s. The model in (9.8) can be estimated using NLS. In practice, it the y_t and x_t may each appear to be PI time series, but in (9.8), for example, the α_s can be set equal to unity. A model similar to (9.8) is fitted to quarterly UK consumption and income series in Birchenhall *et al.* (1989).

The model in (9.8) can be called a periodic cointegration model since the cointegrating vector is $(1, -\kappa_s)$ and varies with the season, and since the adjustment parameter ψ_s is also periodic. To facilitate the classification of various multivariate models, I will denote any model with seasonal variation in the cointegrating relationships as a periodic cointegration (PC) model. The models with only periodic adjustment parameters will be called the periodic error correction model (PECM). A PCM that is an alternative to (9.8) is given by

$$\phi_s(B)\Delta_4 y_t = \gamma_s(B)\Delta_4 x_t + \psi_s(y_{t-1} - \kappa_s x_{t-1}) + \varepsilon_t, \tag{9.9}$$

which reduces to the Davidson *et al.* (1978) model when all $\kappa_s = 1$. This model assumes that the rank of Π in (9.5) is 4. When the Δ_4 filter for either y_t or x_t is replaced by the Δ_1 filter, this PC model assumes that the rank of Π is 7. Notice that (9.9) cannot have the Δ_1 filter for both the y_t and x_t series and that $\kappa_s \neq 1$ for all s. Hence the model in (9.9) requires that for at least one of the time series the Δ_4 filter is needed.

The seasonal cointegration model in Section 5.2 can also be formulated in terms of rank restrictions on the Π matrix in (9.5). When there is cointegration at the zero frequency within a periodic model for y_t and x_t, a useful bivariate model is then

$$\phi_s(B)\Delta_4 y_t = \gamma_s(B)\Delta_4 x_t + \psi_s[S(B)y_{t-1} - \alpha_{12}S(B)x_{t-1}] + \varepsilon_t, \tag{9.10}$$

where $S(B) = (1 + B + B^2 + B^3)$. When this $S(B)$ filter is replaced by $(1 - B + B^2 - B^3)$, the model in (9.10) corresponds to seasonal cointegration at the biannual frequency. As such, the model in (9.10) assumes that the rank of Π is equal to one. For the full seasonal cointegration model as in (5.19) and (5.20), it is assumed that this rank is equal to 4. Notice that when the α_{12} parameter in (9.10) is allowed to vary with the season this model becomes a periodic cointegration model with cointegration at the zero frequency. In the case of the $(1 - B + B^2 - B^3)$ filter, one may obtain a periodic cointegration model with seasonal cointegration at the frequency π (PCSC).

In sum, it appears that various models for a bivariate periodic time series imply a rank reduction of the Π matrix in (9.5). Hence when this rank is known one may evaluate certain models using simple test statistics. For example, if we assume that the rank of Π equals 4, one may compare a full cointegration model, a PCSC model, or a PC model as in (9.9) using *F*-type test statistics, which are asymptotically distributed as a standard *F*-distribution. This naturally follows from the results in Section 8.1 for univariate periodic time series.

The natural question now is how one can determine the rank of the matrix Π in the VQ(1) model in (9.5). Notice that in case of m time series, the Π matrix is of dimension $(4m \times 4m)$. As is indicated in Section 7.5, even for a PVAR(1) model for bivariate time series, the expression of the characteristic equation $|\Lambda_0 - \Lambda_1 z| = 0$ is already quite complicated. Hence for practical purposes it does not seem useful to derive such an explicit expression and to derive the non-linear restrictions on the parameters in a PVAR. For univariate time series this may be useful, as was seen in Chapter 8, but for multivariate time series it does not seem sensible. Furthermore, given the dimension of the Π matrix, it also does not seem useful to apply the Johansen (1991) cointegration method to the VQ(1) model in (9.5). It is by now well known that this method appears to be reasonably reliable in small systems, but that for large systems the empirical results may become less useful.

In this chapter I therefore propose two alternative, somewhat pragmatic, methods to analyse multivariate periodic time series. The first method assumes that the Λ_0 and Λ_1 matrices in (9.2) are like block-diagonal matrices, i.e. that the most relevant parameters for, say, y_t concern lagged y_t variables, while there are only a few parameters that link y_t to x_t. If so, we may first analyse the univariate series for unit root properties and extract the stochastic trend(s) from each of these series. In a second step, we can test for cointegration among these extracted stochastic trends. In a sense, this two-step approach, which is discussed in the next section, corresponds with the Engle–Granger (1987) two-step approach. A second possible approach is to straightaway estimate and test the PC or PECM models in (9.8) or (9.9); in Section 9.3 I will investigate the merits of this method.

9.2 Common Aspects in Periodic Integration

In this section I consider the case of several periodically integrated time series. The aim is to construct a model for the multivariate process. Since such a multivariate model can contain many parameters, it seems useful to investigate whether the univariate time series have certain common properties. The key properties of PI time series are the stochastic trend and periodic parameter variation. Hence in this section I study common trends and common periodicity. When two PI time series have these two properties in common, one can label this 'common periodic integration'.

Consider the availability of two time series, y_t and x_t, which both can be described using PIAR processes. The orders of the periodic AR models do not necessarily have to be equal, nor do the lengths of the autoregressive polynomials in each of the seasons have to be equal.

Definition. Two periodically integrated processes y_t and x_t are cointegrated when the linear combination $y_t - \theta_s x_t$ can be described using a periodically stationary process, where θ_s is a seasonally varying parameter.

The common trend property of two PIAR series implies that there is a single unity solution to the characteristic equation for a VQ model for vector process $\mathbf{Z}_T$.

A natural and simple test procedure seems to be given by evaluating the residuals from the OLS regression

$$y_t = \hat{\mu}_s + \hat{\theta}_s x_t + \hat{u}_t. \tag{9.11}$$

A first step is to estimate a PAR model for $\hat{u}_t$, and as a second step one can apply the LR_i test statistics proposed in Chapter 8. It can be expected that, under the null hypothesis of no cointegration between y_t and x_t, these LR_i statistics follow the asymptotic distribution derived in Phillips and Ouliaris (1990). Limited (unreported) Monte Carlo simulation evidence indicates that the empirical fractiles are very close to the asymptotic fractiles, and hence seems to confirm this conjecture.

Another approach may be to estimate a conditional error correction model such as

$$(1 - \alpha_s B)y_t = \psi_s(y_{t-1} - \theta_{s-1}x_{t-1}) + \varepsilon_t, \tag{9.12}$$

where ψ_s and θ_{s-1} correspond to seasonally varying adjustment and equilibrium parameters and where (9.12) is estimated under the restriction $\alpha_1\alpha_2\alpha_3\alpha_4 = 1$. In Section 9.3 I return to such a model as in (9.12).

An alternative approach is to isolate the stochastic trends from each of the univariate series, and to relate these extracted series in a next step of cointegration. There are several methods of extracting stochastic trends from a set of time series. In this section I analyse the properties of the methods proposed in Box and Tiao (1977) and Gonzalo and Granger (1995).

Extracting stochastic trends using the Box–Tiao method

One simple approach to extracting a stochastic trend from each of the (4×1) vector series $\mathbf{Y}_T$ and $\mathbf{X}_T$ is to use the Box–Tiao (1977) (BT) method. For a vector process of order 1, which is often found useful in practice for univariate time series (see Chapters 7 and 8), this method considers the eigenvalue problem

$$|\lambda I - (\mathbf{W}_T{}'\mathbf{W}_T)^{-1}\mathbf{W}_T{}'\mathbf{W}_{T-1}(\mathbf{W}_{T-1}{}'\mathbf{W}_{T-1})^{-1}\mathbf{W}_{T-1}{}'\mathbf{W}_T| = 0, \tag{9.13}$$

where $\mathbf{W}_T$ is either $\mathbf{Y}_T$ or $\mathbf{X}_T$ when corrected for their respective means. The eigenvector that corresponds to the largest eigenvalue yields the strongest non-stationary linear combination that can be constructed from the four $W_{s,T}$ series (see Box and Tiao 1977 for details). We can consider this linear combination to be the stochastic trend in the multivariate system for $\mathbf{W}_T$. Denoting these stochastic trends for $\mathbf{Y}_T$ and $\mathbf{X}_T$ as Y_T^{BT} and X_T^{BT}, we can compare these annually observed variables in a second step of cointegration analysis.

The obvious question now is which critical values to use for the test statistics in a second cointegration testing step. Table 9.1 reports the empirical fractiles and the empirical size of the Engle–Granger (1987) (EG) and Johansen and Juselius (1990) (JJ) cointegration testing methods, after having extracted the stochastic trend using the BT method. The EG method amounts to calculating the Durbin–Watson statistic (CRDW) and the (augmented) Dickey–Fuller test (CRDF) for the cointegrating regression of Y_T^{BT} on a constant and X_T^{BT}. The standard critical values are taken from MacKinnon (1991) for 25

TABLE 9.1. Empirical performance of cointegration tests in case the DGP consists of two independent PIAR(1) processes[a]

Method[b]		Empirical fractiles(%)			Empirical size(%)[c]		
		5	10	20	5	10	20
EG after BT	CRDW	1.33	1.15	0.93	6.28	12.26	21.70
	CRDF	−3.74	−3.35	−2.88	5.90	11.92	22.10
JJ after BT	$\text{Tr}(r \leq 1)$	8.76	6.97	5.19	6.10	11.10	21.88
	$\text{Tr}(r \leq 0)$	19.46	17.23	14.29	6.20	11.76	21.44
EG after GG	CRDW	1.12	0.96	0.79	2.28	5.72	12.88
	CRDF	−3.27	−2.96	−2.60	2.04	5.50	13.96
	CRDF[d]	−3.45	−3.11	−2.72	3.54	7.80	17.60
JJ after GG	$\text{Tr}(r \leq 1)$	6.89	5.78	4.44	2.20	5.62	15.78
	$\text{Tr}(r \leq 0)$	16.94	14.79	12.56	2.26	5.44	13.30
	$\text{Tr}(r \leq 1)$[d]	8.18	6.65	5.00	4.76	9.76	20.68
	$\text{Tr}(r \leq 0)$[d]	20.25	17.66	14.93	7.14	13.50	24.56

[a] DGP is $y_t = \phi_s y_{t-1} + \varepsilon_t$ and $x_t = \alpha_s x_{t-1} + v_t$, where ε_t, v_t are $N(0,1)$ and $\phi_1 = \alpha_1 = 1.25$, $\phi_2 = \alpha_2 = 0.8$, $\phi_3 = \alpha_3 = 2.0$, $\phi_4 = \alpha_4 = 0.5$. Based on 5,000 replications of effective sample size $n = 100$ or N=25.

[b] The methods and abbreviations are given in the text in Section 9.2. Unless otherwise indicated, the CRDF statistics are based on an auxiliary regression that contains no constant and no lags, and the JJ trace test (Tr) statistics are calculated for a VAR(1) process.

[c] The 'Empirical size' columns report on the rejection frequency when the standard critical values for nonperiodic models are used, which are given in MacKinnon (1991) and Table 8.10.

[d] For the CRDF statistic, the auxiliary regression contains no constant and one lag. The trace test statistics are calculated for a VAR(2) process of Y_T^{GG} and X_T^{GG}.

observations. The JJ method amounts to regressing ΔY_T^{BT}, ΔX_T^{BT}, Y_{T-1}^{BT}, and X_{T-1}^{BT} on a constant and lagged $\Delta Y_T^{BT}$and ΔX_T^{BT} variables, giving the (2×1) vectors of residuals $\mathbf{R}_{0,T}$ and $\mathbf{R}_{1,T}$ and the residual product matrices

$$\mathbf{S}_{ij} = (1/N) \sum_{T=1}^{N} \mathbf{R}_{i,T}\mathbf{R}'_{j,T}, \text{ for } i,j = 0,1. \tag{9.14}$$

The next step is to solve the eigenvalue problem

$$|\lambda \mathbf{S}_{11} - \mathbf{S}_{10}\mathbf{S}_{00}^{-1}\mathbf{S}_{01}| = 0, \tag{9.15}$$

which gives the eigenvalues $\hat{\lambda}_1 \geq \hat{\lambda}_2$ and the corresponding eigenvectors $\hat{\mathbf{v}}_1$ and $\hat{\mathbf{v}}_2$, in this case of only two variables. The two test statistics to check for the presence of cointegration are the maximal eigenvalue test and the trace test. For the case of two series of length 25, I use the critical values in Table 8.10, These fractiles will be used to evaluate the fractiles of the JJ trace test where it is considered for the estimated stochastic trends from PIAR models.

The first panel in Table 9.1 displays the empirical size and fractiles of the CRDW and CRDF as well as of the JJ trace test, to be denoted as $\text{T}r\ (r \leq 1)$ and $\text{T}r\ (r \leq 0)$, where r is the number of cointegrating relationships between

Y_T^{BT} and X_T^{BT}. The results for the maximal eigenvalue test are very similar, and are not reported to save space. The DGP in Table 9.1 consists of two independent PIAR(1) processes. Franses and Paap (1995) provide additional simulation results. As can be observed from the numbers in the first panel of the table, it seems that the empirical rejection frequencies are very close to the nominal sizes even for a sample size of 100 quarterly, i.e. 25 annual, observations. Hence it seems that I may use the standard critical values in a cointegration analysis of Y_T^{BT} and X_T^{BT}.

Extracting stochastic trends using the Gonzalo–Granger method

The Gonzalo–Granger (1995) (GG) method of extracting the stochastic trend exploits the duality between cointegration and stochastic trends (see also Section 2.4). For a (4×1) $\mathbf{W}_T$ series, where $\mathbf{W}_T$ is either $\mathbf{Y}_T$ or $\mathbf{X}_T$, the procedure is as follows. First, we regress $\Delta\mathbf{W}_T$ and $\mathbf{W}_{T-1}$ on a constant and lagged $\Delta\mathbf{W}_T$ variables. This gives the (4×1) residual vectors $\mathbf{R}_{0,T}$ and $\mathbf{R}_{1,T}$ and the (4×4) residual product matrices $\mathbf{S}_{ij}$, $i,j = 0,1$, calculated in the same way as (9.14). The next step is to solve the eigenvalue problem $|\lambda\mathbf{S}_{00} - \mathbf{S}_{01}\mathbf{S}_{11}^{-1}\mathbf{S}_{10}| = 0$, which gives the eigenvalues $\hat{\lambda}_1 > \ldots > \hat{\lambda}_4$ and eigenvectors $\hat{\mathbf{m}}_1, \ldots, \hat{\mathbf{m}}_4$. Given knowledge of three cointegrating relationships between the elements of $\mathbf{W}_T$ for a periodically integrated time series, the stochastic trends here are found to be equal to $\hat{\mathbf{m}}_4'\mathbf{W}_T$. We denote these trends extracted using the GG method as Y_T^{GG} and X_T^{GG}.

Again, there is the question of whether one can use the standard critical values when testing for cointegration between Y_T^{GG} and X_T^{GG}. The second panel of Table 9.1 reports the empirical fractiles and sizes when the DGP consists of two independent PIAR(1) processes. In contrast to the outcomes for the BT method, it seems that the empirical rejection frequency is below the nominal size for the CRDW statistic and for the CRDF test where no lagged variables are included in the auxiliary Dickey–Fuller regression, as well as for the JJ trace tests in the case where a VAR(1) for Y_T^{GG} and X_T^{GG} is assumed to be adequate. However, when I include an additional lag for the CRDF test and consider a VAR(2) instead of a VAR(1) for Y_T^{GG} and X_T^{GG}, the empirical size gets closer to the nominal size.

I perform a similar exercise in case of 400 quarterly, i.e. 100 annual, observations. The results of these simulations are reported in Table 9.2. It can be observed from the results in this table that the empirical size gets closer to the nominal size in larger samples.

Taking the evidence in Tables 9.1 and 9.2 it seems that the BT method appears most useful in small samples. In the empirical analysis below therefore I consider only this method of extracting stochastic trends from several univariate PIAR time series. Before turning to an empirical example, however, I first define the common properties of PIAR time series.

TABLE 9.2. Empirical performance of cointegration tests where the DGP consists of two independent PIAR(1) processes[a]

Method[b]		Empirical fractiles(%)			Empirical size(%)[c]		
		5	10	20	5	10	20
EG after GG	CRDW	0.36	0.30	0.24	3.38	7.62	18.82
	CRDF	–3.30	–3.00	–2.68	4.00	9.16	18.68
JJ after GG	Tr($r\leq1$)	7.79	6.34	4.79	4.22	8.74	19.22
	Tr($r\leq0$)	17.91	15.76	13.32	4.54	7.12	18.96

[a] DGP is $y_t = \phi_s y_{t-1} + \varepsilon_t$ and $x_t = \alpha_s x_{t-1} + v_t$, where ε_t, v_t are $N(0,1)$ and $\phi_1 = 1.106$, $\phi_2 = 0.937$, $\phi_3 = 0.998$, $\phi_4 = 0.967$, $\alpha_1 = 1.014$, $\alpha_2 = 0.966$, $\alpha_3 = 0.963$, $\alpha_4 = 1.060$. Based on 5,000 replications of effective sample size $n = 400$ or $N = 100$.

[b] The methods and abbreviations are given in the text in Section 9.2. Unless otherwise indicated, the CRDF statistics are based on an auxiliary regression that contains no constant and no lags, and the JJ trace test statistics are calculated for a VAR(1) process.

[c] The 'Empirical size' columns report on the rejection frequency when the standard critical values for nonperiodic models are used, which are given in MacKinnon (1991) and Table 8.10.

Definition. The PIAR time series y_t and x_t have common periodicity when there exists a θ for which the linear combination $y_t - \theta x_t$ can be described by a nonperiodic time series model.

Since the univariate time series contain a single stochastic trend, the θ in the linear combination $y_t - \theta x_t$ can be estimated superconsistently (see Engle and Granger 1987). Notice that this definition does not concern the linear combination $y_t - \theta_s x_t$. A simple test procedure for common periodicity is now to estimate a PAR model for the $y_t - \hat{\theta} x_t$ variable, and to check whether the AR part of the model can be restricted to a nonperiodic AR model. Although both models, i.e. the periodic and nonperiodic, may contain a stochastic trend, the asymptotic distribution of the relevant F-test statistic is the standard F-distribution (see Chapter 8).

In the case where the $y_t - \hat{\theta} x_t$ variable can be described using a nonperiodic I(1) process, the accumulation of shocks in different seasons does not have a seasonally varying long-run impact as PIAR processes have (see Section 8.2). In the case of common periodicity, the two time series y_t and x_t in a sense have common slowly changing seasonal patterns.

One approach to representing this common periodicity aspect is given by the unobserved components model

$$\begin{bmatrix} y_t \\ x_t \end{bmatrix} = \begin{bmatrix} \theta w_t \\ w_t \end{bmatrix} + \begin{bmatrix} u_t \\ v_t \end{bmatrix}, \tag{9.16}$$

where w_t is a PIAR process and where u_t and v_t can be a nonperiodic ARIMA type of process. An example of an alternative representation is given by

$$\begin{bmatrix} y_t \\ x_t \end{bmatrix} = \begin{bmatrix} \phi_{1s,1} & \phi_{1s,2} \\ \phi_{1s,3} & \phi_{1s,4} \end{bmatrix} \begin{bmatrix} y_{t-1} \\ x_{t-1} \end{bmatrix} + \begin{bmatrix} \phi_{2s,1} & \phi_{2s,2} \\ \phi_{2s,3} & \phi_{2s,4} \end{bmatrix} \begin{bmatrix} y_{t-2} \\ x_{t-2} \end{bmatrix} + \begin{bmatrix} u_t \\ v_t \end{bmatrix}, \tag{9.17}$$

which is a PVAR(2) process. Given (9.17), one can write

$$y_t - \theta x_t = \lambda_s(y_{t-1} - \theta x_{t-1}) + \pi_s(y_{t-2} - \theta x_{t-2}) + u_t - \theta v_t, \tag{9.18}$$

where

$$\lambda_s = \phi_{1s,1} - \theta\phi_{1s,3} = (-\phi_{1s,2} + \theta\phi_{1s,4}) / \theta \tag{9.19}$$

$$\pi_s = \phi_{2s,1} - \theta\phi_{2s,3} = (-\phi_{2s,2} + \theta\phi_{2s,4}) / \theta. \tag{9.20}$$

The variables y_t and x_t have common periodicity when $\lambda_s = \lambda$ and $\pi_s = \pi$ for all seasons $s = 1,2,3,4$. In the case where all values π_s are equal to zero, i.e. when $y_t - \theta x_t$ can be modelled using a PAR(1) process, we can say that y_t and x_t are in some sense codependent. (See Gourieroux *et al.* (1991) for formal definitions of codependence.) Where all ϕ_s and π_s are equal to zero, we can say that y_t and x_t have a common periodic serial correlation feature. Test statistics for the latter common feature can be derived along similar lines as in Engle and Kozicki (1993). Of course, given the assumption that one has periodically integrated time series, it seems sensible first to analyse possible common trend properties before attempting an analysis of serial correlation features.

Finally, two PIAR time series y_t and x_t are cointegrated and have common changing seasonal fluctuations when the linear combination $y_t - \theta x_t$ can be described by a nonperiodic and stationary time series process. A natural test strategy for this common aspect would seem to be the application of a test for periodicity in the AR parameters when a PAR process is fitted to the estimated residuals $\hat{u}_t$ from (9.11) and, secondly, a test for a unit root along standard lines. In the case of more than two time series, one may use the Box–Tiao method followed by the Johansen method to investigate the number of cointegration relationships. In case of a single cointegration relationship, one may want to regress y_t on a constant and other variables and test whether the residuals are nonperiodic.

Nonperiodic analysis

In practical occasions where PIAR models are found to be appropriate, the estimated values of the α_s in the differencing filter $(1 - \alpha_s B)$ are usually close to unity. As discussed in the previous chapter, an application of tests for seasonal unit roots to such PIAR series in a nonperiodic AR model may then yield evidence for the adequacy of the Δ_4 filter to transform the time series to stationarity. A natural next step in the latter occasion is to investigate nonseasonal cointegration using the $(1 + B + B^2 + B^3)$ filtered time series. Since this is not the appropriate filter to remove the stochastic trend from a PIAR time series, we may expect that this filter may affect the empirical investigation of common stochastic trends in PIAR time series.

To study this conjecture, consider the results of a simulation experiment in Table 9.3. The table displays the rejection frequencies of the two-step method and the nonperiodic cointegration analysis of the $(1 + B + B^2 + B^3)$

TABLE 9.3. Rejection frequencies of the null hypothesis of no cointegration between two PIAR processes[a]

DGP[b]	Nominal size	EG after BT[c]		EG for $(1+B+B^2+B^3)$[d]	
		CRDW	CRDF	CRDW	CRDF
(i)	0.05	0.06	0.05	0.00	0.04
	0.10	0.12	0.11	0.00	0.09
	0.20	0.20	0.21	0.00	0.16
(ii)	0.05	0.97	0.95	0.95	0.36
	0.10	0.99	0.99	0.99	0.48
	0.20	1.00	1.00	1.00	0.61
(iii)	0.05	0.96	0.92	0.05	0.33
	0.10	0.99	0.97	0.19	0.44
	0.20	1.00	0.99	0.53	0.57
(iv)	0.05	0.71	0.60	0.00	0.24
	0.10	0.85	0.75	0.00	0.36
	0.20	0.94	0.88	0.01	0.48
(v)	0.05	0.38	0.29	0.00	0.16
	0.10	0.55	0.44	0.00	0.26
	0.20	0.73	0.62	0.00	0.38

[a] I use the two-step method involving the extraction of stochastic trends via the Box–Tiao method, and a nonperiodic analysis of $(1 + B + B^2 + B^3)$ transformed time series. Based on 5,000 replications of sample size 100.

[b] The DGP is $y_t = \alpha_s y_{t-1} + \varepsilon_t$ with $\alpha_1 = 1.25$, $\alpha_2 = 0.8$, $\alpha_3 = 2$, and $\alpha_4 = 0.5$; ε_t is $N(0,1)$; and (i) $x_t = \gamma_s x_{t-1} + v_t$ with $\gamma_s = \alpha_s$, (ii) $x_t = y_t + v_t$, (iii) $x_t = y_t + v_t/(1-0.5B)$, (iv) $x_t = y_t + v_t/(1-0.8B)$, and (v) $x_t = y_t + v_t/(1-0.9B)$, where v_t is a $N(0,1)$ process.

[c] The Engle–Granger (1987) cointegration analysis applied to the trends extracted using the Box–Tiao (1977) method. The CRDF auxiliary regression contains no lags. The sample size for the second step is N.

[d] The nonperiodic Engle–Granger (1987) cointegration analysis applied to the $(1 + B + B^2 + B^3)$ transformed time series. The lag length in the auxiliary regression for CRDF is determined using the $F_{\text{AR},1\text{–}4}$ test. The sample size in the second step is n.

transformed time series. The first DGP corresponds to the empirical 'size' of the tests, while the next four DGPs correspond to the empirical 'power' of the tests. Clearly, the empirical size of the two-step method is close to the nominal size (see also Table 9.1). Furthermore, the empirical size of the CRDW test for the nonperiodic method is incorrect, while that of the CRDF test seems to be close to the nominal size. The power of the two-step method seems quite reasonable, even in the case when the errors are strongly autocorrelated. On the other hand, the (not corrected for size) power of the CRDW test in nonperiodic analysis is large only when there is no autocorrelation in the cointegration relation, and this power rapidly declines to zero. The power of the CRDF test in nonperiodic analysis is somewhat larger, although not nearly as high as for the two-step method. In sum, the results in Table 9.3 suggest that we find less evidence of cointegration at the nonseasonal frequency in a non-periodic model when the data are generated by cointegrated PIAR processes.

Four UK time series

To illustrate the practical usefulness of a systematic analysis of common properties of periodically integrated time series, I consider a sample from the 11 series analysed in previous chapters. This sample contains real GDP (y_t), real total consumption (c_t), real total investment (i_t), and real total exports (e_t). All data are in logs. In Chapter 8 it was found that all four time series can be described by PIAR models. Hence these macroeconomic variables display changing seasonal patterns over the sample period, and it these changes that are caused by accumulations of shocks. From the estimates of the impact matrix $\Gamma\Phi_0^{-1}$ in Table 8.13, one can observe that, for GDP, consumption, and investment, the values in the first row of this matrix are highest among the four rows. This means that the accumulation of shocks in the first quarter dominates the pattern of the individual time series. Furthermore, for these three series the values in the second column are highest, indicating that the second-quarter observations react the most to shocks. On the other hand, for exports one observes that the values in the first column of $\Gamma\Phi_0^{-1}$ are highest, suggesting that the first quarter reacts the most to accumulations of shocks. In sum, the first quarter seems to be important for all four time series, and therefore I consider these variables for the construction of a multivariate model.

When one is interested in jointly modelling these four variables, it may be useful to investigate whether these series have common properties with respect to the changing seasonal patterns and/or the stochastic trends. A first step may now be to check whether the vector series (y_t,c_t,i_t,e_t) can be described by a PVAR model.

Some diagnostic results for periodic VAR models of orders 1 and 2 are displayed in Table 9.4. The PVAR models contain 4×20 and 4×36 parameters since four seasonal intercept terms are included in each regression. The diagnostics are calculated per equation. It is clear from the results in this table that a PVAR(1) model does not provide an adequate description of the data, and that a PVAR(2) model may be adequate. The ARCH patterns in the i_t and e_t equations are caused by outlying observations in 1963.1, 1963.2, and 1967.4, 1968.1, respectively. When dummy variables are included to remove these outliers, the test values for normality and ARCH become either insignificant or less high. Given that the unrestricted PVAR(2) model contains 36 parameters per equation, it now seems worthwhile investigating the possibilities for simplifying structures within this model.

An initial and possibly insightful linear combination of variables may concern the GDP, consumption, and investment variables. This is because the

TABLE 9.4. Diagnostic test results for periodic vector autoregressive models for GDP, consumption, investment, and exports in the UK

Model		Diagnostics[a]				
		$F_{\mathrm{AR},1-1}$	$F_{\mathrm{AR},1-4}$	$F_{\mathrm{ARCH},1-1}$	$F_{\mathrm{ARCH},1-4}$	JB
PVAR(1)	GDP	6.537**	2.681**	1.713	1.178	4.303
	Consumption	3.233	1.206	1.028	1.050	50.253**
	Investment	4.870**	2.026	4.553**	1.507	8.664**
	Exports	25.865**	8.167**	10.592**	3.299**	24.799**
PVAR(2)	GDP	0.232	2.370	0.830	0.529	0.674
	Consumption	0.105	1.005	0.042	1.471	3.457
	Investment	0.162	0.816	7.254**	2.754**	5.860
	Exports	0.007	1.743	12.294**	3.079**	0.411

[a] These diagnostic measures are calculated for each of the four regressions separately.
** Significant at the 5 per cent level.

estimation results in Table 8.13 suggest that the $\Gamma\Phi_0^{-1}$ matrices for these variables are roughly similar. A regression of y_t on a constant, c_t, and i_t yields

$$\underset{}{y_t} = \underset{(0.125)}{1.143} + \underset{(0.027)}{0.782}c_t + \underset{(0.021)}{0.163}i_t + \hat{u}_t, \tag{9.21}$$

where the calculated standard errors are given in parentheses. These should be interpreted with care since the parameters are estimated superconsistently. A PAR model for $\hat{u}_t$ appears to be a subset PAR model which includes lags 1 and 4. An F-test for the significance of the periodic variation in the AR polynomial obtains a value of 1.025, with a p-value of 0.412. An application of the ADF test (with the inclusion of three lags of the first-order-differenced time series) obtains the insignificant value of –0.912. Hence y_t, c_t, and i_t have common periodicity, though they are not cointegrated. To investigate whether this is due to the exclusion of exports, I next consider the four-variable system.

Before testing for cointegration between the four UK series, I first investigate the small sample properties (similar to Table 9.1) of the two-step method for four variables. Table 9.5 displays several quantiles of the Johansen maximum likelihood method for four variables. This table is needed since I wish to investigate whether one can rely on these critical values in the two-step approach. For that purpose, consider the quantiles in the table for 33 observations (since there are only 33 annual observations to estimate the stochastic trends in the four VQ variables here).

Table 9.6 evaluates the empirical performance of the two-step method when the DGP consists of four independent PIAR(1) processes with parameters that correspond to those in the $\Gamma\Phi_0^{-1}$ matrices in Table 8.13. It is clear from the empirical fractiles, and especially from the reported empirical size, that the small sample empirical performance of my approach seems satisfactory. For the application below therefore I will use the critical values in Table 9.5.

TABLE 9.5. Quantiles of the Johansen cointegration test statistics[a]

m–r	Quantiles (%)						Mean	Var.
	50	80	90	95	97.5	99		
Maximal eigenvalue								
1	2.40	4.84	6.41	8.14	9.72	11.74	3.00	6.90
2	7.76	11.27	13.41	15.52	17.37	19.90	8.41	14.23
3	13.47	17.64	20.13	22.57	24.77	27.65	14.10	20.91
4	19.46	24.42	27.30	30.16	32.82	36.14	20.17	29.96
Trace								
1	2.40	4.84	6.41	8.14	9.72	11.74	3.00	6.90
2	9.61	13.73	16.20	18.42	20.77	23.59	10.27	19.81
3	21.33	27.14	30.50	33.45	36.09	38.82	22.03	39.30
4	37.81	45.25	49.63	53.56	56.98	61.69	38.60	70.01

[a] Sample size is 33. The data-generating process contains no trend, and the constant term is unrestricted. These quantiles are based on 10,000 Monte Carlo replications. The test statistics are computed using the original formulas in Johansen and Juselius (1990). The table corresponds to Table A.2 in Johansen and Juselius (1990). The reader is referred to that paper for more details.

TABLE 9.6. Empirical performance of the Johansen cointegration tests after the stochastic trends have been extracted using the Box–Tiao method where the DGP consists of four independent PIAR(1) processes[a]

Test statistics	4–r	Empirical fractiles(%)			Empirical size(%)		
		5	10	20	5	10	20
λmax	4	30.19	27.48	24.52	0.05	0.11	0.21
	3	22.71	20.34	17.74	0.05	0.11	0.21
	2	15.82	13.83	11.58	0.06	0.11	0.22
	1	8.53	6.86	5.23	0.06	0.12	0.23
Trace	4	53.50	49.69	45.48	0.05	0.10	0.21
	3	33.35	30.64	27.27	0.05	0.10	0.20
	2	18.82	16.62	14.16	0.06	0.11	0.21
	1	8.53	6.86	5.23	0.06	0.12	0.23

[a] The DGPs are $y_{jt} = \phi_{js} y_{jt-1} + \varepsilon_{jt}$, where $\varepsilon_{jt} \sim N(0,1)$ for $j = 1,2,3,4$ and $\phi_{1s} = \{1.011, 0.916, 1.062, 1.017\}$, $\phi_{2s} = \{1.031, 0.918, 1.057, 1.000\}$, $\phi_{3s} = \{1.053, 0.888, 1.071, 0.999\}$, and $\phi_{4s} = \{0.957, 1.022, 1.032, 0.991\}$. The data are generated in this order. This means that for $4 - r = 1$ the DGP is a PIAR(1) with $\phi_{1s} = \{1.011, 0.916, 1.062, 1.017\}$, while for $4–r = 2$ the DGP corresponds to ϕ_{1s} and ϕ_{2s}, and so on. The choice for this DGP is motivated by the estimation results for $\Gamma\Phi_0^{-1}$ in Table 8.13. Based on 5,000 replications of effective sample size $N = 33$.

The Box–Tiao method results in the eigenvalues and eigenvectors as reported in Table 9.7. The eigenvalues clearly support the finding in Chapter 8 that there is only a single unit root in the univariate time series, since the largest eigenvalue is very close to unity, while the next eigenvalues are much closer to zero. Further, the eigenvectors support the tentative observation from Table 8.13 that the first-quarter observations and/or shocks are more important than others, since the first-quarter contribution to the stochastic trend is largest for all four variables.

TABLE 9.7. Extraction of stochastic trends using the Box–Tiao method

Variable	Eigenvalues	Eigenvector for λmax
GDP	0.996, 0.377, 0.166, 0.000	(1, 0.243, 0.886, 0.333)'
Consumption	0.997, 0.345, 0.093, 0.050	(1, –0.057, –0.737, 0.956)'
Investment	0.985, 0.383, 0.152, 0.038	(1, 0.780, 0.462, 0.239)'
Exports	0.996, 0.198, 0.017, 0.001	(1, 0.865, –0.240, 0.017)'

For the (4 × 1) process $\mathbf{V}_{\mathrm{T}}$, which is $(Y_{\mathrm{T}}^{\mathrm{BT}}, C_{\mathrm{T}}^{\mathrm{BT}}, I_{\mathrm{T}}^{\mathrm{BT}}, E_{\mathrm{T}}^{\mathrm{BT}})$, I find a VAR(1) model an adequate description of the data. An application of the Johansen method yields the eigenvalues 0.581, 0.293, 0.185, and 0.012, and the corresponding trace test statistics 0.392, 7.159, 18.601, and 47.299. Comparing these values with the critical values in Table 9.5, it can be concluded that the trace test of 47.299 is significant at about the 10 per cent level, while the eigenvalue 0.581 is significantly different from zero at the 5 per cent level. Hence there seems to be only one cointegrating relation between the four univariate time series. This seems to be verified by the auxiliary cointegrating regression

$$\underset{}{Y_{\mathrm{T}}^{\mathrm{BT}}} = \underset{(0.767)}{8.034} + \underset{(0.140)}{0.761}C_{\mathrm{T}}^{\mathrm{BT}} + \underset{(0.020)}{0.179}I_{\mathrm{T}}^{\mathrm{BT}} + \underset{(0.049)}{0.345}E_{\mathrm{T}}^{\mathrm{BT}}, \quad (9.22)$$

with $R^2 = 0.997$, CRDW = 1.596**, CRDF = –5.014**, where CRDW and CRDF denote the cointegrating regression Durbin–Watson and Dickey–Fuller tests, and ** denotes significant at the 5 per cent level.

Given that there is only one cointegration relationship between y_t, c_t, i_t, and e_t, I perform the following regression:

$$y_t = \mu + \alpha c_t + \beta i_t + \gamma e_t + \varepsilon_t \quad (9.23)$$

for quarterly data. The estimation results are

$$\underset{(0.125)}{\hat{\mu} = 3.236} \quad \underset{(0.025)}{\hat{\alpha} = 0.368} \quad \underset{(0.011)}{\hat{\beta} = 0.179} \quad \underset{(0.012)}{\hat{\gamma} = 0.224},$$

where R^2 is 0.996 and the Durbin–Watson value is 1.305. A PAR(1) model for $\hat{\varepsilon}_t$ in (9.23) cannot be rejected using the diagnostic checks as advocated in Chapter 8. Moreover, the F-test for periodicity in the AR part obtains the insignificant value of 1.473. The Dickey–Fuller test applied to the resulting AR(1) model has a value of –7.442**. The AR parameter is 0.400 with a standard deviation 0.081. Hence GDP, consumption, investment, and exports in the UK have common periodicity and are cointegrated. Finally, note that the AR order of the ε_t process in (9.23) is only 1, while the multivariate model is of order 2. This suggests that the four series seem codependent as well.

Since there is a cointegrating relationship between the four UK variables, we may estimate the PVAR(2) model in periodic differences and with error correction variables $D_{s,t}ecm_{t-2}$, where the ecm_t variable is obtained from (9.23). We denote Δ_p^j as the periodic differencing filter $(1-\alpha_s B)$ for each of the variables, where I use the α_s values as they are given in Table 8.8, and where the index j indicates that the differencing filter varies with the variable $j = y_t, c_t, i_t$, and e_t. The final simplified model is (apart from four seasonal intercept terms)

$$\Delta_p^y y_t = \underset{(0.134)}{-0.549}D_{2,t}\Delta_p^y y_{t-1} \underset{(0.133)}{-0.411}D_{3,t}\Delta_p^y y_{t-1} \underset{(0.287)}{-0.648}D_{1,t}ecm_{t-2} + u_t \tag{9.24}$$

$$\Delta_p^c c_t = \underset{(0.068)}{-0.280}D_{2,t}\Delta_p^e e_{t-1} \underset{(0.050)}{-0.125}D_{3,t}\Delta_p^e e_{t-1} + v_t \tag{9.25}$$

$$\Delta_p^i i_t = \underset{(0.377)}{0.846}D_{1,t}\Delta_p^i i_{t-1} \underset{(0.133)}{-0.352}D_{2,t}\Delta_p^i i_{t-1} + w_t \tag{9.26}$$

$$\begin{aligned}\Delta_p^e e_t = {} & \underset{(0.175)}{-0.709}D_{2,t}\Delta_p^e e_{t-1} \underset{(0.128)}{-0.398}D_{3,t}\Delta_p^e e_{t-1} \underset{(0.160)}{-0.611}D_{4,t}\Delta_p^e e_{t-1} \\ & \underset{(0.462)}{+1.380}D_{2,t}ecm_{t-2} + z_t.\end{aligned} \tag{9.27}$$

It is clear from these estimation results that the highly parameterized PVAR model can be reduced to a reasonably simple model, with error correction in the first quarter for GDP and in the second quarter for exports. The model in (9.24)–(9.27) is also estimated using NLS under the various restrictions that $\alpha_1\alpha_2\alpha_3\alpha_4 = 1$. The tests for the hypothesis that these $\alpha_s = 1$ for some or all variables result in rejections of the null hypotheses. Hence no equation in (9.24)–(9.27) can be reduced to an equation for a $(1 - B)$ transformed time series.

The empirical results reported in this section may not come as a complete surprise. Given the accounting definitions that ensure relationships between the macroeconomic variables considered, one may expect the presence of simplifying structures for this set of variables. On the whole, the presented outcomes seem to confirm that the proposed test strategy for common aspects in periodically integrated time series has some empirical power. A disadvantage of the proposed empirical method in this section is that it may not be easy to assign a direct economic interpretation to the multivariate model in (9.24)–(9.27). Hence it may be sensible to interpret this method as an approach to summarizing a set of time series, each with quite complicated seasonal and trend patterns, into a single multivariate statistical model. When the purpose is to obtain for such data an econometric model that has a somewhat clearer economic interpretation, it seems useful to analyse the periodic cointegration models like (9.9). In the next section I discuss a empirical modelling strategy for such models.

9.3 Periodic Cointegration

In this section I focus on the representation, estimation, and testing issues of periodic cointegration models, i.e. error correction models where the cointegration parameters and the short-run adjustment parameters are allowed to vary with the season. In the first subsection I motivate such a PC model using somewhat more economic arguments and compare it with the full seasonal cointegration model presented in Section 5.2. I then investigate two empirical modelling strategies, and apply these to the quarterly income and consumption data for Sweden. In the final subsection, I study the impact of a nonperiodic analysis of these data.

Representation

One of the main attractions of error correction models (ECMs) in general is their connection with linear-quadratic adjustment costs (LQAC) models (see Nickell 1985 and Gregory 1994). Such models can be applied to many economic problems, such as the demand for and supply of labour and money (see Sargent 1978; Kennan 1979; Dolado *et al.* 1991; and Engsted and Haldrup 1992). In this section I briefly survey some aspects of ECMs and LQAC models, and discuss how such models can be modified for the analysis of seasonally observed time series.

Consider an observed economic time series y_t with its unobserved target value y_t^*, where this target is related to a vector of forcing (explanatory) variables X_t as follows:

$$y_t^* = \theta' X_t + u_t, \tag{9.28}$$

where u_t is a white noise process. Assume that at time t economic agents choose the sequence $y_{t+j}, j = 0,1,2,\ldots$, in order to minimize the expected intertemporal loss function

$$L_t = E\{\sum_{j=0}^{\infty} \rho^j [\delta(y_{t+j} - y_{t+j}^*)^2 + (y_{t+j} - y_{t+j-1})^2] \mid I_t\}, \tag{9.29}$$

where I_t is the information set available at time t, $\rho \in (0,1)$ is a discount factor, and $\delta > 0$ is a weighting factor reflecting adjustment costs (see e.g. Sargent 1978; Gregory 1994).

Assuming that the explanatory variables can be modelled by a random walk, i.e. that $\Delta_1 X_t$ is a vector white noise process, it can be shown (see Gregory 1994) that the solution of the intertemporal optimization problem can be represented by the following error correction model:

$$\Delta_1 y_t = \lambda(y_{t-1} - \theta' X_{t-1}) + \beta' \Delta_1 X_t + \varepsilon_t, \tag{9.30}$$

where ε_t is a standard white noise process. More generally, if $\Delta_1 X_t$ is some stationary VAR process, the resulting ECM will be

$$\gamma(B)\Delta_1 y_t = \lambda(y_{t-1} - \theta' X_{t-1}) + \beta(B)' \Delta_1 X_t + \varepsilon_t. \tag{9.31}$$

The parameters λ, $\gamma(B)$, and $\beta(B)$ are functions of ρ, δ, and θ in (9.28) and (9.29) and of the parameters in the VAR model for $\Delta_1 X_t$. In particular, if δ approaches zero then λ does too, because if there is no cost associated with being out of equilibrium there will be no incentive for adjustment towards an equilibrium. Alternatively, if $\delta > 0$, then $\lambda < 0$ and the model in (9.31) is an ECM in which the variables are cointegrated through $y_{t-1} - \theta' X_{t-1}$.

In the case of seasonal time series with seasonal unit roots, one may consider the seasonal cointegration model as given in Section 5.2. For the so-called full seasonal cointegration model, i.e. when there is cointegration at the seasonal frequencies π and $\pi/2$ and at the nonseasonal frequency, Osborn (1993) shows that its simplest version corresponds to

$$\Delta_4 y_t = \sum_{i=1}^{4} \lambda_i (y_{t-i} - \theta_i x_{t-i}) + \varepsilon_t. \tag{9.32}$$

Hence, in terms of (9.28), the seasonal cointegration model implies four different target relationships and adjustment parameters, each associated with a different lag. From an economic point of view, the model in (9.32) assumes that the degree of adjustment towards equilibrium depends on the lag of the disequilibrium error.

In the form of (9.30), a simple version of a periodic cointegration model as in (9.9) is

$$\Delta_4 y_t = \lambda_s (y_{t-4} - \theta_s' X_{t-4}) + \beta' \Delta_4 X_t + \varepsilon_t. \tag{9.33}$$

There are four different target relations in this model, and, in contrast to (9.32), the parameters of these relations vary not with the lag but with the season. In the context of the LQAC model in (9.28)–(9.29), the PCM in (9.33) can be interpreted as follows. First, the seasonally varying adjustment parameter λ_s is a reflection of adjustment costs varying over the quarters. Thus, in some seasons it can be worse to be out of equilibrium. This may be the case, for example, with employment in typically seasonal industries as tourism and construction. Secondly, the long-run parameter θ_s and the associated target relationship (9.28) are not constant. This reflects seasonally varying preferences; for example, a particular consumption bundle may not generate the same level of utility in summer as in winter. Notice that this corresponds with the univariate consumption model in Osborn (1988). Finally, and similar to the periodic integration model, an important and possibly useful feature of the PC model is that the impact of shocks differs throughout the year. An implication of such variation is that the effects of policy measures, in both the short run and the long run, can depend on the timing of these measures. In Section 9.4 I will investigate using Monte Carlo simulation whether a PC model can generate time series which in SA form may result in seasonality in business cycle peaks and troughs.

In the next subsection I discuss estimation and testing issues of PC models. In this discussion I consider so-called full PC models, i.e. models like (9.33) with all $\lambda_s \neq 0$, and partial PC models; i.e., some of the λ_s in (9.33) can be equal to zero. Of course, when all $\lambda_s = 0$, the hypothesis of no cointegration cannot be rejected.

Estimation and testing

Statistical inference on periodic cointegration models is discussed in great detail in Boswijk and Franses (1995). In this subsection I confine myself to presenting the main results that are necessary for a sensible empirical use of a PC model.

To be more specific, consider the following PC model for y_t, i.e.

$$\Delta_4 y_t = \lambda_s(y_{t-4} - \theta_s' X_{t-4}) + \sum_{j=1}^{p} \gamma_j \Delta_4 y_{t-j} + \sum_{i=0}^{p} \beta_i' \Delta_4 X_{t-i} + \varepsilon_t, \tag{9.34}$$

and assume that $\Delta_4 X_t$ is generated by some VAR process, that $\Delta_4 y_t$ does not contain any unit roots in its AR polynomial, and that X_t is weakly exogenous. The latter assumption implies that the Brownian motions corresponding to X_t and the cumulative sum of ε_t are independent. This requires that in a model for X_t the coefficients of the error-correcting terms $(y_{t-4} - \theta_s' X_{t-4})$ are all equal to zero. In the model in (9.34) the $\Delta_4 X_t$ variables can be replaced by $\Delta_1 X_t$ variables. Notice that (9.34) together with the three assumptions puts several restrictions on multivariate models such as (9.2). In practice, therefore, it is relevant to investigate whether at least the assumption of weak exogeneity holds. (See Boswijk (1992) for a simple test procedure.) Finally, it should be mentioned that all forthcoming results apply equally to models such as (9.34) where the lag polynomials for $\Delta_4 y_t$ and $\Delta_4 X_t$ contain seasonally varying parameters.

A simple method to test for cointegration

A simple empirical model specification strategy for the PC model in (9.34) follows the well-known two-step strategy given in Engle and Granger (1987). An additional step is to test whether the various cointegration and adjustment parameters are indeed periodically varying. Consider two quarterly observed time series y_t and x_t, and their vector time series $\mathbf{Y}_\mathrm{T} = (Y_{1,\mathrm{T}}, Y_{2,\mathrm{T}}, Y_{3,\mathrm{T}}, Y_{4,\mathrm{T}})$ and $\mathbf{X}_\mathrm{T} = (X_{1,\mathrm{T}}, X_{2,\mathrm{T}}, X_{3,\mathrm{T}}, X_{4,\mathrm{T}})$. Periodic cointegration as in (9.34) amounts to four cointegration relationships between y_t and x_t, i.e. one cointegration relation per quarter. It is clear that these cointegration relations can be written as $Y_{s,\mathrm{T}} - \theta_s X_{s,\mathrm{T}}$ for $s = 1,2,3,4$. A first step in specifying models like (9.34) may then be to regress $Y_{s,\mathrm{T}}$ on a constant and $X_{s,\mathrm{T}}$ for each of the seasons. In the case of cointegration between $Y_{s,\mathrm{T}}$ and $X_{s,\mathrm{T}}$, the residuals of one or more of these four regressions should be stationary time series. Of course, it may be that one will obtain stationary residuals for only a few seasons. In this case a partial PC model appears. When all four residual series are stationary, one has a full PC model.

To test for cointegration between $Y_{s,\mathrm{T}}$ and $X_{s,\mathrm{T}}$, one may choose to check the value of the Durbin–Watson statistic (CRDW_s) for each season s and the Dickey–Fuller test (CRDF_s). The asymptotic distributions of these test statistics can be derived from the asymptotic distributions where the parameters do not vary over the seasons. This is caused by the orthogonality of the regressors $D_{s,t}y_{t-4}$ and $D_{s,t}x_{t-4}$ in (9.34). For example, the CRDF_s test should have the same distribution as the well-known cointegration Dickey–Fuller test in the case of two nonperiodic variables (see Engle and Granger 1987). To verify these conjectures on asymptotics, especially for small samples, consider the simulated critical values for these test statistics in Table 9.8 for the cases where x_t should be transformed using a Δ_4 filter or a Δ_1 filter. One can infer from the critical values in this table and in MacKinnon (1991) for 25 observations that the critical values for these two cases seem to be similar.

In the case where one obtains evidence that there is indeed cointegration in each season, a useful next step is to check whether the estimated parameters in the cointegration vectors, as well as the adjustment parameters, vary over the seasons. In other words, one may want to test the hypotheses $\theta_s = \theta$ and $\lambda_s = \lambda$ in (9.34). The test for $\theta_s = \theta$ can be performed by comparing the residual sums of squares of the four regressions of $Y_{s,\mathrm{T}}$ on a constant and $X_{s,\mathrm{T}}$ with the *RSS* of the regression of y_t on four seasonal dummies and x_t. Assuming cointegration, one can construct an *F*-test for the hypothesis $\theta_s = \theta$, which follows a standard *F*-distribution under the null hypothesis. This result follows from theorem 3.1 in Johansen (1991). This test will be denoted as $F(\theta_s = \theta)$. Furthermore, in the case of cointegration, the *F*-test for the hypothesis that $\lambda_s = \lambda$ also follows

TABLE 9.8. Critical values of periodic cointegration test statistics for a sample of 25 years of quarterly observations[a]

Test statistic	Significance level			
	0.20	0.10	0.05	0.01
$\Delta_4 x_t = v_t$				
CRDW_s	0.86	1.08	1.26	1.63
CRDF_s	−2.70	−3.13	−3.50	−4.27
$\Delta_1 x_t = v_t$				
CRDW_s	0.88	1.09	1.27	1.63
CRDF_s	−2.71	−3.12	−3.48	−4.31

[a] Based on 20,000 Monte Carlo replications. The CRDW_s for s = 1,2,3,4, as well as the Dickey–Fuller *t*-test statistic CRDF_s for the null hypothesis of no cointegration are found by estimating the model, $Y_{s,\mathrm{T}} = \mu_s + \tau_s X_{s,\mathrm{T}} + \zeta_{s,\mathrm{T}}$, for $s = 1,2,3,4$, where $\Delta_4 y_t = \varepsilon_t$ and $\Delta_4 x_t = v_t$ or $\Delta_1 x_t = v_t$, and where $\varepsilon_t, v_t \sim N(0,1)$.

a standard F-distribution, since $y_{t-4} - \hat{\theta}_s x_{t-4}$ are stationary variables (see Engle and Granger 1987). This test will be denoted as $F(\lambda_s = \lambda)$. Similar test statistics can be constructed in the case of a partial instead of a full PC model, although one should then make sure that the F-tests for restrictions on the θ_s and λ_s are calculated only for those seasons where one obtains cointegrating relationships. Otherwise these F-tests do not follow standard F-distributions.

Finally, a test for the weak exogeneity of x_t for the cointegration relations can be performed via testing the significance of the periodic error correction variables in a model for $\Delta_4 x_t$. Again, since these error correction variables are stationary variables, the relevant F-test follows a standard F-distribution under the null hypothesis (see Boswijk 1992).

An application of this simple empirical estimation strategy is given in Franses and Kloek (1995), where it is applied to consumption and income data for Austria. Below I will apply it to Swedish consumption and income data, and will compare it with the empirical results that are obtained using the method that is discussed next.

An extension of the Boswijk method

As can be observed from (9.34), periodic cointegration requires that the adjustment parameters λ_s are strictly smaller than zero. Here I will discuss a class of Wald tests for the null hypothesis of no cointegration versus the alternative of periodic cointegration. Two versions of this test are considered: the Wald$_s$ statistic for periodic cointegration in season s, and the joint Wald statistic. Both versions are extensions of the Boswijk (1994) cointegration test, which has been developed for nonseasonal data.

Consider again model (9.34), and define w_t as the vector containing the various differenced variables in this model, such that (9.34) can be written as

$$\Delta_4 y_t = \sum_{s=1}^{4} (\delta_{1s} D_{s,t} y_{t-4} + \delta_{2s}' D_{s,t} X_{t-4}) + \pi' w_t + \varepsilon_t, \quad (9.35)$$

where $\delta_{1s} = \lambda_s$ and $\delta_{2s} = -\lambda_s \theta_s'$. Notice that from this definition it follows that $\lambda_s = 0$ implies that $\delta_{1s} = \delta_{2s} = 0$. The null and alternative hypotheses for the Wald$_s$ tests are, for some particular s,

$$H_{0s}: \delta_s = 0 \text{ versus } H_{1s}: \delta_s \neq 0, \text{ for some } s, \quad (9.36)$$

whereas for the Wald test these hypotheses are

$$H_0: \delta_s = 0 \text{ for all } s \text{ versus } H_1: \delta_s \neq 0 \text{ for at least one } s, \quad (9.37)$$

where s is either 1, 2, 3, or 4. Notice that the Wald$_s$ statistics are designed to test for partial periodic cointegration, whereas the Wald statistic concerns the full periodic cointegration hypothesis. Of course, one may expect both tests to have some power against both alternative hypotheses.

Let $\hat{\delta}_s = (\hat{\delta}_{1s}, \hat{\delta}_{2s})$ denote the OLS estimator of δ_s, and let $\hat{V}(\hat{\delta}_{ss})$ denote the OLS covariance matrix estimator. Similarly, define $\hat{\delta}_s = (\hat{\delta}_{s1}, \hat{\delta}_{s2}, \hat{\delta}_{s3}, \hat{\delta}_{s4})$ and its estimated covariance matrix $\hat{V}(\hat{\delta}_s)$. The two Wald statistics are now given by

$$\text{Wald}_s = \hat{\delta}_s'(\hat{V}(\hat{\delta}_s))^{-1}\hat{\delta}_s = (n-l)((RSS_{0s} - RSS_1)/RSS_1) \tag{9.38}$$

and

$$\text{Wald} = \hat{\delta}'(\hat{V}(\hat{\delta}))^{-1}\hat{\delta} = (n-l)(RSS_0 - RSS_1)/RSS_1), \tag{9.39}$$

where l is the number of estimated parameters in (9.35) and RSS_{0s}, RSS_0, and RSS_1 denote the OLS residual sums of squares under H_{0s} in (9.36), H_0 in (9.37), and H_1 in both expressions.

The model in (9.35) does not contain deterministic regressors. However, if the long-run relations require an intercept, one should add seasonal dummy variables to the PC model; i.e., the Wald tests should be based on

$$\Delta_4 y_t = \sum_{s=1}^{4} \mu_s D_{s,t} + \sum_{s=1}^{4} (\delta_{1s} D_{s,t} y_{t-4} + \delta_{2s}' D_{s,t} X_{t-4}) + \pi' w_t + \varepsilon_t. \tag{9.40}$$

Finally, if the variables contain a drift term, the asymptotic distributions of the Wald tests change (see Park and Phillips 1988, 1989). In order to obtain distributions that are invariant to the presence of drifts, four linear trends should be added to the regressors; i.e., (9.34) becomes

$$\Delta_4 y_t = \sum_{s=1}^{4} \mu_s D_{s,t} + \sum_{s=1}^{4} \tau_s D_{s,t} t + \sum_{s=1}^{4} (\delta_{1s} D_{s,t} y_{t-4} + \delta_{2s}' D_{s,t} X_{t-4}) + \pi' w_t + \varepsilon_t. \tag{9.41}$$

Notice that this also allows for a trend term to appear in the cointegration relations.

The asymptotic distributions of the Wald test statistics in (9.38) and (9.39) are nonstandard and are derived in Boswijk and Franses (1995); to save space, I do not present these distributions in any detail, and refer the interested reader to that paper. For practical purposes, however, I give the asymptotic critical values in the case of two to four variables, i.e. one to three weakly exogenous variables, in Table 9.9

Where one obtains evidence for the presence of cointegration in all or some seasons, it is of particular interest to test for the following parameter restrictions:

$$H_0(\lambda_s = \lambda):\ \lambda_s = \lambda \quad \text{for all } s = 1,2,3,4 \tag{9.42}$$
$$H_0(\theta_s = \theta):\ \theta_s = \theta \quad \text{for all } s = 1,2,3,4 \tag{9.43}$$
$$H_0(\delta_s = \delta):\ \delta_s = \delta \quad \text{for all } s = 1,2,3,4 \tag{9.44}$$

Each of these hypotheses may be tested using an F-type test statistic. For the hypotheses $H_0(\lambda_s = \lambda)$ and $H_0(\delta_s = \delta)$, these F-tests are the classical F-tests since the model is linear under the null and alternative hypotheses. These F-tests are denoted as $F(\lambda_s = \lambda)$ and $F(\delta_s = \delta)$. For $H(\theta_s = \theta)$, one may use the likelihood ratio based test statistic $F(\theta_s = \theta) = [(n-l)/h][(RSS_1 - RSS_0)/RSS_0]$, where RSS_0 and RSS_1 are the residual sums of squares under $H_0(\theta_s = \theta)$ and a NLS regression under the alternative. Under weak exogeneity and given cointegration, these three F-statistics are all distributed as the standard F-distribution. As discussed above, a test for weak exogeneity can be performed by adding the cointegrating variables $\hat{\phi}_{t-4} = D_{s,t}(y_{t-4} - \hat{\theta}_s X_{t-4})$ to a model for $\Delta_4 X_t$, i.e. by estimating

TABLE 9.9. Asymptotic critical values of the Wald_s and Wald test statistics for periodic cointegration[a]

No. of variables	Significance level Wald_s			Wald		
	0.20	0.10	0.05	0.20	0.10	0.05
Regression contains no constants and no trends						
2	4.80	6.48	8.10	16.17	19.09	21.65
3	7.40	9.38	11.18	25.26	28.73	31.75
4	9.87	12.10	14.20	34.02	38.03	41.50
Regression contains constants and no trends						
2	7.49	9.50	11.36	25.34	28.75	31.82
3	9.92	12.18	14.24	34.13	38.07	41.51
4	12.29	14.79	16.99	42.85	47.22	51.06
Regression contains constants and trends						
2	10.13	12.38	14.39	35.00	38.97	42.49
3	12.45	14.89	17.11	43.50	47.92	51.73
4	14.78	17.39	19.78	51.93	56.72	60.78

[a] Based on the regression model (9.35), (9.40), or (9.41),where the number of lags is set equal to zero. The quantiles are obtained through Monte Carlo simulations using 50,000 replications, where Brownian motions are approximated by Gaussian random walks of 500 observations.

$$\alpha(B)\Delta_4 z_t = \beta(B)\Delta_4 y_{t-1} + \kappa_s \hat{\phi}_{t-4} + v_t, \tag{9.45}$$

where $\alpha(B)$ and $\beta(B)$ are matrix lag polynomials and v_t is a vector white noise process. In Boswijk (1994) it is shown that, given cointegration, the LR test for $\kappa_s = 0$ for all s follows an asymptotic $\chi^2(4)$ distribution in the case of a full PCM. Where the null hypothesis of weak exogeneity is rejected, one may turn to alternative estimators or models. Boswijk and Franses (1995) deal with a few of such alternative approaches.

A summary of an empirical modelling procedure

An empirical modelling procedure for a periodic cointegration model involves four steps. The first is an analysis of univariate time series with respect to their unit root periodicity properties. It would be useful to select those variables for X_t for which one can be reasonably confident that there are no cointegration relations among the X_t variables. The second step is to specify a conditional error correction model such as in (9.34) to estimate the parameters and to evaluate its empirical relevance using diagnostic measures. In case of no misspecification, one would calculate the Wald_s and Wald statistics (and possibly the Engle–Granger type test statistics: see Table 9.8). Where the null hypothesis of no cointegration can be rejected, the third step is to test the hypotheses formulated in (9.42)–(9.44). Finally, in the fourth step one tests for weak exogeneity using (9.45).

A Monte Carlo investigation

To investigate the finite sample performance of the Wald and F-tests above, I report on a limited simulation experiment. The DGP in this experiment is the bivariate time series process,

$$\Delta_4 y_t = \sum_{s=1}^{4} \lambda_s D_{s,t}(y_{t-4} - \theta_s x_{t-4}) + \gamma \Delta_4 y_{t-1} + \beta \Delta_4 x_t + \varepsilon_t, \tag{9.46}$$

$$\Delta_q x_t = \sum_{s=1}^{4} \kappa_s D_{s,t}(y_{t-4} - \theta_s x_{t-4}) + \eta_t, \tag{9.47}$$

where ε_t and η_t are independent and identically distributed (i.i.d.) $N(0,1)$ variables and q in (9.47) can be equal to 1 or 4. In all cases I take $\beta = 0.5$ and the number of effective observations n as 100 or 200 (i.e. 25 and 50 years). I consider seven different cases. The number of replications is 10,000. In each replication the model in (9.46) is estimated and analysed using the various Wald and F-tests. Cases A–C are used to assess the size of the cointegration tests. In case B, x_t is a random walk process. In case C, there are short-run dynamics in $\Delta_4 y_t$, and this may lead to lower significance levels of the Wald$_s$ and Wald test statistics. The remaining cases D–G are used to study the power of the cointegration tests, and the size and power of the various tests for parameter variation in the cointegration and/or adjustment term. Case E is the only case where weak exogeneity is violated; this may lead to a size distortion of the tests for seasonal parameter variation.

The rejection frequencies at the nominal level of 5 per cent for the Wald$_s$ and Wald tests are presented in Table 9.10. From the rejection frequencies in this table for cases A and B, we can see that these are quite close to the nominal size, whereas in case C the empirical size is somewhat smaller. This corresponds to the theoretical results in Boswijk and Franses (1995), where it is shown that short-run dynamics may affect the asymptotic distributions. For cases D–G, we can see that the power of the tests increases with the sample size, and with the absolute values of the λ_s and θ_s parameters. It can also be observed that power of the joint Wald test is much larger than that of the individual Wald$_s$ tests. Note that this result is not unexpected since data are generated from a full PC model.

Table 9.11 reports the rejection frequencies of the F-tests for periodic variation, i.e. the F-tests for the hypotheses in (9.42)–(9.44). From these results we can see that in cases D and E the empirical size is close to the nominal size, but that the $F(\theta_s = \theta)$ and $F(\delta_s = \delta)$ tests suffer from size distortions in small samples. When the sample size increases, the empirical size gets closer to the nominal size. The power of the tests in cases F and G seems quite satisfactory. In summary, it seems that the extension of the Boswijk (1994) method to investigate

TABLE 9.10. Rejection frequencies of the Wald_s and Wald statistics[a]

Case[b]	n	Wald_1	Wald_2	Wald_3	Wald_4	Wald
Empirical size						
A	100	0.053	0.049	0.053	0.052	0.062
	200	0.053	0.048	0.051	0.052	0.056
B	100	0.054	0.051	0.054	0.052	0.063
	200	0.052	0.050	0.052	0.051	0.060
C	100	0.051	0.045	0.048	0.047	0.050
	200	0.045	0.042	0.043	0.047	0.041
Empirical power						
D	100	0.571	0.566	0.567	0.569	0.982
	200	0.964	0.962	0.964	0.964	1.000
E	100	0.455	0.456	0.456	0.464	0.879
	200	0.892	0.889	0.892	0.890	1.000
F	100	0.493	0.567	0.671	0.767	0.993
	200	0.932	0.962	0.984	0.995	1.000
G	100	0.129	0.399	0.723	0.910	0.989
	200	0.341	0.868	0.992	1.000	1.000

[a] The DGP is $\Delta_4 y_t = \Sigma_{s=1}^4 \lambda_s D_{s,t}(y_{t-4} - \tau_s x_{t-4}) + \gamma\Delta_4 y_{t-1} + \beta\Delta_4 x_t + \varepsilon_t$, $\Delta_q x_t = \Sigma_{s=1}^4 \kappa_s D_{s,t}(y_{t-4} - \theta_s x_{t-4}) + \eta_t$, with ε_t and η_t are i.i.d. $N(0,1)$. Based on 10,000 Monte Carlo replications. Nominal size is 5 per cent.

[b] The cases are: A: $\lambda_s = \lambda = 0$; B: $\lambda_s = \lambda = 0$ and q = 1; C: $\lambda_s = \lambda = 0$, q = 4 and $\gamma = 0.3$; D: $\lambda_s = \lambda = -0.5$, $\theta_s = \theta = 1$; E: case D with $\kappa_s = \kappa = 0.3$; F: $\lambda_s = \lambda = -0.5$, $\theta_1 = 0.8$, $\theta_2 = \theta_4 = 1$, $\theta_3 = 1.2$; and G: $\lambda_1 = -0.2$, $\lambda_2 = -0.4$, $\lambda_3 = -0.6$, $\lambda_4 = -0.8$, and $\theta_s = \theta = 1$. Unless otherwise indicated, $\gamma = 0$, $q = 4$, and $\kappa = 0$.

TABLE 9.11. Rejection frequencies of the F-statistics for parameter restrictions in periodic cointegration models[a]

Case[b]	n	$F(\lambda_s{=}\lambda)$	$F(\tau_s{=}\tau)$	$F(\delta_s{=}\delta)$
D	100	0.040	0.076	0.065
	200	0.046	0.069	0.063
E	100	0.040	0.077	0.070
	200	0.048	0.079	0.073
F	100	0.042	0.353	0.318
	200	0.049	0.841	0.807
G	100	0.420	0.092	0.437
	200	0.857	0.085	0.838

[a] Nominal size is 5 per cent.

[b] See note b in Table 9.10 for the DGPs.

periodic cointegration seems to work well, although for small samples one may need to treat some F-tests for periodic variation with some care.

Consumption and income in Sweden

A periodic error correction model can be expected to be most useful for the description of economic variables when economic agents may have seasonally varying preferences or when they face seasonally varying adjustment costs. (See Osborn (1988), where it is derived that a PIAR(1) model is useful to describe UK consumption nondurables; see also Chapter 8.) In Osborn (1988) it is also shown that this PIAR(1) model may be extended by incorporating lagged income. In the case of periodicity in the consumption variable, one useful extension of the univariate PIAR model is a PC model. In this subsection I investigate, for nondurables consumption and income data for Sweden, whether such a PC model yields a concise description of the quarterly time series. First, I start with a univariate analysis. Then I investigate periodic cointegration. Finally, I consider cointegration in a nonperiodic seasonal cointegration model and in seasonally adjusted data. It should be mentioned that all empirical models below are checked using the usual diagnostic measures. To save space, I do not report all details of the results of these diagnostic tests.

Univariate analysis

To select an appropriate univariate model for Swedish consumption and income, I start with a general PAR(6) model, which is subsequently simplified when F-tests indicate this option and when diagnostic measures indicate that the model cannot be rejected by the data. Consumption is denoted as c_t and income as y_t. For c_t the F_{PAR} test in the PAR(6) model obtains a value of 2.525**. It appears that the PAR(6) model can be reduced to the subset PAR(1,4) model, i.e.

$$c_t = \mu_s + \phi_{1s}c_{t-1} + \phi_{4s}c_{t-4} + \varepsilon_t. \tag{9.48}$$

The OLS estimation results are given in Table 9.12. As can be observed from these estimation results, the null hypothesis of no periodicity cannot be rejected. To test for a unit root in this c_t series, rewrite (9.48) as $\Phi_0\mathbf{C}_T = \mu + \Phi_1\mathbf{C}_{T-1} + \varepsilon_T$, with

$$\phi_0 = \begin{bmatrix} 1 & 0 & 0 & 0 \\ -\phi_{12} & 1 & 0 & 0 \\ 0 & -\phi_{13} & 1 & 0 \\ 0 & 0 & -\phi_{14} & 1 \end{bmatrix} \quad \text{and} \quad \phi_1 = \begin{bmatrix} \phi_{41} & 0 & 0 & \phi_{11} \\ 0 & \phi_{42} & 0 & 0 \\ 0 & 0 & \phi_{43} & 0 \\ 0 & 0 & 0 & \phi_{44} \end{bmatrix}. \tag{9.49}$$

TABLE 9.12. Parameter estimates for a subset PAR(1,4) model for quarterly consumption nondurables in Sweden[a]

	Quarter			
	1	2	3	4
$\hat{\mu}_s$	–0.211	0.322	0.096	0.278
	(0.129)	(0.084)	(0.074)	(0.083)
ϕ_{1s}	0.468	0.499	0.558	0.504
	(0.164)	(0.159)	(0.200)	(0.152)
ϕ_{4s}	0.623	0.359	0.393	0.389
	(0.118)	(0.184)	(0.196)	(0.160)

Diagnostics

$F_{\text{AR},1\text{–}1} = 1.261$ $F_{\text{ARCH},1\text{–}1} = 0.211$ $F_{Per\text{AR},1\text{–}1} = 1.188$

$F_{\text{AR},1\text{–}4} = 2.399$ $F_{\text{ARCH},1\text{–}4} = 0.536$ $JB = 2.942$

$F_{\text{PAR}} = 3.784^{**}$

[a] $c_t = \mu_s + \phi_{1s}c_{t-1} + \phi_{4s}c_{t-4} + \varepsilon_t$, where OLS standard errors are given in parentheses.

** Significant at the 5 per cent significance level.

The corresponding characteristic equation is

$$|\Phi_0 - \Phi_1 z| = (1 - \phi_{41}z)(1 - \phi_{42}z)(1 - \phi_{43}z)(1 - \phi_{44}z) - \phi_{11}\phi_{12}\phi_{13}\phi_{14}z = 0, \qquad (9.50)$$

and hence the restriction that yields a stochastic trend is

$$(1 - \phi_{41})(1 - \phi_{42})(1 - \phi_{43})(1 - \phi_{44}) - \phi_{11}\phi_{12}\phi_{13}\phi_{14} = 0. \qquad (9.51)$$

This restriction can be tested with the LR_1 or $LR_{1\tau}$ test in Chapter 8. When the estimated parameters $\hat{\phi}_{is}$ $(i = 1,4)$ are substituted in (9.50), the roots of the characteristic equation are the reciprocals of 0.936, 0.244, and 0.325 ± 0.211i. Hence it seems that there may be only a single unit root. The $LR_{1\tau}$ test obtains a value of –2.543, which is not significant at the 10 per cent level. When deterministic trend terms are included in (9.48), the $LR_{2\tau}$ test obtains the insignificant value of –2.766.

For the disposable income variable, I again start with a PAR(6) model. This model cannot be rejected by the data. However, the F_{PAR}-test statistic has an insignificant value of 0.663, which implies that a nonperiodic model is adequate. The next step is to test for seasonal unit roots using the HEGY test method. When two lags of $\Delta_4 y_t$ are added to the auxiliary HEGY regression, in addition to four seasonal dummies and one linear trend, the $t(\pi_1)$-test has a value of –1.500, the $t(\pi_2)$ has a value of –2.588, and the $F(\pi_3,\pi_4)$-test obtains a value of 5.761. Comparing these outcomes with the critical values in Table 5.1, I find that y_t is appropriately transformed using the Δ_4 filter. In fact, some (unreported) estimation results indicate that an AR(2) model for the $\Delta_4 y_t$ time series is empirically adequate. Taking the univariate results together, I can conclude that the total number of stochastic trends is equal to five. In the next empirical analysis, I investigate whether there are some long-run links between the two variables.

Periodic cointegration analysis

After some (unreported) experimentation with different lag lengths and various forms of parameter variation in the short-run dynamics, I decided to analyse the following PC model for Swedish consumption and income:

$$\Delta_4 c_t = \mu_s + \beta \Delta_4 y_t + \gamma \Delta_4 c_{t-1} + \lambda_s (c_{t-4} - \theta_s y_{t-4}) + \varepsilon_t. \qquad (9.52)$$

Unrestricted estimation of this model for 99 effective observations (1964.2–1988.4), which involves 14 parameters, yields estimated residuals for which the usual diagnostics do not indicate any misspecification. To investigate the presence of cointegration, I calculated the Engle–Granger-based statistics and the Boswijk–Wald tests and present the results in Table 9.13. The Engle–Granger-based test statistics indicate that the null hypothesis of no cointegration is rejected for the second and fourth quarters. The results for the Wald tests suggest that this hypothesis cannot be rejected for each of these quarters, or for all quarters. The Wald test statistics take the largest values in the second and fourth quarters. Although the evidence for cointegration may be viewed as not very strong, I proceed with the estimation of a PC model, where it is assumed that there may be cointegration in the second and fourth quarters. The estimation results for this partial model are

TABLE 9.13. Testing for periodic cointegration among nondurables consumption and disposable income in Sweden[a]

Quarter	Engle–Granger method[b]			Boswijk method[c]
	θ_s	CRDW_s	CRDF_s	Wald
1	0.885	0.703	−1.841	2.253
2	0.940	1.057	−3.526**	8.433
3	0.951	0.550	−1.792	0.744
4	0.661	1.380**	−3.551**	8.343
All				19.280

[a] Based on model (9.52)

[b] The θ_s refers to the estimated parameter for $Y_{s,T}$ of the regression of $C_{s,T}$ on 1 and $Y_{s,T}$ per quarter s. Estimated standard errors are given in parentheses and should be treated with care since θ_s is a superconsistent estimator (see Engle and Granger 1987). The CRDW_s is the cointegrating regression Durbin–Watson statistic, and the CRDF_s refers to the Dickey–Fuller t-ratio.

[c] The Wald test statistics are given in (9.38) and (9.39).

** Significant at the 5 per cent level.

* Significant at the 10 per cent level.

$$\Delta_4 c_t = \underset{(0.003)}{0.008} + \underset{(0.055)}{0.213}\Delta_4 y_t + \underset{(0.090)}{0.224}\Delta_4 c_{t-1} - \underset{(0.109)}{0.311} D_{2,t}(c_{t-4} - \underset{(0.261)}{0.049} - \underset{(0.107)}{0.868} y_{t-4})$$
$$- \underset{(0.160)}{0.448} D_{4,t}(c_{t-4} - \underset{(0.170)}{0.517} - \underset{(0.062)}{0.614} y_{t-4}) + \hat{\varepsilon}_t, \qquad (9.53)$$

where $R^2 = 0.325$ and where the (unreported) diagnostics do not indicate misspecification. The $F(\lambda_s = \lambda)$-test obtains an insignificant value of 0.507, the $F(\theta_s = \theta)$-test obtains the insignificant value of 2.663, and the joint test $F(\delta_s = \delta)$ has the value of 2.253. Hence each null hypothesis in (9.42)–(9.44) cannot be rejected. A test for the equality of all parameters in the ECM terms (including the two intercept terms) has a value of 3.938**. Our final estimated model therefore reads

$$\Delta_4 c_t = \underset{(0.002)}{0.008} + \underset{(0.055)}{0.196}\Delta_4 y_t + \underset{(0.091)}{0.238}\Delta_4 c_{t-1} - \underset{(0.080)}{0.260} D_{2,t}(c_{t-4} - \underset{(0.210)}{0.489} - \underset{(0.086)}{0.687} y_{t-4})$$
$$- \underset{(0.080)}{0.260} D_{4,t}(c_{t-4} - \underset{(0.236)}{0.313} - \underset{(0.086)}{0.687} y_{t-4}) + \hat{\varepsilon}_t, \qquad (9.54)$$

with diagnostics

$$F_{\text{AR},1\text{-}1} = 0.088, \quad JB = 3.732 \quad F_{\text{ARCH},1\text{-}1} = 0.249$$
$$F_{\text{AR},1\text{-}4} = 1.820, \qquad F_{\text{ARCH},1\text{-}4} = 0.409.$$

To test for weak exogeneity of y_t for the long-run parameters, I extract the error correction variables from (9.54) and add these to the univariate AR(2) model for $\Delta_4 y_t$. An F-test statistic for their joint significance is 2.985, which is not significant at the 5 per cent level, although it is at the 10 per cent level.

In sum, using a periodic cointegration model, one can adequately describe the quarterly consumption and income series for Sweden. Notice that the model in (9.54) contains only seven parameters. Because the series are in logs, the final PC model implies two long-run targets in terms of (9.29), i.e. $C = 1.631\, Y^{0.687}$ in the second quarter and $C = 1.368\, Y^{0.687}$ in the fourth quarter. This means that a change in the disposable income in the fourth quarter has a smaller long-run impact than it has in the second quarter. Finally, notice from (9.54) that the adjustment parameters are quite significant.

Nonperiodic analysis

Given the above finding that there appears to be cointegration between income and consumption in some quarters and not in other quarters, it now seems of interest to study the effect of analysing the same data using a nonperiodic cointegration model. This can be done using the seasonal cointegration model for the unadjusted series and with a standard cointegration model for SA time series.

We start with an analysis of the seasonal cointegration model. Before turning to this analysis, I examine the consumption series within a univariate HEGY framework to test for seasonal unit roots. The estimated HEGY auxiliary regression, which included four seasonal dummies, a trend, and $\Delta_4 c_{t-3}$, yields $t(\pi_1) = -2.235$, $t(\pi_2) = -1.714$, and $F(\pi_3, \pi_4) = 7.189^{**}$. Hence the appropriate filter for this consumption variable is $(1-B^2)$. Given that income appears to require the Δ_4 differencing filter, I proceed with testing whether c_t and y_t have their nonseasonal unit root and/or their seasonal unit at the biannual frequency in common.

The first seasonal cointegration method considered is the Engle *et al.* (1993) (EGHL) method. Here we start with a regression of $(1 + B + B^2 + B^3)c_t$ on a constant, three seasonal dummies, a trend, and $(1 + B + B^2 + B^3)y_t$. The residuals $\hat{u}_t$ are evaluated using a standard ADF regression (see (5.21)). With two lags of $\Delta_1 \hat{u}_t$ in this regression, this ADF test obtains the significant value of -3.595^{**}. When excluding the trend variable in the first regression, we find that the ADF is -2.516, which is insignificant. Hence there appears to be zero frequency cointegration between c_t and y_t around a linear trend. A test for seasonal cointegration at the biannual frequency amounts to regressing $(1 - B + B^2 - B^3)c_t$ on four seasonal dummies and $(1 - B + B^2 - B^3)y_t$, and evaluating the estimated residuals $\hat{v}_t$ in the ADF regression (5.22). This ADF, which includes four lags of $(1 + B)\hat{v}_t$, obtains an insignificant value of -1.709. Hence there is no seasonal cointegration at the frequency π.

These empirical results roughly correspond with the test results obtained using the seasonal cointegration approach advocated in Lee (1992) (see also Section 5.2). When estimating (5.24), where four lags of each Δ_4-differenced variable are needed, and calculating the relevant eigenvalues according to the Johansen (1991) method, I obtain the trace test statistic values 2.914 and 8.447 for the Π_1 matrix and 1.754 and 12.962 for the Π_2 matrix. Comparing these results with the critical values in Table 5.6, I conclude that this full system method yields no evidence for cointegration at the zero frequency nor at the biannual frequency.

Seasonally adjusted data

The univariate simulation and empirical results in Chapters 7 and 8 indicate that seasonally adjusting periodic time series with unit roots has an impact on the short-run dynamics of the time series, while it seems to preserve the stochastic trend properties. Hence in the context of periodic cointegration we may expect seasonally adjusting periodically cointegrated variables to weaken the evidence for cointegration since the short-run adjustment parameters may be affected.

To investigate this conjecture, I apply the linear Census X-11 seasonal adjustment filter, as it is given in Table 4.1, to the consumption and income series for Sweden. To forecast and backcast the observations, the method as advocated in Ooms (1994) is used, which considers a nonperiodic AR model with a trend and four seasonal dummies. These data are denoted by $\hat{c}_t^{ns}$ and $\hat{y}_t^{ns}$.

Univariate analysis of the $\hat{c}_t^{ns}$ and $\hat{y}_t^{ns}$ series results in ADF test values of

-2.284 and -1.531 in auxiliary regressions which include two and three lags of Δ_1 transformed time series, respectively. When a PAR(3) model is estimated for $\hat{c}_t^{ns}$, the F_{PAR}-test obtains the insignificant value of 0.587. Hence the application of the SA filter does not yield significant periodic parameter variation. When the standard Boswijk approach is applied (see Section 2.4), a nonperiodic version of (9.54) is obtained, i.e.

$$\underset{}{\Delta_1\hat{c}_t^{ns}} = \underset{(0.030)}{0.036} - \underset{(0.090)}{0.375}\Delta_1\hat{c}_{t-1}^{ns} + \underset{(0.051)}{0.164}\Delta_1\hat{y}_t^{ns} - \underset{(0.055)}{0.111}(\hat{c}_{t-1}^{ns} - \underset{(0.128)}{0.746}\hat{y}_{t-1}^{ns}) + \hat{\varepsilon}_t, \tag{9.55}$$

which cannot be rejected by the data. The Wald test for cointegration in this nonperiodic ECM obtains the insignificant value of 5.854. When the ECM is enlarged with a linear trend term, the Wald test has the insignificant value of 10.792. Notice from (9.55) that the key difference from (9.54) is that the absolute value of the adjustment parameter drops from 0.260 to 0.111.

In summary, the analysis of SA data shows that a linear adjustment filter can affect cointegration inference. The empirical results above suggest that it seems useful to analyse periodic data using a periodic model.

9.4 Seasonal Variation in Business Cycles

In this final section, I turn again to an investigation of the impact of the nonperiodic analysis of data that are generated from periodic cointegration models. In Chapter 8 I have already illustrated for univariate PIAR time series that nonperiodic analysis of such data may lead to the finding of seasonal variation in business cycle peaks and troughs and to a positive correlation between certain 'seasonal' and 'business' cycles. In this section I consider two Monte Carlo simulation experiments to check whether the univariate results extend to multivariate periodic models with stochastic trends.

First, consider the auxiliary regression (3.7), which is advocated in Beaulieu *et al.* (1992) and Miron (1994), and which is used in these studies to analyse the correlation between 'seasonal' and 'business' cycles. The regression is

$$\Delta_1 y_t = \delta_1 D_{1,t} + \delta_2 D_{2,t} + \delta_3 D_{3,t} + \delta_4 D_{4,t} + v_t. \tag{9.56}$$

The 'seasonal' cycle is assumed to be the variance of the fit of regression (9.56), and the 'business' cycle is assumed to be the variance of the $\hat{v}_t$. In Beaulieu *et al.* (1992) it is shown that, for several countries and macroeconomic variables, the two variances are positively correlated. In the previous chapters we have seen that (9.56) is unlikely to amount to an adequate data description. In Chapter 8 I found that similar results as in the above studies can be found using PIAR models with increasing variation in the periodic parameters in the periodic differencing filter.

To investigate whether periodic cointegration models can generate similar positive correlation between the variance of the fit and of the residuals from (9.46), I generate observations from the PC model:

$$\Delta_4 y_t = \alpha_s(y_{t-4} - \beta_s x_{t-4}) + 0.5\Delta_4 y_{t-1} + \varepsilon_t \quad (9.57)$$

and

$$\Delta_1 x_t = 0.2 + 1.2\Delta_1 x_{t-1} - 0.4\Delta_1 x_{t-2} + \phi_t, \quad (9.58)$$

where ε_t and ϕ_t are i.i.d. $N(0,1)$ variables. The x_t variable displays cyclical behaviour around a stochastic trend.

Table 9.14 reports the variance of the fit and of the residuals from the regression (9.56) for different choices of α_s and β_s. Since I generate the data using the same ε_t and ϕ_t variables, the results in this table can be evaluated across the various values of α_s and β_s. The outcomes clearly indicate that when the variation in α_s or β_s increases the variances of the fit and the residuals increase as well. Hence a positive correlation between these two latter variances may be caused by more variation in the periodic cointegration parameters.

The second investigation into the outcomes of a nonperiodic analysis of periodically cointegrated time series concerns the determination of business

TABLE 9.14. Some estimation results from the auxiliary regression $\Delta_1 y_t = \delta_1 D_{1,t} + \delta_2 D_{2,t} + \delta_3 D_{3,t} + \delta_4 D_{4,t} + v_t$, when the data are generated by the periodic cointegration model[a]

α_3^b	σ^2_{fit}	σ^2_v	β_3^c	σ^2_{fit}	σ^2_v
–0.20	1.281	1.941	1.00	0.450	2.286
–0.25	1.720	2.398	1.05	4.012	3.576
–0.30	3.190	2.980	1.10	8.289	5.745
–0.35	4.341	3.512	1.15	12.572	8.115
–0.40	5.240	3.968	1.20	16.857	10.553
–0.45	5.961	4.357	1.25	21.142	13.020
–0.50	6.553	4.687	1.30	25.428	15.502
–0.55	7.048	4.970	1.35	29.713	17.993
–0.60	7.470	5.215	1.40	34.000	20.490
–0.65	7.834	5.427	1.45	38.285	22.991
–0.70	8.152	5.614	1.50	42.571	25.495
–0.75	8.432	5.779	1.55	46.857	28.000
–0.80	8.681	5.927	1.60	51.143	30.507
–0.85	8.902	6.059	1.65	55.429	33.016
–0.90	9.103	6.177	1.70	59.715	35.525

[a] The periodic cointegration model: $\Delta_4 y_t = \alpha_s(y_{t-4} - \beta_s x_{t-4}) + 0.5\Delta_4 y_{t-1} + \varepsilon_t$ and $\Delta_1 x_t = 0.2 + 1.2\Delta_1 x_{t-1} - 0.4\Delta_1 t_{t-2} + \phi_t$, with ε_t and ϕ_t i.i.d $N(0,1)$. The estimation results are based on a single simulation run, where the ε_t and ϕ_t processes are the same for each parameter combination. The effective sample size is 124 observations. The auxiliary regression is the same as advocated in Miron (1994). σ^2_{fit} is the variance of $(\Delta_1 y_t)$ from the auxiliary regression and σ^2_v is the estimated variance of v_t.

[b] The α_1, α_2 and α_4 values are set equal to –0.2.

[c] The β_1, β_2, and β_4 values are set equal to 1.

cycle turning points using SA data. Ghysels (1994) convincingly argues that the distribution of such peaks and troughs is not constant throughout the year, and also that the duration of contractions and expansions varies with the season in which such a period starts. Again, I simulate data from a PC model such as (9.57) and (9.58) to study whether a PC model may be useful to describe unadjusted data when the SA data are found to display seasonal variation in peaks and troughs. The results are reported in Table 9.15. The simulation experiment amounts to generating an effective sample of 2,000 SA data, where again I use the linear SA method in Table 4.1.

The first panel of Table 9.15 shows that, when the data are generated from a nonperiodic error correction model, the distribution of peaks and troughs and the median of the business cycle period length do not display obvious seasonal variation. However, when the α_s parameter in (9.57) is allowed to vary with the season, as occurs in the second panel of the table, we can observe marked changes in the distribution of peaks and troughs and in the length of the cycle. From the third panel it can be seen that such changes seem smaller when the β_s parameter is allowed to vary. Hence it appears that the adjustment parameter in the PC model can have the most impact on seasonal variation in peaks and troughs.

TABLE 9.15. Seasonality in business cycles when data are generated by a periodic cointegration model and analysed in seasonally adjusted form[a]

Parameters			Quarters			
			1	2	3	4
$\alpha=-0.5, \beta=1$	Contraction	No. of peaks	37	31	33	23
		Median length	6	6	6	6
	Expansion	No. of troughs	23	29	32	40
		Median length	6	7	6	6
$\alpha_1=-0.2, \alpha_2=-0.4$	Contraction	No. of peaks	38	11	27	53
$\alpha_3=-0.6, \alpha_4=-0.8$		Median length	3	6	5	4
$\beta=1$	Expansion	No. of troughs	11	8	9	103
		Median length	5	8	4	4
$\alpha=-0.5$	Contraction	No. of peaks	3	32	33	54
$\beta_1=0.8, \beta_2=0.9$		Median length	5	5	6	7
$\beta_3=1.0, \beta_4=1.1$	Expansion	No. of troughs	37	29	39	17
		Median length	6	6	5	6

[a] The DGP is $\Delta_4 y_t = \alpha_s(y_{t-4} - \beta_s x_{t-4}) + 0.5\,\Delta_4 y_{t-1} + \varepsilon_t$ and $\Delta_1 x_t = 0.2 + 1.2\Delta_1 x_{t-1} - 0.55\Delta_1 x_{t-2} + \phi_t$, with ε_t and ϕ_t i.i.d $N(0,1)$. Effective sample length is 2,000 quarterly observations. The cells are the frequency that a certain quarter is found to be a peak or a trough and the median of the length of a contraction or expansion. The data are seasonally adjusted using the linear Census X-11 filter in Table 4.1. A contraction is defined as two consecutive quarters for which $\Delta_1 \hat{y}_t^{ns}$ is negative. The results are based on a single simulation experiment. The data are all generated using the same ε_t and ϕ_t processes.

Conclusion

In this chapter I have considered a few specific periodic models for multivariate time series. Such specific structures appear necessary in practice since general PVAR models are difficult to analyse with respect to the presence of (common) stochastic trends. The presented empirical results are encouraging in the sense that reasonably simple multivariate models can be fitted to sets of time series with complicated univariate properties. The periodic cointegration model may have links with some economic theoretical considerations. Finally, it appears that nonperiodic analysis of periodically cointegrated variables results in less evidence in favour of cointegration.

10

Conclusion

In this book I have considered the class of periodic time series models with stochastic trends to describe and forecast seasonally observed time series. I have argued that this class is particularly useful for several macroeconomic time series since such models can describe variables for which (i) seasonal patterns change slowly over time, (ii) seasonality is a dominant source of variation in a time series, and (iii) the seasonal fluctuations depend on the business cycle and stochastic trend fluctuations. In Chapter 3 I showed that properties (i) and (ii) are apparent for several example series. In the next chapter I reviewed commonly used seasonal adjustment methods. In Chapter 5 I reviewed recent developments in nonperiodic models with nonseasonal and seasonal unit roots. Seasonal adjustment methods and seasonal unit root models share the property that it is assumed that seasonal and nonseasonal movements can be meaningfully separated. In Chapter 6, however, I used several examples to conjecture that for some macroeconomic variables it may be difficult, if not impossible, to separate seasonal from nonseasonal movements. This led me to propose the class of periodic models with stochastic trends, which can take account of the various documented empirical regularities.

In Chapter 7, I surveyed important aspects of periodic time series models. I proposed a simple model selection strategy, and applied it to the example series. For all these quarterly series, which concern variables such as industrial production in the USA, total consumption in the UK, unemployment in Canada, and real GNP in Germany, strong evidence is found for periodic dynamics. Using extensive Monte Carlo experiments, I showed that the model selection strategy performs well, and that, for example, the probability of fitting periodic models while nonperiodic models are adequate is small. Furthermore, I investigated the effects of seasonal adjustment of periodic time series. I found that such seasonally adjusted series show more persistence than unadjusted periodic series. Additionally, it appears that nonperiodic models for periodic time series need many lags to whiten the error process.

In Chapter 8, I focused on periodic integration for univariate time series. Loosely speaking, periodic integration amounts to using a differencing filter to remove the stochastic trend that varies with the season. I proposed statistics to test for periodic integration. An application of the tests to our sample series leads to the conclusion that many macroeconomic time series may be periodically integrated of order 1, i.e. that each series is generated by only a single stochastic trend. A possible drawback of periodic integration is that the differencing filter has to be estimated from the data. However, the simulation and empirical results show that the forecasting gain from periodic integration relative to alternative data descriptions can be large. Moreover, a nonperiodic model for periodically integrated time series may suggest the presence of a seasonal unit root. In this chapter I also showed that a property of periodic integration is that it allows seasonal fluctuations to depend on the stochastic trend. This implies that shocks in certain seasons have different long-run impacts from shocks in other seasons. This, in turn, may lead to periodic patterns in, say, business cycle turning points, even after seasonal adjustment filters have been applied.

In Chapter 9 I proposed multivariate extensions of the univariate periodic integration concept. I showed that a standard extension of a vector autoregression that allows for periodic parameters may be difficult to analyse for the presence of stochastic trends. In the rest of the chapter therefore I considered two possible routes which may yield useful empirical models. The first assumes that a multivariate periodic model for nonstationary time series can be constructed in two steps: the first step amounts to an extraction of stochastic trends from the univariate series; the second correlates the estimated stochastic trends in a cointegration analysis. I proposed several methods and evaluated these using extensive simulation experiments. The second route assumes the possible adequacy of a so-called periodic cointegration model. This model allows equilibrium parameters and adjustment parameters to vary across the seasons. Several specification strategies are proposed and applied to some sample series. An analysis of the effects of a nonperiodic analysis, possibly with seasonally adjusted time series, showed that one may then find less evidence in favour of cointegration. Hence, again, a nonperiodic analysis yields too many unit roots.

In Chapters 8 and 9 I argued that univariate and multivariate periodic time series models with stochastic trends can have links with theoretical economic models. For example, the extension of the life cycle–permanent income hypothesis in Osborn (1988), where consumers are assumed to have seasonally varying utility functions, implies a periodic autoregression of order 1 with a unit root. However, it is my view that additional research is needed to embed the empirically successful periodic integration and cointegration models into a theoretical framework. Recently there have appeared several economic studies which explicitly have focused on seasonality, and which have sometimes considered periodic parameter variation. For example, Miron and Zeldes (1988) and Krane (1993) consider seasonality in production-smoothing models of inventories. Faig (1989) considers seasonal fluctuations in money demand. Miron (1986) and Osborn (1988) study how the life cycle–permanent income hypothesis can be modified to allow for the marked seasonal fluctuations in consumption. Todd (1990), Chatterjee and Ravikumar (1992), Hansen and Sargent (1993), and Braun and Evans (1995) propose modifications of (variants of) real business cycle models to cope with apparent seasonal fluctuations in macroeconomic variables such as consumption and production. Future research should indicate whether the class of models can be fitted into a more formal economic framework.

There are also some topics for further econometric research. The first considers the number of parameters. In Chapters 8 and 9 I showed that the periodic integration and cointegration models in a sense encompass nonperiodic models with seasonal unit roots, since these can describe the same as well as additional phenomena. Of course, such encompassing often involves the estimation of additional parameters, although it is observed that the increase in the number of parameters can be quite small. All the sample series concerned quarterly time series variables. I have some (unreported) experience with monthly time series with periodic patterns, and it appears that for some specific cases such models can become cumbersome to analyse empirically. One can imagine that for periodic vector autoregressions for multivariate time series such problems become even larger. Hence further research may be dedicated to proposing useful empirical modelling strategies for monthly time series.

A second topic concerns nonlinearity. Recent research summarized in, e.g., Granger and Teräsvirta (1993) shows that several economic time series display nonlinear features such as threshold and limit cycle behaviour. Empirical models that appear useful are, for example, smooth transition regression models and neural network models. It seems important to investigate how periodicity in seasonal time series can be incorporated in such nonlinear models.

I can summarize this book by making two main conclusions. First, periodic models for economic time series with stochastic trends can adequately describe and forecast macroeconomic variables. Such models can also generate observed empirical regularities, even when these regularities are measured using inadequate empirical models. Second, one should preferably not construct econometric models for seasonally adjusted time series. This is because seasonally adjusted data suggest too much persistence of shocks and can hide important long-run and short-run correlations between variables; moreover, the empirical results in this book show that it is not at all difficult to construct reasonably simple models for seasonally unadjusted data.

Data Appendix

This appendix lists the sources of the data that are used in the empirical applications throughout the various chapters. The data themselves are also available upon request from the author. All data are analysed in logs, except when otherwise indicated.

The total industrial production index (1985 = 100) for the USA (1960.1–1991.4) is obtained from the OECD *Main Economic Indicators*. The unemployment in Canada (×1,000, 1960.1–1987.4) is also obtained from the OECD *Main Economic Indicators*. This time series is analysed without any prior transformation. The source of real GNP in Germany (1960.1–1990.4) is Wolters (1992). These three sources include seasonally unadjusted and adjusted data.

The source of total investment in the UK (1955.1–1988.4) at 1985 prices is given in Osborn (1990). The same applies for gross domestic product at 1985 prices, total consumption at 1985 prices, nondurables consumption at 1985 prices, exports of goods and services at 1985 prices, imports of goods and services at 1985 prices, public investment at 1985 prices, and workforce, consisting of workforce in employment and the unemployed for the UK, and all for the sample 1955.1–1988.4.

Finally, the source of real per capita nondurables consumption and real per capita disposable income in Sweden for 1963.1–1988.4 is Assarsson (1991). The seasonally adjusted data are created by applying the linear Census X-11 filter given in Table 4.1, where the forecasts and backcasts are generated as described in Ooms (1994).

References

Abeysinghe, T. (1994), 'Deterministic Seasonal Models and Spurious Regressions', *Journal of Econometrics*, 61: 259–72.

Abraham, B. and A. Chuang (1989), 'Outlier Detection and Time Series Modeling', *Technometrics*, 31: 241–8.

—— and J. Ledolter (1983), *Statistical Methods for Forecasting*. New York: John Wiley.

—— and N. Yatawara (1988), 'A Score Test for Detection of Time Series Outliers', *Journal of Time Series Analysis*, 9: 109–19.

Adams, G.J. and G. C. Goodwin (1995), 'Parameter Estimation for Periodic ARMA Models', *Journal of Time Series Analysis*, 16: 127–45.

Agiakloglou, C. and P. Newbold (1992), 'Empirical Evidence on Dickey–Fuller Type Tests', *Journal of Time Series Analysis*, 13: 471–83.

Akaike, H. (1969), 'Fitting Autoregressive Models for Prediction', *Annals of the Institute of Statistical Mathematics*, 21: 243–7.

—— (1974), 'A New Look at the Statistical Model Identification', *IEEE Transactions on Automatic Control*, AC-19: 716–23.

Andel, J. (1983), 'Statistical Analysis of Periodic Autoregression', *Aplikace Matematiky*, 28: 364–85.

Anderson, P. L. and A. V. Vecchia (1993), 'Asymptotic Results for Periodic Autoregressive Moving-Average Models', *Journal of Time Series Analysis*, 14: 1–18.

Anderson, T. W. (1971), *The Statistical Analysis of Time Series*. New York: John Wiley.

—— and A. M. Walker (1964), 'On the Asymptotic Distribution of the Autocorrelations of a Sample from a Linear Stochastic Process', *Annals of Mathematical Statistics*, 35: 1296–1303.

Andrews, D. W. K. (1993), 'Tests for Parameter Instability and Structural Change with Unknown Change Point', *Econometrica*, 61: 821–56.

Assarsson, B. (1991), 'The Stochastic Behaviour of Durables and Nondurables Consumption in Sweden', Working Paper 1991:21, Department of Economics, Uppsala University.

Baillie, R. T. (1996), Long Memory Processes and Fractional Integration in Econometrics, *Journal of Econometrics*.

Banerjee, A., J. Dolado, J. W. Galbraith, and D. F. Hendry (1993), *Co-integration, Error Correction, and the Econometric Analysis of Non-stationary Data*. Oxford: Oxford University Press.

Barsky, R. B. and J. A. Miron (1989), 'The Seasonal Cycle and the Business Cycle', *Journal of Political Economy*, 97: 503–35.

Beaulieu, J. J. and J. A. Miron (1993), 'Seasonal Unit Roots in Aggregate US Data', *Journal of Econometrics*, 55: 305–28.

—— J. K. MacKie–Mason, and J. A. Miron (1992), 'Why Do Countries and Industries with Large Seasonal Cycles Also Have Large Business Cycles?' *Quarterly Journal of Economics*, 107: 621–56.

Bell, W. R. (1984), 'Signal Extraction for Nonstationary Time Series', *Annals of Statistics*, 12: 646–64.

—— (1987), 'A Note on Overdifferencing and the Equivalence of Seasonal Time Series Models with Monthly Means and Models with $(0,1,1)_{12}$ Seasonal Parts when $\theta = 1$', *Journal of Business and Economic Statistics*, 5: 383–7.

—— and S. C. Hillmer (1984), 'Issues Involved with the Seasonal Adjustment of Economic Time Series' (with discussion), *Journal of Business and Economic Statistics*, 2: 291–320.

Bentarzi, M. and M. Hallin (1994*a*), 'On the Invertibility of Periodic Moving-Average Models', *Journal of Time Series Analysis*, 15: 263–8.

—— —— (1994*b*), 'Locally Optimal Tests against Periodical Autoregression: Parametric and Nonparametric Approaches', unpublished manuscript, Free University Brussels.

Bera, A. K. and C. M. Jarque (1982), 'Model Specification Tests: a Simultaneous Approach', *Journal of Econometrics*, 20: 59–82.

Bewley, R., D. Orden, M. Yang, and L. A. Fisher (1994), 'Comparison of Box–Tiao and Johansen Canonical Estimators of Cointegrating Vectors in VEC(1) Models', *Journal of Econometrics*, 64: 3–28.

Bhargava, A. (1990), 'An Econometric Analysis of the US Postwar GNP', *Journal of Population Economics*, 3: 147–56.

Birchenhall, C. R., R. C. Bladen-Hovell, A. P. L. Chui, D. R. Osborn, and J. P. Smith (1989), 'A Seasonal Model of Consumption', *Economic Journal*, 99: 837–43.

Bloomfield, P., H. L. Hurd, and R. B. Lund (1994), 'Periodic Correlation in Stratospheric Ozone Data', *Journal of Time Series Analysis*, 15: 127–50.

Boldin, M. D. (1994), 'Dating Turning Points in the Business Cycle', *Journal of Business*, 67: 97–131.

Bollerslev, T. (1986), 'Generalized Autoregressive Conditional Heteroskedasticity', *Journal of Econometrics*, 31: 307–27.

Boswijk, H. P. (1992), *Cointegration, Identification and Exogeneity*. Amsterdam: Thesis.

—— (1994), 'Testing for an Unstable Root in Conditional and Structural Error Correction Models', *Journal of Econometrics*, 63: 37–60.

—— (1995), 'Asymptotic Theory for Integrated Processes', unpublished manuscript', University of Amsterdam.

—— and P. H. Franses (1992), 'Dynamic Specification and Cointegration', *Oxford Bulletin of Economics and Statistics*, 54: 369–81.

—— —— (1995), 'Periodic Cointegration: Representation and Inference', *Review of Economics and Statistics*, 77: 436–54.

—— —— (1996), 'Unit Roots in Periodic Autoregressions', *Journal of Time Series Analysis*.

—— —— and N. Haldrup (1995), 'Multiple Unit Roots in Periodic Autoregression', unpublished manuscript, Erasmus University Rotterdam.

Bowerman, B. L., A. B. Koehler, and D. J. Pack (1990), 'Forecasting Time Series with Increasing Seasonal Variation', *Journal of Forecasting*, 9: 419–36.

Box, G. E. P. and G. M. Jenkins (1970), *Time Series Analysis; Forecasting and Control*. San Francisco: Holden-Day.

—— and D. A. Pierce (1970), 'Distribution of Residual Autocorrelation in Autoregressive Integrated Moving Average Time Series Models', *Journal of the American Statistical Association*, 65: 1509–26.

—— and G. C. Tiao (1977), 'A Canonical Analysis of Multiple Time Series', *Biometrika*, 64: 355–65.

Braun, R. A. and C. L. Evans (1995), 'Seasonality and Equilibrium Business Cycle Theories', *Journal of Economic Dynamics and Control*, 19: 503–32.

Breitung, J. (1994), 'Some Simple Tests of the Moving-Average Unit Root Hypothesis', *Journal of Time Series Analysis*, 15: 351–70.

—— and P. H. Franses (1995), 'Impulse Response Functions for Periodic Integration', Discussion Paper, Institute of Statistics and Econometrics, Humboldt University Berlin.

Burridge, P. and K. F. Wallis (1985), 'Calculating the Variance of Seasonally Adjusted Series', *Journal of the American Statistical Association*, 80: 541–52.

—— —— (1990), 'Seasonal Adjustment and Kalman Filtering: Extension to Periodic Variances', *Journal of Forecasting*, 9: 109–18.

Campbell, J. Y. and P. Perron (1991), 'Pitfalls and Opportunities: What Macroeconomists Should Know about Unit Roots', in O. J. Blanchard and S. Fisher (eds.), *NBER Macroeconomics Annual*. Boston: MIT Press.

Canova, F. and E. Ghysels (1994), 'Changes in Seasonal Patterns: Are They Cyclical?' *Journal of Economic Dynamics and Control*, 18: 1143–71.

—— and B. E. Hansen (1995), 'Are Seasonal Patterns Constant over Time? A Test for Seasonal Stability', *Journal of Business and Economic Statistics*, 13: 237–52.

Chatterjee, S. and B. Ravikumar (1992), 'A Neoclassical Model of Seasonal Fluctuations', *Journal of Monetary Economics*, 29: 59–86.

Chen, C. and L. Liu (1993), 'Joint Estimation of Model Parameters and Outlier Effects in Time Series', *Journal of the American Statistical Association*, 88: 284–97.

Chow, G. C. (1960), 'Tests of Equality Between Sets of Coefficients in Two Linear Regressions', *Econometrica*, 28: 591–605.

Cipra, T. (1985), 'Periodic Moving Average Processes', *Aplikace Matematiky*, 30: 218–29.

Cleveland, W. P. and G. C. Tiao (1976), 'Decomposition of Seasonal Time Series: a Model for the X-11 Program', *Journal of the American Statistical Association*, 71: 581–7.

—— —— (1979), 'Modelling Seasonal Time Series', *Revue Economique Appliquee*, 32: 107–29.

Davidson, J. E. H., D. F. Hendry, F. Srba, and S. Yeo (1978), 'Econometric Modelling of the Aggregate Time-Series Relationship Between Consumers' Expenditure and Income in the United Kingdom', *Economic Journal*, 88: 661–92.

Davisson, L. D. (1965), 'The Predictive Error of Stationary Gaussian Time Series of Unknown Variance', *IEEE Transactions on Information Theory*, IT-11: 527–37.

De Gooijer, J. J., B. Abraham, A. Gould, and L. Robinson (1985), 'Methods for Determining the Order of an Autoregressive–Moving Average Process: a Survey', *International Statistical Review*, 85: 301–29.

Dickey, D. A. and W. A. Fuller (1979), 'Distribution of the Estimators for Autoregressive Time Series with a Unit Root', *Journal of the American Statistical Association*, 74: 427–31.

—— —— (1981), 'Likelihood Ratio Statistics for Autoregressive Time Series with a Unit Root', *Econometrica*, 49: 1057–72.

—— D. P. Hasza, and W. A. Fuller (1984), 'Testing for Unit Roots in Seasonal Time Series', *Journal of the American Statistical Association*, 79: 355–67.

Dickey, D. A., W. R. Bell and R. B. Miller (1986), 'Unit Roots in Time Series Models: Tests and Implications', *American Statistician*, 40: 12–26.

Dolado, J. J., J. W. Galbraith, and A. Banerjee (1991), 'Estimating Euler Equations with Integrated Series', *International Economic Review*, 32: 919–36.

Engle, R. F. (1982), 'Autoregressive Conditional Heteroscedasticity with Estimates of the Variance of UK Inflation', *Econometrica*, 50: 987–1008.

—— and C. W. J. Granger (1987), 'Co-integration and Error Correction: Representation, Estimation, and Testing', *Econometrica*, 55: 251–76.

—— and S. Kozicki (1993), 'Testing for Common Features' (with discussion), *Journal of Business and Economic Statistics*, 11: 369–95.

—— D. F. Hendry, and J. F. Richard (1983), 'Exogeneity', *Econometrica*, 51: 277–304.

—— C. W. J. Granger, and J. J. Hallman (1989), 'Merging Short- and Long-Run Forecasts: an Application of Seasonal Cointegration to Monthly Electricity Sales Forecasting', *Journal of Econometrics*, 40: 45–62.

—— —— S. Hylleberg, and H. S. Lee (1993), 'Seasonal Cointegration: the Japanese Consumption Function', *Journal of Econometrics*, 55: 275–98.

Engsted, T. and N. Haldrup (1992), 'Testing Quadratic Adjustment Cost Models in Cointegrated VAR Models'. Institute of Economics Memo 1992–9, Aarhus University.

Ericsson, N. R., D. F. Hendry, and H. A. Tran (1994), 'Cointegration, Seasonality, Encompassing, and the Demand for Money in the United Kingdom', in C. P. Hargreaves (ed.), *Nonstationary Time Series Analysis and Cointegration*. Oxford: Oxford University Press.

Faig, M. (1989), 'Seasonal Fluctuations and the Demand for Money', *Quarterly Journal of Economics*, 99: 847–61.

Fox, A. J. (1972), 'Outliers in Time Series', *Journal of the Royal Statistical Society* B, 34: 350–63.

Franses, P. H. (1991*a*), *Model Selection and Seasonality in Time Series*. Amsterdam: Thesis.

—— (1991*b*), 'Seasonality, Nonstationarity and the Forecasting of Monthly Time Series', *International Journal of Forecasting*, 7: 199–208.

—— (1993), 'Periodically Integrated Subset Autoregressions for Dutch Industrial Production and Money Stock', *Journal of Forecasting*, 12: 601–13.

—— (1994), 'A Multivariate Approach to Modeling Univariate Seasonal Time Series', *Journal of Econometrics*, 63: 133–51.

—— (1995*a*), 'A Differencing Test', *Econometric Reviews*, 14: 183–93.

—— (1995*b*), 'On Periodic Autoregressions and Structural Breaks in Seasonal Time Series', *Environmetrics*, 6: 451–5.

—— and N. Haldrup (1994), 'The Effects of Additive Outliers on Tests for Unit Roots and Cointegration', *Journal of Business and Economic Statistics*, 12: 471–8.

—— and B. Hobijn (1995), 'Critical Values for Unit Root Tests in Seasonal Time Series', *Journal of Applied Statistics*.

—— and T. Kloek (1995), 'A Periodic Cointegration Model of Quarterly Consumption', *Applied Stochastic Models and Data Analysis*, 11: 159–66.

—— and A. B. Koehler (1994), 'Model Selection Strategies for Time Series with Increasing Seasonal Variation', revised version of Econometric Institute Report 9308, Erasmus University Rotterdam.

—— and M. McAleer (1995*a*), 'Testing for Unit Roots and Non-linear Transformations', Econometric Institute Report 9507, Erasmus University Rotterdam.

—— —— (1995*b*), 'Testing Nested and Nonnested Periodically Integrated Autoregressive Models', Center Discussion Paper 9510, Tilburg University.

—— and M. Ooms (1995), 'Forecasting Changing Seasonal Components in US and German Unemployment using Periodic Correlations', Tinbergen Institute Discussion Paper 95–77, Erasmus University Rotterdam.

—— and R. Paap (1995), 'Seasonality and Stochastic Trends in German Consumption and Income, 1960.1–1987.4', *Empirical Economics*, 20: 109–32.

—— and G. Romijn (1993), 'Periodic Integration in Quarterly UK Macroeconomic Variables', *International Journal of Forecasting*, 9: 467–76.

—— and T. J. Vogelsang (1995), 'Testing for Seasonal Unit Roots in the Presence of Changing Seasonal Means', Econometric Institute Report 9532, Erasmus University Rotterdam.

—— S. Hylleberg, and H. S. Lee (1995), 'Spurious Deterministic Seasonality', *Economics Letters*, 48: 249–56.

Fuller, W. A. (1976), *Introduction to Statistical Time Series*. New York: John Wiley.

Gersovitz, M. and J. G. MacKinnon (1978), 'Seasonality in Regression: an Application of Smoothness Priors', *Journal of the American Statistical Association*, 73: 264–73.

Ghysels, E. (1988), 'A Study towards a Dynamic Theory of Seasonality for Economic Time Series', *Journal of American Statistical Association*, 83: 168–72.

—— (1990), 'Unit Root Tests and the Statistical Pitfalls of Seasonal Adjustment: the Case of US Postwar Real Gross National Product', *Journal of Business and Economic Statistics*, 8: 145–52.

—— (1993), 'A Time Series Model with Periodic Stochastic Regime Switching', unpublished manuscript, CRDE, University of Montreal.

—— (1994), 'On the Periodic Structure of the Business Cycle', *Journal of Business and Economic Statistics*, 12: 289–98.

—— and A. Hall (1993*a*), 'Testing Periodicity in Some Linear Macroeconomic Models', unpublished manuscript, CRDE, University of Montreal.

—— —— (1993*b*), 'On Periodic Time Series and Testing the Unit Root Hypothesis', CRDE Discussion Paper 2693, University of Montreal.

—— and M. Nerlove (1988), 'Seasonality in Surveys: a Comparison of Belgian, French and Business Survey Tests', *European Economic Review*, 32: 81–99.

—— and P. Perron (1993), 'The Effect of Seasonal Adjustment Filters on Tests for a Unit Root', *Journal of Econometrics*, 55: 57–98.

—— A. Hall, and H. S. Lee (1994*a*), 'On Periodic Structures and Testing for Seasonal Unit Roots', unpublished manuscript, University of Montreal.

—— H. S. Lee, and J. Noh (1994*b*), 'Testing for Unit Roots in Seasonal Time Series', *Journal of Econometrics*, 62: 415–42.

—— C. W. J. Granger, and P. L. Siklos (1995), 'Is Seasonal Adjustment a Linear or Nonlinear Data-Filtering Process?' Unpublished manuscript, CRDE University of Montreal.

Gladyshev, E. G. (1961), 'Periodically Correlated Random Sequences', *Soviet Mathematics*, 2: 385–8.

Godfrey, L. G. (1979), 'Testing the Adequacy of a Time Series Model', *Biometrika*, 66: 67–72.

Gonzalo, J. and C. W. J. Granger (1995), 'Estimation of Common Long-Memory Components in Cointegrated Systems', *Journal of Business and Economic Statistics*, 13: 27–36.

Gourieroux, C., A. Monfort, and E. Renault (1991), 'A General Framework for Factor Models', Discussion Paper 9107, INSEE, Paris.

Granger, C. W. J. (1986), 'Developments in the Study of Cointegrated Variables', *Oxford Bulletin of Economics and Statistics*, 48: 213–28.

—— and J. Hallman (1991), 'Nonlinear Transformations of Integrated Time Series', *Journal of Time Series Analysis*, 12: 207–24.

—— and R. Joyeux (1980), 'An Introduction to Long-Memory Time Series Models and Fractional Differencing', *Journal of Time Series Analysis*, 1: 15–39.

—— and P. Newbold (1974). 'Spurious Regressions in Econometrics', *Journal of Econometrics*, 2: 111–20.

—— —— (1986), *Forecasting Economic Time Series*, 2nd edn. San Diego: Academic Press.

—— and T. Teräsvirta (1993), *Modelling Nonlinear Economic Relationships*. Oxford: Oxford University Press.

—— M. L. King, and H. White (1995), 'Comments on Testing Economic Theories and the Use of Model Selection Criteria', *Journal of Econometrics*, 67: 173–88.

Gregory, A. W. (1994), 'Testing for Cointegration in Linear Quadratic Models', *Journal of Business and Economic Statistics*, 12: 347–60.

Grether, D. M. and M. Nerlove (1970), 'Some Properties of Optimal Seasonal Adjustment', *Econometrica*, 38: 682–703.

Haldrup, N. (1994*a*), 'The Asymptotics of Single Equation Cointegration Regression Models with I(1) and I(2) Variables', *Journal of Econometrics*, 63: 153–81.

—— (1994*b*), 'Semi-parametric Tests for Double Unit Roots', *Journal of Business and Economic Statistics*, 12: 109–22.

Hall, A. D. and M. McAleer (1989), 'A Monte Carlo Study of Some Tests of Model Adequacy in Time Series Analysis', *Journal of Business and Economic Statistics*, 7: 95–106.

Hall, R. E. (1978), 'Stochastic Implications of the Life Cycle–Permanent Income Hypothesis', *Journal of Political Economy*, 86: 971–87.

Hamilton, J. D. (1989), 'A New Approach to the Economic Analysis of Nonstationary Time Series and the Business Cycle', *Econometrica*, 57: 357–84.

—— (1994), *Time Series Analysis*. Princeton: Princeton University Press.

Hannan, E. J. (1970), *Multiple Time Series*. New York: John Wiley.

Hansen, L. P. and T. J. Sargent (1993), 'Seasonality and Approximation Errors in Rational Expectations Models', *Journal of Econometrics*, 55: 21–56.

Harvey, A. C. (1984), 'A Unified View of Statistical Forecasting Procedures', *Journal of Forecasting*, 3: 245–75.

—— (1989), *Forecasting, Structural Time Series Models and the Kalman Filter*. Cambridge: Cambridge University Press.

—— and S. J. Koopman (1992), 'Diagnostic Checking of Unobserved-Components Time Series Models', *Journal of Business and Economic Statistics*, 10: 377–89.

Hasza, D. P. and W. A. Fuller (1982), 'Testing for Nonstationary Parameter Specifications in Seasonal Time Series Models', *Annals of Statistics*, 10: 1209–16.

Hillmer, S. C. and G. C. Tiao (1982), 'An ARIMA-Model-Based Approach to Seasonal Adjustment', *Journal of the American Statistical Association*, 77: 63–70.

Hosking, J. R. M. (1981), 'Fractional Differencing', *Biometrika*, 68: 165–76.

Hurd, H. L. and N. L. Gerr (1991), 'Graphical Methods for Determining the Presence of Periodic Correlation', *Journal of Time Series Analysis*, 12: 337–50.

Hylleberg, S. (1986), *Seasonality in Regression*. Orlando, Fla.: Academic Press.
—— (ed.) (1992), *Modelling Seasonality*. Oxford: Oxford University Press.
—— (1994), 'Modelling Seasonal Variation', in C. P. Hargreaves (ed.), *Nonstationary Time Series Analysis and Cointegration*. Oxford: Oxford University Press.
—— (1995), 'Tests for Seasonal Unit Roots: General to Specific or Specific to General?' *Journal of Econometrics*, 69: 5–26.
—— and G. E. Mizon (1989*a*), 'A Note on the Distribution of the Least Squares Estimator of a Random Walk With Drift', *Economic Letters*, 29: 225–30.
—— —— (1989*b*), 'Cointegration and Error Correction Mechanisms', *Economic Journal*, 99: 113–25.
—— R. F. Engle, C. W. J. Granger and B. S. Yoo (1990), 'Seasonal Integration and Cointegration', *Journal of Econometrics*, 44: 215–38.
—— C. Jørgensen and N. K. Sørensen (1993), 'Seasonality in Macroeconomic Time Series', *Empirical Economics*, 18: 321–35.
Jaeger, A. and R. M. Kunst (1990), 'Seasonal Adjustment and Measuring Persistence in Output', *Journal of Applied Econometrics*, 5: 47–58.
Johansen, S. (1988), 'Statistical Analysis of Cointegration Vectors', *Journal of Economic Dynamics and Control*, 12: 231–54.
—— (1989), *Likelihood Based Inference of Cointegration. Theory and Applications*. Bologne: Centro Interuniversitario di Econometria.
—— (1991), 'Estimation and Hypothesis Testing of Cointegration Vectors in Gaussian Vector Autoregressive Models', *Econometrica*, 59: 1551–80.
—— (1992*a*), 'Cointegration in Partial Systems and the Efficiency of Single-Equation Analysis', *Journal of Econometrics*, 52: 389–402.
—— (1992*b*), 'A Representation of Vector Autoregressive Processes Integrated of Order 2', *Econometric Theory*, 8: 188–202.
—— and K. Juselius (1990), 'Maximum Likelihood Estimation and Inference on Cointegration—with Applications to the Demand for Money', *Oxford Bulletin of Economics and Statistics*, 52: 169–210.
Jones, R. H. and W. M. Brelsford (1967). 'Time Series with Periodic Structure', *Biometrika*, 54: 403–7.
Kennan, J. (1979), 'The Estimation of Partial Adjustment Model with Rational Expectations', *Econometrica*, 47: 1441–55.
Kleibergen, F. and P. H. Franses (1995), 'Direct Cointegration Testing in Periodic VAR Models', Econometric Institute Report 9518, Erasmus University Rotterdam.
Krane, S. D. (1993), 'Induced Seasonality and Production-Smoothing Models of Inventory Behavior', *Journal of Econometrics*, 55: 135–68.
Kuiper, W. E., P. H. Franses, and T. Kloek (1993), 'Testing Rational Expectations in Agricultural Markets using Periodic Models', Econometric Institute Report 9312, Erasmus University Rotterdam.
Kunst, R. M. (1993), 'Seasonal Cointegration in Macroeconomic Systems: Case Studies for Small and Large European Countries', *Review of Economics and Statistics*, 75: 325–30.
Laroque, G. (1977), 'Analyse d'une méthode de désaissonnalisation: le programme X-11 du Bureau of Census, Version Trimestrielle', *Annales de l'INSEE*, 28: 105–27.
Lee, H. S. (1992), 'Maximum Likelihood Inference on Cointegration and Seasonal Cointegration', *Journal of Econometrics*, 54: 351–65.

—— and P. L. Siklos (1992), 'Seasonality in Time Series: Money–Income Causality in US Data Revisited', unpublished manuscript, Tulane University, New Orleans.

Ljung, G. M. and G. E. P. Box (1978), 'On a Measure of Lack of Fit in Time Series Models', *Biometrika*, 65: 297–303.

Lomnicki, Z. A. (1961), 'Tests for Departure from Normality in the Case of Linear Stochastic Processes', *Metrika*, 4: 27–62.

Lucas, A. (1995), 'An Outlier Robust Unit Root Test With Application to the Extended Nelson–Plosser Data', *Journal of Econometrics*, 66: 153–73.

Lütkepohl, H. (1991), *Introduction to Multiple Time Series Analysis*. Berlin: Springer–Verlag.

MacKinnon, J. G. (1991), 'Critical Values for Co-Integration Tests', in R. F. Engle and C. W. J. Granger (eds.), *Long-Run Economic Relationships*. Oxford: Oxford University Press.

Maravall, A. (1985), 'On Structural Time Series Models and the Characterization of Components', *Journal of Business and Economic Statistics*, 3: 350–5.

—— (1995), 'Unobserved Components in Economic Time Series', *Handbook of Applied Econometrics*.

—— and D. A. Pierce (1987), 'A Prototypical Seasonal Adjustment Model', *Journal of Time Series Analysis*, 8: 177–93.

McClave, J. T. (1975), 'Subset Autoregression', *Technometrics*, 17: 312–20.

McLeod, A. I. (1993), 'Parsimony, Model Adequacy and Periodic Correlation in Time Series Forecasting', *International Statistical Review*, 61: 387–93.

—— (1994), 'Diagnostic Checking of Periodic Autoregression Models with Application', *Journal of Time Series Analysis*, 15: 221–33.

Miron, J. A. (1986), 'Seasonal Fluctuations and the Life Cycle–Permanent Income Model of Consumption', *Journal of Political Economy*, 94: 1258–79.

—— (1994), 'The Economics of Seasonal Cycles', in C. A. Sims (ed.), *Advances in Econometrics, Sixth World Congress of the Econometric Society*. Cambridge: Cambridge University Press.

—— and S. P. Zeldes (1988), 'Seasonality, Cost Shocks, and the Production Smoothing Model of Inventories', *Econometrica*, 56: 877–908.

Nankervis, J. C. and N. E. Savin (1987), 'Finite Sample Distributions of *t* and *F* Statistics in an AR(1) Model with an Exogenous Variable', *Econometric Theory*, 3: 387–408.

Neftçi, S. N. (1984), 'Are Economic Time Series Asymmetric over the Business Cycle?', *Journal of Political Economy*, 92: 307–28.

Nelson, C. R. (1973), *Applied Time Series Analysis for Managerial Forecasting*. San Francisco: Holden Day.

Nerlove, M., D. M. Grether, and J. L. Carvalho (1979), *Analysis of Economic Time Series*. New York: Academic Press.

Nickell, S. (1985), 'Error Correction, Partial Adjustment and All That: an Expository Note', *Oxford Bulletin of Economics and Statistics*, 47: 119–29.

Noakes, D. J., A. I. McLeod, and K. W. Hipel (1985), 'Forecasting Monthly Riverflow Time Series', *International Journal of Forecasting*, 1: 179–90.

Ooms, M. (1994), *Empirical Vector Autoregressive Modeling*. Berlin: Springer-Verlag.

Osborn, D. R. (1988), 'Seasonality and Habit Persistence in a Life-Cycle Model of Consumption', *Journal of Applied Econometrics*, 3: 255–66.

—— (1990), 'A Survey of Seasonality in UK Macroeconomic Variables', *International Journal of Forecasting*, 6: 327–36.

—— (1991), 'The Implications of Periodically Varying Coefficients for Seasonal Time-Series Processes', *Journal of Econometrics*, 48: 373–84.

—— (1993), 'Comment on Engle *et al.* [1993]', *Journal of Econometrics*, 55: 299–303.

—— A. P. L. Chui, J. P. Smith, and C. R. Birchenhall (1988), 'Seasonality and the Order of Integration for Consumption', *Oxford Bulletin of Economics and Statistics*, 50: 361–77.

—— and J. P. Smith (1989), 'The Performance of Periodic Autoregressive Models in Forecasting Seasonal UK Consumption', *Journal of Business and Economic Statistics*, 7: 117–27.

Osterwald–Lenum, M. (1992), 'A Note with Quantiles of the Asymptotic Distribution of the Maximum Likelihood Cointegration Rank Test Statistics: Four Cases', *Oxford Bulletin of Economics and Statistics*, 54: 461–72.

Otto, G. and T. Wirjanto (1990), 'Seasonal Unit Root Tests on Canadian Macroeconomic Time Series', *Economics Letters*, 34: 117–20.

Paap, R. (1995), 'Trends in Periodic Autoregressions', *Tinbergen Institute PhD Research Bulletin*, 7: 1–12.

Pagano, M. (1978), 'On Periodic and Multiple Autoregressions'. *Annals of Statistics*, 6: 1310–17.

Pantula, S. G. (1989), 'Testing for Unit Roots in Time Series Data', *Econometric Theory*, 5: 256–71.

Park, J. Y. and P. C. B. Phillips (1988), 'Statistical Inference in Regressions with Integrated Processes: Part I', *Econometric Theory*, 4: 468–97.

—— —— (1989), 'Statistical Inference in Regressions with Integrated Processes: Part II', *Econometric Theory*, 5: 95–131.

Parzen, E. and M. Pagano (1979), 'An Approach to Modeling Seasonally Stationary Time Series', *Journal of Econometrics*, 9: 137–53.

Peña, D. (1990), 'Influential Observations in Time Series', *Journal of Business and Economic Statistics*, 8: 235–41.

Perron, P. (1989), 'The Great Crash, the Oil Price Shock, and the Unit Root Hypothesis', *Econometrica*, 57: 1361–1401.

—— (1990), 'Testing for a Unit Root in a Time Series with a Changing Mean', *Journal of Business and Economic Statistics*, 8: 153–62.

—— and T. J. Vogelsang (1992), 'Nonstationarity and Level Shifts with an Application to Purchasing Power Parity', *Journal of Business and Economic Statistics*, 10: 301–20.

Phillips, P. C. B. (1987), 'Time Series Regression with a Unit Root', *Econometrica*, 55: 277–301.

—— and S. Ouliaris (1990), 'Asymptotic Properties of Residual Based Tests for Cointegration', *Econometrica*, 58: 165–93.

Pierce, D. A. (1980), 'Data Revisions in Moving Average Seasonal Adjustment Procedures, *Journal of Econometrics*, 14: 95–114.

Plosser, C. I. and G. W. Schwert (1977), 'Estimation of a Non-invertible Moving Average Process: the Case of Overdifferencing', *Journal of Econometrics*, 6: 199–224.

Porter-Hudak, S. (1990), 'An Application of the Seasonally Fractionally Differenced Model to the Monetary Aggregates', *Journal of the American Statistical Association*, 85: 338–44.

Poskitt, D. S. and A. R. Tremayne (1980), 'Testing the Specification of a Fitted Autoregressive-Moving Average Model', *Biometrika*, 67: 359–63.

Ray, B. (1993), 'Long-range Forecasting of IBM Product Revenues using a Fractionally Differenced ARMA Model', *International Journal of Forecasting*, 9: 255–69.

Rissanen, J. (1978), 'Modelling by Shortest Data Description', *Automatica*, 14: 465–71.

Said, S. E. and D. A. Dickey (1984), 'Testing for Unit Roots in Autoregressive-Moving Average Models of Unknown Order', *Biometrika*, 71: 599–607.

—— —— (1985), 'Hypothesis Testing in ARIMA (p,1,q) Models', *Journal of the American Statistical Association*, 80: 369–74.

Sakai, H. (1982), 'Circular Lattice Filtering using Pagano's Method', *IEEE Transactions on Acoustics, Speech and Signal Processing*, 30: 279–87.

Sargent, T. J. (1978), 'Estimation of Dynamic Labour Demand Schedules under Rational Expectations', *Journal of Political Economy*, 86: 1009–44.

Schwarz, G. (1978), 'Estimating the Dimension of a Model', *Annals of Statistics*, 6: 461–4.

Shiskin, J. and H. Eisenpress (1957), 'Seasonal Adjustment by Electronic Computer Methods', *Journal of the American Statistical Association*, 52: 415–49.

—— A. H. Young, and J. C. Musgrave (1967), 'The X-11 Variant of the Census Method II Seasonal Adjustment Program', Technical Report 15, Bureau of the Census, US Department of Commerce, Washington DC.

Sims, C. A. (1974), 'Seasonality in Regression', *Journal of the American Statistical Association*, 69: 618–27.

Stock, J. and M. W. Watson (1989), 'Testing for Common Trends', *Journal of the American Statistical Association*, 83: 1097–1107.

Teräsvirta, T. and I. Mellin (1986), 'Model Selection Criteria and Model Selection Tests in Regression Models', *Scandinavian Journal of Statistics*, 13: 159–71.

Tiao, G. C. and M. R. Grupe (1980), 'Hidden Periodic Autoregressive-Moving Average Models in Time Series Data', *Biometrika*, 67: 365–73.

Todd, R. (1990), 'Periodic Linear–Quadratic Methods for Modeling Seasonality', *Journal of Economic Dynamics and Control*, 14: 763–95.

Troutman, B. M. (1979), 'Some Results in Periodic Autoregression', *Biometrika*, 66: 219–28.

Tsay, R. (1988), 'Outliers, Level Shifts, and Variance Changes in Time Series', *Journal of Forecasting*, 7: 1–20.

—— (1993), 'Testing for Noninvertible Models with Applications', *Journal of Business and Economic Statistics*, 11: 225–33.

Ula, T. A. (1993), 'Forecasting of Multivariate Periodic Autoregressive Moving-Average Processes', *Journal of Time Series Analysis*, 14: 645–57.

Vecchia, A. V. (1985), 'Maximum Likelihood Estimation for Periodic Autoregressive Moving Average Models', *Technometrics*, 27: 375–84.

—— and R. Ballerini (1991), 'Testing for Periodic Autocorrelations in Seasonal Time Series Data', *Biometrika*, 78: 53–63.

Wallis, K. F. (1974), 'Seasonal Adjustment and Relations between Variables', *Journal of the American Statistical Association*, 69: 18–31.

Watson, M. W. (1994), 'Business-Cycle Durations and Postwar Stabilization of the US Economy', *American Economic Review*, 84: 24–46.

Wisniewski, J. (1934), 'Interdependence of Cyclical and Seasonal Variation', *Econometrica*, 2: 176–85.

Wolters, J. (1992), 'Persistence and Seasonality in Output and Employment of the Federal Republic of Germany', *Recherches Economiques de Louvain*, 58: 421–39.

Author Index

Abeysinghe, T., 64, 215
Abraham, B., 10–11, 42, 215–17
Adams, G. J., 104, 215
Agiakloglou, C., 63, 66, 215
Akaike, H., 13, 215
Andel, J., 93, 103, 215
Anderson, P. L., 103, 215
Anderson, T. W., 4, 8, 23, 47, 215
Andrews, D. W. K., 12, 165, 215
Assarsson, B., 214, 215

Baillie, R. T., 23, 215
Ballerini, R., 101, 114, 224
Banerjee, A., 20, 215, 218
Barsky, R. B., 38, 215
Beaulieu, J. J., 38, 64, 69, 171, 207, 215
Bell, W. R., 47, 49, 53, 216, 218
Bentarzi, M., 98, 101, 216
Bera, A. K., 10, 216
Bewley, R., 30, 216
Bhargava, A., 57, 216
Birchenhall, C. R., 188, 216, 223
Bladen–Hovell, R. C., 216
Bloomfield, P., 101, 216
Boldin, M. D., 85, 216
Bollerslev, T., 11, 216
Boswijk, H. P., 20, 29–31, 104, 126, 129–30, 132–3, 135, 154, 178, 195, 197–200, 216
Bowerman, B. L., 62, 79, 216
Box, G. E. P., 4, 7–9, 30, 32, 41–2, 61–2, 182, 187, 216–17, 222
Braun, R. A., 212, 217
Breitung, J., 23, 156, 217
Brelsford, W. M., 93, 221
Burridge, P., 54, 102, 217

Campbell, J. Y., 21, 217
Canova, F., 73, 80, 84, 86–7, 217
Carvalho, J. L., 222
Chatterjee, S., 212, 217
Chen, C., 10, 217
Chow, G. C., 12, 217
Chuang, A., 10, 215
Chui, A. P. L., 216, 223
Cipra, T., 94, 98, 103, 217
Cleveland, W. P., 53, 93, 217

Davidson, J. E. H., 180, 217
Davisson, L. D., 13, 217
De Gooijer, J. J., 13, 217
Dickey, D. A., 19–21, 63–4, 66, 73, 217–18, 224
Dolado, J., 193, 215, 218

Eisenpress, H., 51, 55, 80, 224
Engle, R. F., 11–12, 24–5, 29, 66, 74–5, 148, 153, 181–2, 185–7, 195–7, 204, 206, 218, 221
Engsted, T., 193, 218
Ericsson, N. R., 59, 218
Evans, C. L., 212, 217

Faig, M., 212, 218
Fisher, L. A., 216
Fox, A. J., 10, 218
Franses, P. H., 12, 22–3, 31, 47, 55, 62–4, 66–7, 70–1, 73, 81, 93, 102, 104, 114, 126, 129–30, 132–3, 135, 137, 148–9, 154, 156, 164, 175, 177–8, 184, 195, 197–200, 216–18, 221
Fuller, W. A., 4, 19, 20–1, 62, 130, 133, 138, 217–20

Galbraith, J. W., 215–18
Gerr, N. L., 101, 220
Gersovitz, M., 92, 94, 219
Ghysels, E., 51, 57–9, 66–7, 69, 80, 84–8, 90–1, 93, 102, 135–6, 142, 209, 217, 219
Gladyshev, E. G., 93, 95, 219
Godfrey, L. G., 10, 219
Gonzalo, J., 30, 156, 182, 184, 219
Goodwin, G. C., 104, 215
Gould, A., 217
Gourieroux, C., 186, 220
Granger, C. W. J., 4, 8, 12–14, 22, 24–5, 30, 42, 64, 74–5, 128, 153, 156, 181–2, 184–5, 187, 195–7, 204, 213, 218–21
Gregory, A. W., 193, 220
Grether, D. M., 51, 220, 222
Grupe, M. R., 95, 117, 122, 224

Haldrup, N., 22, 193, 216, 218, 220
Hall, A., 102, 135–6, 219
Hall, A. D., 10, 220
Hall, R. E., 141–2, 220
Hallin, M., 98, 101, 216
Hallman, J. J., 12, 218, 220
Hamilton, J. D., 4, 84, 87, 93, 220
Hannan, E. J., 16, 220

Hansen, B. E., 73, 217
Hansen, L. P., 212, 220
Harvey, A. C., 8–9, 48, 54, 220
Hasza, D. P., 62, 217–18, 220
Hendry, D. F., 215, 217–18
Hillmer, S. C., 49, 53, 216, 220
Hipel, K. W., 222
Hobijn, B., 66–7, 71, 218
Hosking, J. R. M., 22, 220
Hurd, H. L., 101, 216, 220
Hylleberg, S., 2, 18, 21, 24, 32–3, 35, 39, 41, 49, 51–2, 54, 64, 66–7, 69, 73, 78–9, 218–19, 221

Jaeger, A., 57, 221
Jarque, C. M., 10, 216
Jenkins, G. M., 4, 7–8, 32, 41–2, 61–2, 216
Johansen, S., 26–9, 101, 132, 142, 153, 181–2, 190, 196, 206, 221
Jones, R. H., 93, 221
Jørgensen, C., 221
Joyeux, R., 22, 220
Juselius, K., 26, 28, 182, 190, 221

Kennan, J., 193, 221
King, M. L., 220
Kleibergen, F., 178, 221
Kloek, T., 197, 218, 221
Koehler, A. B., 62–3, 216, 218
Koopman, S. J., 9, 220
Kozicki, S., 186, 218
Krane, S. D., 212, 221
Kuiper, W. E., 92, 221
Kunst, R. M., 57, 75, 221

Laroque, G., 51–2, 221
Ledolter, J., 42, 215
Lee, H. S., 75–6, 206, 218–19, 221
Liu, L., 10, 217
Ljung, G. M., 9, 222
Lomnicki, Z. A., 10, 222
Lucas, A., 11, 22, 222
Lund, R. B., 216
Lütkepohl, H., 16–17, 95–6, 104, 123, 157, 178, 222

MacKie–Mason, J. K., 215
MacKinnon, J. G., 25, 92, 94, 182–3, 185, 196, 219, 222
Maravall, A., 49, 51–3, 59, 222
McAleer, M., 10, 12, 137, 164, 218–20
McClave, J. T., 8, 222
McLeod, A. I., 92, 94, 101, 222
Mellin, I., 13, 224
Miller, R. B., 218
Miron, J. A., 38, 40, 64, 69, 80, 171, 207–8, 212, 215, 222
Mizon, G. E., 18, 21, 24, 220
Monfort, A., 220
Musgrave, J. C., 224

Nankervis, J. C., 22, 222
Neftçi, S. N., 87, 222
Nelson, C. R., 42, 222
Nerlove, M., 42, 51, 88, 90, 219–20, 222
Newbold, P., 4, 8, 14, 42, 63–4, 66, 215, 220
Nickell, S., 193, 222
Noakes, D. J., 94, 222
Noh, J., 219

Ooms, M., 16, 52–3, 55, 81, 175, 206, 214, 219, 222
Orden, D., 216
Osborn, D. R., 61–3, 67, 69, 77, 92–3, 95, 99, 117–18, 128, 141, 144, 167, 176, 194, 202, 212, 214, 216, 222–3
Osterwald–Lenum, M., 28, 132, 138, 143, 223
Otto, G., 69, 223
Ouliaris, S., 25, 182, 223

Paap, R., 93, 152, 177, 184, 219, 223
Pack, D. J., 216
Pagano, M., 93, 103, 223
Pantula, S. G., 22, 223
Park, J. Y., 132, 198, 223
Parzen, E., 93, 223
Peña, D., 11, 223
Perron, P., 21–2, 51, 58, 69, 70, 114, 217, 219, 223
Phillips, P. C. B., 20, 25, 132, 182, 198, 223
Pierce, D. A., 9, 53–4, 216, 222–3
Plosser, C. I., 23, 223
Porter–Hudak, S., 73, 223
Poskitt, D. S., 10, 223

Ravikumar, B., 212, 217
Ray, B., 73, 224
Renault, E., 220
Richard, J. F., 218
Rissanen, J., 14, 224
Robinson, L., 217
Romijn, G., 148, 219

Said, S. E., 20–1, 224
Sakai, H., 101, 224
Sargent, T. J., 193, 212, 220, 224
Savin, N. E., 22, 222
Schwarz, G., 13, 224
Schwert, G. W., 23, 223
Shiskin, J., 51, 55, 80, 224
Siklos, P. L., 75–6, 219, 222
Sims, C. A., 59, 224
Smith, J. P., 93, 216, 223
Sørensen, N. K., 221
Srba, F., 217
Stock, J., 30, 224

Teräsvirta, T., 12–13, 213, 220, 224
Tiao, G. C., 30, 53, 93, 95, 117, 122, 182, 187, 217, 220, 224
Todd, R., 92, 212, 224
Tran, H. A., 218
Tremayne, A. R., 10, 224
Troutman, B. M., 93, 103, 223
Tsay, R., 11, 23, 224

Ula, T. A., 123, 224

Vecchia, A. V., 93–4, 101, 103, 114, 215, 224
Vogelsang, T. J., 70–1, 114, 219, 223

Wallis, K. F., 54, 59, 102, 217, 224
Walker, A. M., 23, 47, 215
Watson, M. W., 30, 84, 85, 224
White, H., 220
Wirjanto, T., 69, 223
Wisniewski, J., 79, 224
Wolters, J., 214, 224

Yang, M., 216
Yatawara, N., 11, 215
Yeo, S., 217
Yoo, B. S., 221
Young, A. H., 224

Zeldes, S. P., 212, 222

Subject Index

Accumulation of shocks, 152–6, 162
Adjoint matrix, 17
Adjustment parameters, 26
Airline model, 42–3
Autocorrelation function [ACF]:
  definition, 6–8
  nonperiodic ACF, 140
  of residuals, 9
  of vector process, 16
  periodic ACF, 101, 114
Autocovariance, 6
Autoregressive conditional heteroskedasticity (ARCH), 11
Autoregressive model (AR), 5
Autoregressive moving average model (ARMA), 5–6

Backward shift operator, 5, 35, 97
Brownian motion, 20, 130, 132

Canonical correlation, 27
Characteristic equation:
  of multivariate time series model, 15
  of periodic autoregression, 98, 203
Codependence, 186, 192
Cointegration:
  between elements of $\mathbf{Y}_T$ process, 127, 142
  Boswijk method, 29–30
  Engle–Granger method, 25
  Johansen method, 26–8
  linear restrictions on parameters, 28
  long-run equilibrium, 24
  nonseasonal cointegration, 75
  parameters, 26
  spurious cointegration, 31
  testing for, 25–30
Consumer confidence:
  index, 80, 88–90
  seasonal adjustment of, 88
Cycles:
  asymmetry of, 86–7
  business cycles and seasonal cycles, 170–1
  business cycle turning-points, 84, 209
  cyclical behaviour, 37
  cyclical variation, 79
  expansions and recessions, 79, 84, 209
  peaks and troughs, 84, 171–5, 207–9

Data transformations:
  general, 117
  log transformation, 12, 79
Diagnostic measures, 9–12
Differencing:
  appropriate filter, 61
  double filter, 42, 52
  first order filter, 18, 146
  notation, 35
  periodically varying filter, 104, 126, 128, 140
  seasonal filter, 41

Eigenvalues:
  of canonical correlation matrix, 27, 182–5, 191
  of $\Phi_0^{-1}\,\Phi_1$, 113
Eigenvectors, 27, 182–5, 191
Error correction:
  conditional, 29
  model 26, 179, 182, 193
  parameters, 74
Exogeneity:
  properties, 59
  weak exogeneity, 29–30, 198–9
Expectations:
  time-varying expectations, 90

Forecasting, 14–15, 157–63
Forward shift operator, 50
Frequency:
  annual, 74
  biannual, 74
  nonseasonal (zero), 62, 100
  seasonal, 74

Growth rates:
  annual, 41
  quarterly, 35

h-step forecast, 14

Identification, 7, 41
Implied univariate models, 17
Impulse response function, 156
Integrated:
  fractional, 22, 73
  of order 1, I(1), 18
  of order 2, I(2), 22, 100
  seasonal, 61, 63, 100
Invertibility, 6, 19

Lagrange multiplier principle, 10
Linearity, 12
Linear-quadratic adjustment cost, 193
Local-level model, 8

Mean shifts, 22, 61, 163–5
Misspecified homogeneous model, 118, 122, 167–8
Model selection criteria:
  AIC, 13, 16, 105
  FPE, 13
  MAPE, 15
  RMSE, 14, 161–3
  SC, 13, 16, 105
Moving average [MA]:
  asymmetric filter, 51
  model, 5
  symmetric linear centred filter, 50, 53, 58

Noninvertibility, 47, 52
Nonlinearity, 12
Nonperiodic AR model, 93, 186–8, 205–7
Nonstationarity, 6
Normality, 10

Orthogonality:
  assumption, 56, 60
  definition, 54
  regression, 55
  testing the assumption, 80
Outliers:
  additive, 10, 71
  innovation, 11, 70
  removal of, 51
Overdifferencing, 22, 64, 150

Parameter constancy, 12
Parameter restrictions in PAR models, 133–4
Partial autocorrelation function, 6–8
Periodic autoregression [PAR]:
  for an I(1) time series [PARI], 133, 136
  general, 92–3
  model selection strategy, 133
  multivariate representation of univariate PAR, 95–8
  order selection, 105–8
Periodic cointegration [PC]:
  Boswijk test method, 197–9
  Engle–Granger test method, 195–7
  full PC model, 195
  model, 178, 180
  partial PC model, 195
  with seasonal cointegration, 180
Periodic error correction model:
  multivariate, 180
  univariate, 147
Periodicity:
  common, 181, 185–6
  spurious, 114
  testing for, 102
  time series, 33
  variation in AR parameters, 104
Periodic integration [PI]:
  comparison with seasonal integration, 150
  definition, 99–100, 128
  differencing filter $(1-\alpha_s B)$, 128–9, 172
  PI autoregression [PIAR], 133
Periodic models:
  estimation of, 103
  Markov regime-switching model, 84, 93
  multivariate periodic autoregression, 93, 123
  periodic moving average, 92, 150
  periodic VAR, 124, 177, 179–80
  periodic ARMA, 92, 94
  regression model, 81
Portmanteau test, 9

Recursive estimation, 69
Reduced rank regression, 27
Robustness:
  estimation, 11, 22
  of seasonal adjustment, 54

Seasonal adjustment:
  and periodic time series, 95, 117, 120–2, 169–70, 206
  Census X-11, 49, 51–3
  evaluation criteria of, 54–9
  model-dependent approach, 49, 53–4
  (not) seasonally adjusted [(N)SA] data, 49–50
Seasonal cointegration:
  full cointegration model, 77, 194
  full system method, 75–6, 206
  multivariate model, 61
  residual based method, 74–5
Seasonal constants, 34
Seasonal dummy variables, 34, 39
Seasonal heteroskedasticity, 94, 102, 118
Seasonal mean shifts, 69, 114
Seasonal time series:
  as vector process, 34
  definition, 32
Seasonal time series models:
  estimated SARIMA model, 42–6
  multiplicative SARIMA, 41
  seasonal ARIMA [SARIMA], 32, 41–6, 53
  seasonal unobserved component models, 32, 48
Seasonality:
  amount of seasonal variation, 38
  changing, 33
  constant, 73
  definition, 33
  deterministic, 42

Seasonality: (*cont.*)
  increasing, 62, 151
  in utility, 92, 194
  multiplicative, 50
  stochastic, 42
  seemingly nonstationary, 43–4
Stationarity:
  covariance-stationarity, 6, 19
  difference-stationarity, 18
  long-memory, 23, 57
  periodic, 129, 135, 181
  trend-stationarity, 18
Structural break, 12, 102
Structural models:
  for seasonal time series, 48, 54
  unobserved components, 8
Subset:
  autoregression, 8
  moving average, 7
  periodic autoregression, 94, 149, 202

Temporal aggregation, 166
Transfer function models, 17
Trend:
  breaking trend, 22
  common stochastic trend, 30, 181–5
  deterministic, 18, 150–1
  seasonal deterministic, 132, 152
  seasonal stochastic, 72
  stochastic, 18, 23, 30–1, 156

Underdifferencing, 64
Unit roots:
  in MA polynomial, 23, 47
  multiple unit roots, 142
  no mean reversion, 18
  nonseasonal, 52, 57, 99
  permanent effect of shocks, 18
  persistence of shocks, 58
  random walk, 19
  seasonal, 52–3, 59, 61, 99, 129, 144
  testing for, 19–23
Unobserved components:
  irregular component, 50, 53
  nonseasonal component, 50
  seasonal component, 50, 53, 55
  trend and cycles, 50, 53

Vector time series:
  of quarters [VQ], 95–6, 142, 145–6
  vector ARMA, 15, 159
  vector MA, 150, 152
  vector white noise, 15

White noise process, 4, 7, 39
Wiener process, 20